STP 1357

Time Dependent and Nonlinear Effects in Polymers and Composites

Richard A. Schapery and C. T. Sun, editors

ASTM Stock Number: STP 1357

ASTM
100 Barr Harbor Drive
West Conshohocken, PA 19428-2959
Printed in the U.S.A.

Library of Congress Cataloging-in-Publication Data

Schapery, Richard Allan.
    Time dependent and nonlinear effects in polymers and composites / Richard A.
    Schapery and C. T. Sun, editors.
        p. cm. — (STP ; 1357)
    "ASTM Stock Number: STP1357."
    Includes bibliographical references and index.
    ISBN 0-8031-2601-8
        1. Polymers—Mechanical properties.   2. Composite materials—Mechanical properties.   3.
    Nonlinear mechanics.   I. Sun, C. T. (Chang Tsan), 1928–   II. Title.   III. ASTM special
    technical publication ; 1357.

    TA455.P58 S34 2000
    620.1'9204292—dc21

                                                                              99-056137

Photocopy Rights

Authorization to photocopy items for internal, personal, or educational classroom use, or the internal, personal, or educational classroom use of specific clients, is granted by the American Society for Testing and Materials (ASTM) provided that the appropriate fee is paid to the Copyright Clearance Center, 222 Rosewood Drive, Danvers, MA 01923, Tel: 508-750-8400; online: http://www.copyright.com/.

Peer Review Policy

Each paper published in this volume was evaluated by two peer reviewers and at least one editor. The authors addressed all of the reviewers' comments to the satisfaction of both the technical editor(s) and the ASTM Committee on Publications.

The quality of the papers in this publication reflects not only the obvious efforts of the authors and the technical editor(s), but also the work of the peer reviewers. In keeping with long standing publication practices, ASTM maintains the anonymity of the peer reviewers. The ASTM Committee on Publications acknowledges with appreciation their dedication and contribution of time and effort on behalf of ASTM.

Printed in Philadelphia, PA
January 2000

Foreword

The Symposium on *Time Dependent and Nonlinear Effects in Polymers and Composites* was held on May 4–5, 1998 in Atlanta, Georgia. ASTM Committee D-30 on Composite Materials sponsored the Symposium. Richard Schapery, the University of Texas at Austin, and C. T. Sun, Purdue University, presided as symposium co-chairmen and are editors of this publication.

Contents

Overview

Polymeric composites exhibit appreciable time-dependent and nonlinear mechanical behavior in many structural applications. Improved fundamental understanding of and predictive models for this behavior over service lifetimes in realistic environments are needed for composites to gain wider acceptance and to achieve a significantly increased level of structural efficiency and reliability in commercial and military applications.

In advanced fibrous composites, the fiber phase is basically linearly elastic and shows little nonlinear and time-dependent behavior, if any. Thus, the nonelastic properties originate primarily from the matrix phase, including the polymer-fiber interphase zone. It is evident that to understand and to model the nonelastic behavior of composites, one must understand and be able to model this behavior in polymers.

Understanding the in situ behavior of the polymeric matrix and its interaction with fibers is an important part of developing improved predictive models. Because there is now a considerable amount of research activity worldwide on time-dependent and nonlinear effects in polymers and composites, it is desirable to describe and assess recent developments and their practical significance as well as to identify important unsolved fundamental problems. For these reasons, the two related disciplines, polymers and composites, were brought together at the *Symposium on Time-dependent and Nonlinear Effects in Polymers and Composites.* One of the main objectives of this symposium was to join the effort of specialists in these two disciplines to advance research in this important area of polymeric composites.

This volume contains eighteen papers presented at the Symposium. They are grouped under two subheadings, *Polymers* and *Composites.* Under *Polymers,* the primary topics are chemical and physical aging, nonlinear viscoelasticity and viscoplasticity. A number of topics are addressed by the papers under *Composites,* including the effect of physical aging on time-dependent behavior of composites, multiaxial nonlinear effects, compressive behavior, nonlinear viscoelasticity and viscoplasticity, failure and failure mechanisms, hygrothermal effects, durability, and accelerated strength testing.

We want to express our sincere thanks to all those who made the symposium and this STP possible. The excellent contributions of the authors, reviewers, presenters, session chairs, ASTM staff, and sponsoring technical committee are deeply appreciated.

Richard A. Schapery

The University of Texas, Austin, Texas;
symposium cochairman and editor

C. T. Sun

Purdue University, West Lafayette, Indiana;
symposium cochairman and coeditor

Polymers

Luis C. Tsuji,[1] *Hugh L. McManus,*[1] *and Kenneth J. Bowles*[2]

Mechanical Properties of Degraded PMR-15 Resin

REFERENCE: Tsuji, L. C., McManus, H. L., and Bowles, K. J., "**Mechanical Properties of Degraded PMR-15 Resin,**" *Time Dependent and Nonlinear Effects in Polymers and Composites, ASTM STP 1357,* R. A. Schapery and C. T. Sun, Eds., American Society for Testing and Materials, West Conshohocken, PA, 2000, pp. 3–17.

ABSTRACT: Thermo-oxidative aging produces a non-uniform degradation state in PMR-15 resin. A surface layer, usually attributed to oxidative degradation, forms. This surface layer has different properties from the inner material. A set of material tests was designed to separate the properties of the oxidized surface layer from the properties of interior material. Test specimens were aged at 316°C in either air or nitrogen, for durations of up to 800 h. The thickness of the oxidized surface layer in air aged specimens, and the shrinkage and coefficient of thermal expansion (CTE) of nitrogen aged specimens were measured directly. Four-point-bend tests were performed to determine modulus of both the oxidized surface layer and the interior material. Bimaterial strip specimens consisting of oxidized surface material and unoxidized interior material were constructed and used to determine surface layer shrinkage and CTE. Results confirm that the surface layer and core materials have substantially different properties.

KEYWORDS: testing, aging, degradation, polymer matrix, material properties, test methods

Background

Polymer matrix composite materials are being increasingly considered for use in environments that challenge their durability. Such applications include turbine engine structures and high speed aircraft skins. In these environments the materials are exposed to high temperature and to oxygen, both of which contribute to the degradation of the polymer matrix.

Significant progress has been made in the understanding of the aging effects on the thermo-oxidative stability of polymer matrix composites [1–3]. Much of this work has focused on PMR-15 resin as a representative material. Thermo-oxidative aging produces a non-uniform degradation state in PMR-15 resin. While thermal degradation occurs throughout the material, oxidative degradation occurs only where oxygen diffuses into the material. This produces an oxidized surface layer that has different properties from the unoxidized inner material. Current models of coupled diffusion-reaction mechanism [4–6] attempt to capture the behavior of this chemical degradation. However, there is a need for data on the material properties of the degraded material(s), in order to link the diffusion-reaction models of chemical degradation to thermo-mechanical models [7].

Air aged material specimens have an oxidized surface layer and an unoxidized inner material which have different material properties. Tests performed on such specimens as if

[1] Graduate Student and Associate Professor, respectively, Department of Aeronautics and Astronautics, Massachusetts Institute of Technology, 77 Massachusetts Avenue, Cambridge, MA 02139.
[2] Senior Material Engineer, NASA Lewis Research Center, 21000 Brookpark Road, Cleveland, OH 44135.

they were homogeneous produce results that are difficult to interpret and may have little meaning. Tests must be carefully designed to separate the properties of the surface layer from the properties of the inner material.

Problem

The problem addressed here is to separate the properties of the oxidized surface layer from the properties of the unoxidized inner material, in PMR-15 resin specimens aged in air for various times at 316°C. Specifically, the thickness of the oxidized surface layer and the modulus, CTE, and shrinkage of both surface and inner material are determined as functions of aging duration.

Approach

The approach to this problem is primarily experimental. The main challenge of this project, and the problem that drove the experimental design, is how to determine the shrinkage and CTE of the oxidized surface layer. The shrinkage is small, and the changes are slight. Typical measurements of length are confounded with the effects of surface erosion on the ends, and the stress and strain interactions of the surface layer and the unoxidized inner material. In order to determine the surface layer shrinkage and the surface layer CTE, a new test based on a bimaterial strip model was designed and used.

The new test involved the manufacturing of specimens that resembled a bimaterial strip, with the oxidized surface material on one side, and unoxidized inner material on the other side. These specimens were manufactured simply by slicing a piece of an aged specimen lengthwise through the thickness. When one side contracts or expands relative to the other, the specimen curves. By utilizing a bimaterial strip model, the amount of curvature and the change in curvature with temperature can be used to determine the surface layer shrinkage and the CTE. An illustration of the curvature specimen is shown (Fig. 1).

In order to calculate the surface layer shrinkage and CTE from the curvature and change in curvature with temperature, several other properties must be known. These include surface layer thickness, modulus of both the oxidized surface layer and unoxidized material, shrinkage of unoxidized aged material, and CTE of unoxidized aged material. These properties are determined from other tests.

Surface layer thickness is measured simply by polishing a cross section of a specimen and measuring the surface layer. The modulus of unoxidized aged material is determined by a standard four-point bend test (Fig. 2). The modulus of the oxidized surface layer is derived using basic beam theory from a bend test of an air aged specimen, as surface layer thickness and the modulus of the unoxidized material are known. Figure 2 also illustrates the bend test of an air-aged specimen. Shrinkage of unoxidized aged material is measured simply by measuring the length of a specimen both before and after aging. The CTE of unoxidized aged material is determined from a typical test using thermo-mechanical analysis (Fig. 3).

Experimental Procedure

Overview

Specimens were cut from plaques of PMR-15 resin. Each specimen was then measured to determine initial dimensions. After initial measurements were completed, the specimens were aged in an oven at 316°C. Half of the specimens were aged in air in the oven, and half were aged in nitrogen in a nitrogen chamber placed inside the oven. Once specimens were

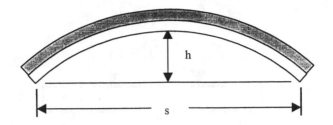

Test Gives:	Measurements:	Required inputs:
◆ Oxidized surface layer shrinkage ◆ CTE	◆ Temperature ◆ Height of arc h and ◆ Span of arc s or ◆ Temperature ◆ Coordinates of three points on arc	◆ Layer thicknesses ◆ CTE of unoxidized material ◆ Moduli of both layers ◆ Shrinkage of unoxidized material

FIG. 1—*Curve test schematic.*

removed from the aging ovens, their dimensions were remeasured. They were then tested to determine modulus by using four point bend tests. After bend testing, the specimens were cut again into smaller specimens for bimaterial strip curve tests, thermo-mechanical analysis (TMA), and for measuring surface layer thickness. Typically five specimens were used per test.

Specimen Preparation

These experiments used 152.4 mm by 152.4 mm PMR-15 neat resin plaques from the same batch as the resin plaques used by Kamvouris and Roberts [8] so that comparisons could be made. The plaques were made from HyComp 100, a PMR-15 powder supplied by HyComp, Inc. The powder was compression molded using an automated heated press with vacuum. Both temperature and pressure were ramped gradually, to a 2-h hold at 316°C and 9.3 MPa. The material was not given a separate post cure before aging.

Specimens were cut from the plaques using a water-cooled micro-machining diamond saw. The specimens were cut to the dimensions specified for four-point bend testing in the ASTM Standard Test Method for Flexural Properties of Unreinforced and Reinforced Plastics and Electrical Insulating Materials (D 790M). Specimens measured approximately 54 mm by 10 mm by 2.5 mm. After aging, dimensional measurements, and bend testing were completed, the specimens were further cut into smaller pieces for use in other tests (Fig. 4). Long thin specimens approximately 53 × 3 × 1 mm were cut for use in bimaterial strip curvature tests. Specimens approximately 26 × 3 × 2.5 mm were cut for determining layer thickness. Specimens approximately 5 × 3 × 2.5 mm were cut for use in thermo-mechanical analysis.

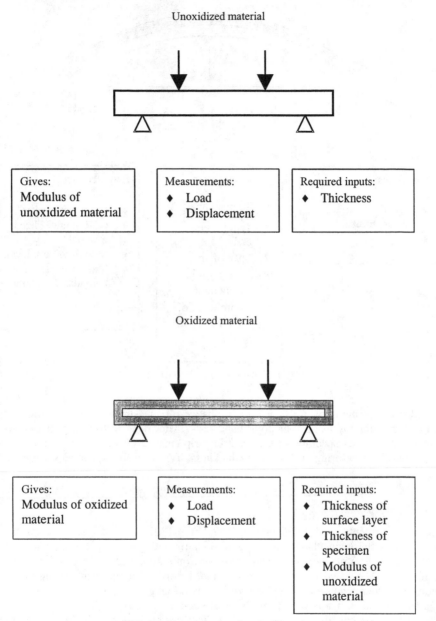

FIG. 2—*Four-point bend test schematics.*

Specimens were stored in sealed plastic bags inside a dessicator, and were dried before each measurement for at least 30 min in a 120°C oven, except for TMA specimens, which were dried for two or more hours.

Specimens were aged in a Blue M oven at 315°C. Specimens aged in air were placed on a tray in the oven, while specimens aged in nitrogen were placed inside a nitrogen chamber

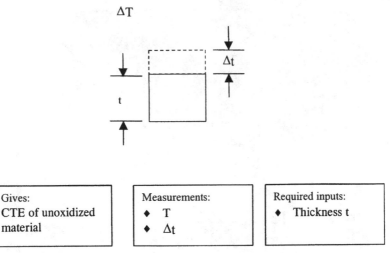

FIG. 3—*CTE test schematic.*

inside the oven. Specimen groups were aged 24 h, 48 h, 96 h, 168 h, 240 h, 336 h, 465 h, 633 h, and 801 h. One group was not aged.

Tests

Nitrogen Aged Specimen Shrinkage—Shrinkage of nitrogen aged specimens was determined from direct measurements of the length of the bend test specimens before and after aging. Measurements were made using a traveling measuring microscope, which measured to 0.001 mm. The coordinates of each corner were recorded for both the top and bottom of each specimen. This produced four measurements of length for each specimen. The shrinkage could then be determined by comparing the length of each specimen before aging to the length after.

Bend Tests—An Instron 4505 load frame with a 4500 controller was used for the bend tests. Steel four point bend fixtures with ceramic rollers were used. The load span to support span ratio was two. The bend tests conformed to ASTM D 790M, with one exception. Because of difficulties obtaining an accurate means of measuring center point deflection, load frame crosshead deflection was used instead. To compensate for slack and compliance of the load train, a test with a rigid bar in the fixture was performed to determine the compliance of the system. This was then used to correct the load versus displacement data for the tested specimens. Specimens were not tested to failure.

Thermo-Mechanical Analysis—Thermo-mechanical analysis was used to determine the CTE of nitrogen aged specimens. A TA Instruments model 2940 with a weighted expansion probe was used to measure the expansion of the small specimens as temperature was increased. The probe was weighted with a 5 g mass, applying a force of 0.05 N to the sample. CTE was determined from the slope of expansion versus temperature, over the range of temperature from room temperature to 316°C.

Surface Layer Thickness—Surface layer thickness was determined from specimens cut from the original aged specimens (Fig. 4). These specimens were mounted in epoxy and the cross section of the specimen was polished, to enable viewing of the surface layer with microscopy. Photomicrographs were then taken of the specimens. The thickness of the sur-

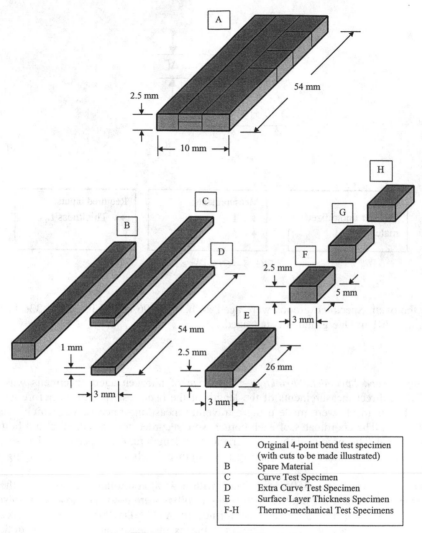

A	Original 4-point bend test specimen (with cuts to be made illustrated)
B	Spare Material
C	Curve Test Specimen
D	Extra Curve Test Specimen
E	Surface Layer Thickness Specimen
F-H	Thermo-mechanical Test Specimens

FIG. 4—*Specimen cutting illustration.*

face layer was then determined from measurements taken from the photomicrographs. This method of examination was known to produce measurable results due to the experience of Bowles, using similar material and aging conditions [9].

Curvature Tests—Specimen curvature at room temperature was determined by measuring the location of three points on the edge of the curvature specimens, using a traveling measuring microscope that measured to 0.001 mm. Measurements were taken on both sides of each specimen. To make measurements at elevated temperatures, specimens were placed in the oven of the Instron 4505 test machine. Measurements were made using a cathetometer that looked through the oven window at the specimens. Specimens were placed on small stands in the oven, and measurements were taken of the height and span of the arc (Fig. 1).

Analysis

Assumptions

The following analysis rests on a few basic assumptions. First is that the surface layer has uniform properties throughout the layer. This assumption is supported by the work of Cunningham, who showed both experimentally and with models that the surface degradation occurs on a sharp front, with little apparent gradient in degradation level within the surface layer [10]. Second is the assumption that the moduli of the surface and interior layers have the same temperature dependence. This assumption is as yet unsupported, but it affects only the surface layer CTE calculations in this work. Third is the assumption that viscoelastic relaxation is negligible. Relaxation, if it occurred, would be much more significant in the more highly stressed surface layer. Relaxation in this layer would cause the unsymmetrical beams in the curvature tests to have curvature that decreased with time, and hence the measured shrinkage would decrease with time. This was not observed. Finally, it is assumed that the nitrogen-aged specimens and the apparently unoxidized core material in the air aged specimens have the same properties.

Nitrogen Aged Specimen Shrinkage

The shrinkage of a nitrogen aged specimen, which represents the shrinkage of unoxidized aged resin, is simply as follows:

$$\varepsilon_{sc} = \frac{(l_{aged} - l_0)}{l_0} - (\alpha_{c(aged)} - \alpha_{c(o)})\Delta T \tag{1}$$

where ε_{sc} is the shrinkage strain of the unoxidized material, l_{aged} is the aged length, and l_0 is the unaged length. The second term compensates for the small difference between the aged and unaged core material CTE's, $\alpha_{c(aged)}$ and $\alpha_{c(o)}$, with ΔT being the difference between the aging temperature and room temperature.

Bend Tests

The modulus of nitrogen aged specimens, E_c, was calculated according to ASTM D790M

$$E_c = \frac{0.17L^3m}{wt^3} \tag{2}$$

where L is the support span, w is the specimen width, t is the specimen thickness, and m is the slope of the tangent to the initial straight-line portion of the load-deflection curve. For the specimens tested, the load-deflection curves were very linear.

The modulus of the oxidized surface layer, E_s, was determined by using basic beam theory

$$E_s = \frac{1}{I_s}\left(\frac{L^3m}{96} - E_cI_c\right) \tag{3}$$

where L is the support span, m is the slope of the tangent to the initial straight line portion of the load deflection curve, and:

$$I_c = \frac{(w - 2t_s)(t - 2t_s)^3}{12} \tag{4}$$

$$I_s = \frac{wt^3}{12} - I_c \tag{5}$$

where t is the total thickness, and t_s is the thickness of the surface layer.

Thermo-Mechanical Analysis of Nitrogen Specimens

The CTE of the unoxidized core material was determined from thermo-mechanical analysis of nitrogen aged specimens. The TMA testing machine determined the slope of dimensional change versus temperature. From there calculating the CTE was simply

$$\alpha_c = \frac{m_{CTE}}{t} \tag{6}$$

where α_c is the coefficient of thermal expansion and m_{CTE} is the slope of dimension change versus temperature.

Surface Layer Thickness

Surface layer thickness was determined by optical microscope measurement of prepared cross sections. Calculations were only to convert scale.

Curvature Tests

The curvature tests are based on a model of a bimaterial beam (Fig. 1). Since the beam is unrestrained, the moment balance is of the form

$$M = 0 = \int \sigma z \, dz \tag{7}$$

where the stress includes a thermal expansion term and a shrinkage term

$$\sigma = E(\varepsilon - \alpha \Delta T - \varepsilon_s) \tag{8}$$

Using Bernoulli-Euler beam theory, the strain term can be expressed as

$$\varepsilon = kz + \varepsilon_0 \tag{9}$$

where k is the curvature and ε_0 is the strain at the neutral axis. Thus stress can be expressed as

$$\sigma = E(kz + \varepsilon_0 - \alpha \Delta T - \varepsilon_s) \tag{10}$$

and the moment is then of the form

$$M = \int E(kz + \varepsilon_0 - \alpha\Delta T - \varepsilon_s)z\,dz \qquad (11)$$

It would be desirable to transfer coordinates to the neutral axis before integrating. The location of the neutral axis (Fig. 5) is

$$z^* = \frac{E_s\left(t_c + \dfrac{t_s}{2}\right)t_s + E_c\left(\dfrac{t_c}{2}\right)t_c}{E_s t_s + E_c t_c} \qquad (12)$$

where t_s is the thickness of the surface layer and t_c is the thickness of the unoxidized core material. Establishing the new coordinate system we have

$$\zeta_1 = t_s + t_c - z^*$$

$$\zeta_2 = t_c - z^* \qquad (13)$$

$$\zeta_3 = -z^*$$

Thus the moment balance equation becomes

$$M = 0 = \int_{\zeta_2}^{\zeta_1} E_s(k\zeta + \varepsilon_0 - \alpha_s\Delta T - \varepsilon_{ss})\zeta\,d\zeta + \int_{\zeta_3}^{\zeta_2} E_c(k\zeta + \varepsilon_0 - \alpha_c\Delta T - \varepsilon_{sc})\zeta\,d\zeta \qquad (14)$$

where α_s and α_c are the surface layer and core CTE's respectively, and ε_{ss} and ε_{sc} are the surface layer and core shrinkage strains. Curvature k can be determined from measurements as

$$k = \frac{8h}{s^2} \qquad (15)$$

where h is the height of the arc and s is the span. Plotting values of k versus temperature and fitting a line to the data produces a graph (Fig. 6), where k' is the slope of k versus T and k'' is the curvature at the cure temperature of 316°C, where ΔT is zero.

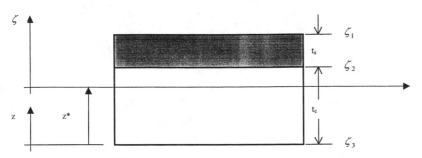

FIG. 5—*Neutral axis and coordinate system.*

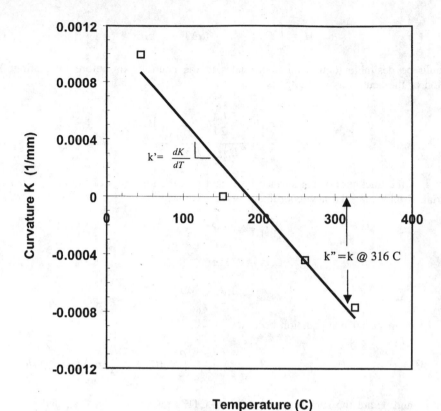

FIG. 6—*Sample graph of curvature versus temperature, showing* k' *and* k''.

Thus k can be expressed as

$$k = k'\Delta T + k'' \tag{16}$$

Integrating and simplifying the moment balance equation leads to

$$M = 0 = \tfrac{1}{3} aE_s k + \tfrac{1}{3} bE_c k - \tfrac{1}{2} cE_s \varepsilon_{ss} - \tfrac{1}{2} dE_c \varepsilon_{sc} - [\tfrac{1}{2} cE_s \alpha_s + \tfrac{1}{2} dE_c \alpha_c]\Delta T \tag{17}$$

where

$$a = \zeta_1^3 - \zeta_2^3$$

$$b = \zeta_2^3 - \zeta_3^3$$

$$c = \zeta_1^2 - \zeta_2^2 \tag{18}$$

$$d = \zeta_2^2 - \zeta_3^2$$

and, as a consequence of the choice of axis

$$cE_s = -dE_c \qquad (19)$$

For $\Delta T = 0$, we can find

$$M = 0 = \tfrac{1}{3} aE_s k'' + \tfrac{1}{3} bE_c k'' - \tfrac{1}{2} cE_s \varepsilon_{ss} - \tfrac{1}{2} dE_c \varepsilon_{sc} \qquad (20)$$

Solving Eq 20 for surface layer shrinkage, and using Eq 19

$$\varepsilon_{ss} = \frac{\tfrac{2}{3} aE_s k'' + \tfrac{2}{3} bE_c k''}{cE_s} + \varepsilon_{sc} \qquad (21)$$

Now for arbitrary ΔT it is possible to determine the coefficient of thermal expansion of the surface layer material. Inserting Eq 19 into Eq 17 and solving

$$\alpha_s = \frac{\tfrac{2}{3} aE_s k + \tfrac{2}{3} bE_c k}{cE_s \Delta T} - \frac{\varepsilon_{ss}}{\Delta T} + \frac{\varepsilon_{sc}}{\Delta T} + \alpha_c \qquad (22)$$

It is convenient to present shrinkage strain ε_s as a percent shrinkage P_s; this can be calculated simply from

$$P_s = -100\varepsilon_{ss} \qquad (23)$$

Results

Format for Plotting

Data for each specimen was reduced individually. Results from replicate specimens (typically five per condition) were then averaged, and the standard deviations were calculated. Symbols in the following plots indicate the mean, and error bars represent plus-minus one standard deviation.

Surface Layer Thickness

Surface layer thickness (Fig. 7) increases with aging duration, but the rate of growth decreases. The curve very closely resembles similar measurements made by Bowles [9].

Modulus

The measured value of the stiffness for the nitrogen aged specimens (Fig. 8) remained rather constant regardless of aging duration. The calculated value for the stiffness of the oxidized surface layer indicates a significant increase in the modulus over that of unoxidized material. However, the modulus of oxidized material also remains rather constant with aging duration.

Surface Layer Shrinkage

Shrinkage of nitrogen aged specimens and calculated shrinkage of the oxidized surface layer are shown (Fig. 9). Values of shrinkage strain were derived using Eqs 1 and 21, then

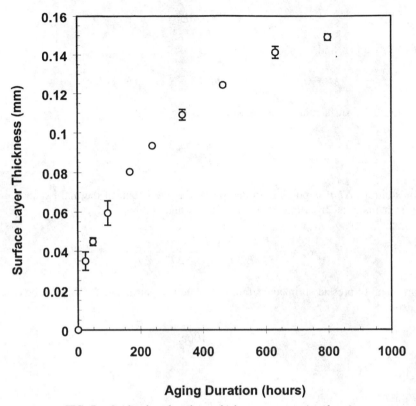

FIG. 7—*Oxidized surface layer thickness versus aging duration.*

converted to percent shrinkage for plotting using Eq 23. The plots indicate that oxidized surface layer shrinkage is significantly greater than that of the inner unoxidized material. Shrinkage increases as aging duration increases although the unoxidized nitrogen aged specimen shrinkage appears to approach a limit as aging duration increases. Oxidized surface layer shrinkage does not have a similar limiting behavior.

Coefficient of Thermal Expansion

Thermo-mechanical testing was used to determine CTE of nitrogen aged specimens. The CTE of the oxidized surface layer was derived from test results using Eq 22. A plot of CTE versus aging duration is shown (Fig. 10) for nitrogen aged specimens and for the calculated value of the oxidized surface layer. The data indicates that the CTE of material aged in nitrogen decreases slightly as time increases. The calculated CTE of the oxidized surface layer decreases more markedly.

Discussion and Conclusion

This set of experiments successfully provides separate properties for the surface layer and the inner material of aged PMR-15 resin. These properties will be valuable for both greater

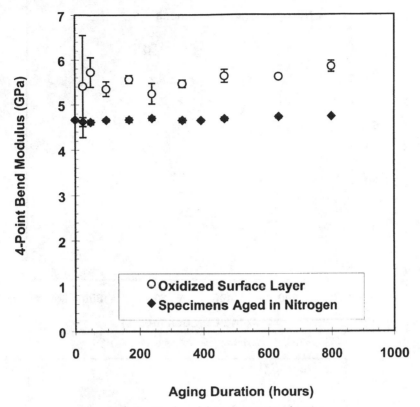

FIG. 8—*Bend modulus versus aging duration.*

understanding of aged material behavior, and also as data that can be used to complete models of thermo-oxidative degradation.

The behavior of the resin remains time-dependent throughout the 800+ hour period studied. The surface layer continues to grow throughout the period. The properties of the core material seem to stabilize after a few hundred hours. In contrast, in the surface layer, the modulus quickly stabilizes (after a substantial increase), but the CTE continues to decrease and the shrinkage continues to increase throughout the period studied. This suggests several different mechanisms are at work, with differing rates. We will not speculate here as to what these mechanisms are, except to repeat the observation that relaxation would cause measured shrinkage to decrease with time, and hence is unlikely as a primary mechanism.

The core and surface layer material prove to be notably different. This difference explains some of the complex observed behavior of thermo-oxidatively aged specimens. For example, it was observed that curvature specimens at room temperature curved towards the interior material, when it was initially assumed that the specimens would curve toward the shrunken surface layer. Curvature specimens at the aging temperature curve toward the surface layer, as expected. The changes in CTE's revealed by this set of experiments explain this behavior. The increased stiffness and shrinkage seen in the surface layer also helps to explain surface cracking observed in other investigations in air-aged specimens. The shrinkage of the surface is restrained by the core, resulting in tensile stresses, which are only enhanced by the material's increased modulus.

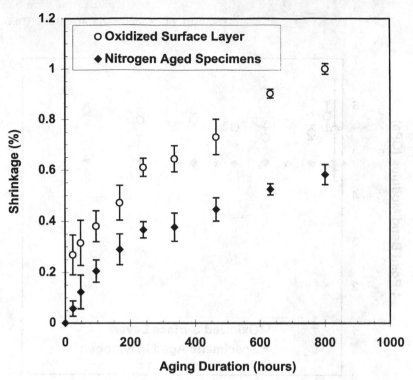

FIG. 9—*Shrinkage versus aging duration.*

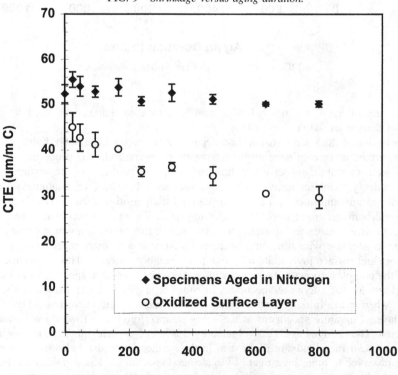

FIG. 10—*CTE versus aging duration.*

References

[1] Nam, J. D. and Seferis, J. C., "Anisotropic Thermo-oxidative Stability of Carbon Fiber Reinforced Polymeric Composites," *SAMPE Quarterly,* May 1992.

[2] Bowles, K. J. and Meyers, A., "Specimen Geometry Effects on Graphite/PMR-15 Composites During Thermo-Oxidative Aging," NASA Technical Memorandum 87204, Cleveland, OH, 1986.

[3] Bowles, K. J., Roberts, G. D., and Kamvouris, J. E., "Long-Term Isothermal Aging Effects on Carbon Fabric-Reinforced PMR-15 Composites: Compression Strength," NASA Technical Memorandum 107129, 1995.

[4] McManus, H. L. and Cunningham, R., "Coupled Materials and Mechanics Analyses of Durability Tests for High Temperature Polymer Matrix Composites," *ASTM STP 1302: High Temperature and Environmental Effects on Polymeric Composites,* 1997, pp. 1–17.

[5] Cunningham, R. and McManus, H. L., "Coupled Diffusion-Reaction Models for Predicting the Distribution of Degradation in Polymer Matrix Composites," *Symposium on Composite Materials,* ASME International Mechanical Engineering Congress and Exposition, Atlanta, GA, Nov. 1996.

[6] McManus, H. L., Foch, B., and Cunningham, R., "Mechanism-Based Modeling of Long-Term Degradation," DURACOSYS 97: Progress in Durability Analysis of Composite Systems, Blacksburg VA, Sept. 1997, and submitted to the *Journal of Composites Technology and Research,* March 1998.

[7] McManus, H. L. and Charmis, C. C., "Stress and Damage in Polymer Matrix Composite Materials Due to Material Degradation at High Temperatures," NASA Technical Memorandum 4682, 1996.

[8] Kamvouris, J. E., Roberts, G. D., Pereira, J. M., and Rabzak, C., "Physical and Chemical Aging Effects in PMR-15 Neat Resin," *ASTM Second Symposium on High Temperature and Environmental Effects on Polymeric Composites,* 1995, Norfolk, VA, preprint.

[9] Bowles, K. J., Jayne, D., and Leonhardt, T. A., "Isothermal Aging Effects on PMR-15 Resin," *SAMPE Quarterly,* Vol. 24, No. 2, 1993, pp. 2–9.

[10] Cunningham, R. A., Massachusetts Institute of Technology, S. M. Thesis, 1997, pp. 138–146.

Gregory B. McKenna[1] and Sindee L. Simon[2]

Time Dependent Volume and Enthalpy Responses in Polymers

REFERENCE: McKenna, G. B. and Simon, S. L., **"Time Dependent Volume and Enthalpy Responses in Polymers,"** *Time Dependent and Nonlinear Effects in Polymers and Composites, ASTM STP 1357,* R. A. Schapery and C. T. Sun, Eds., American Society for Testing and Materials, West Conshohocken, PA, 2000, pp. 18–46.

ABSTRACT: Most users of polymeric materials have a good sense that the glass transition event is kinetic in nature, i.e., it depends on cooling or heating rate in conventional experiments, and that the glassy state is a non-equilibrium state. However, it is often not appreciated that the structural recovery which occurs as a glass attempts to reach equilibrium is non-linear (e.g., the rate of volume recovery depends on the instantaneous volume). The non-linear viscoelastic nature of structural recovery can lead to surprising behaviors in certain kinds of measurements. It results in, for example, the asymmetry of approach in up and down-temperature jumps. The features of enthalpy and volume recovery, including the sub-T_g peaks and excess enthalpy overshoots in differential scanning calorimetry, are well-described by models of structural recovery developed in the 1970's. The purpose of the present work is to describe structural recovery and physical aging and their impacts on material performance and measurement of material properties. In addition, we present new results from calculations using the structural recovery models in which we demonstrate that new analytical tools, such as Temperature Modulated Differential Scanning Calorimetry (TMDSC), need to be used with caution when glass-forming systems are studied because of the nonlinear viscoelastic nature of structural recovery.

KEYWORDS: structural recovery, glass transition, kinetics, volume recovery, enthalpy recovery, nonlinear viscoelasticity, temperature modulated DSC, DSC, TMDSC

It is well known that upon cooling of glass forming liquids, including polymers, there is a point in temperature space at which the thermodynamic properties (volume, enthalpy, etc.) fall out of equilibrium [1]. This point is generally referred to as the glass transition temperature T_g and is shown schematically in Fig. 1 where we show the specific volume v and enthalpy H behaviors versus temperature. Importantly, there is a significant kinetic aspect to the magnitude of T_g and this is commonly manifested as a cooling rate dependence of the glass transition [2]. This is also depicted in Fig. 1 where q is the cooling rate. The impact of cooling rate on the value of T_g is generally of the order of 3 K per logarithmic decade of rate, leading to a decrease of the glass transition temperature measured on cooling as the cooling rate decreases [1–3].

[1] Professor, Department of Chemical Engineering, Texas Tech University, Lubbock, TX 79409; previously Group Leader, Structure and Mechanics Group, Polymers Division, National Institute of Standards and Technology, Gaithersburg, MD 20899.

[2] Associate Professor, Department of Chemical Engineering, Texas Tech University, Lubbock, TX 79409; previously Assistant Professor, Department of Chemical Engineering, University of Pittsburgh, Pittsburgh, PA 15261.

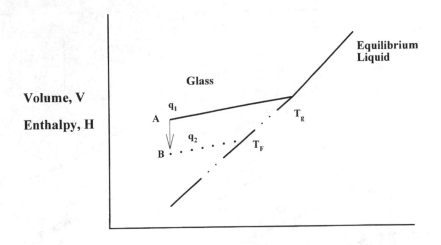

FIG. 1—*Schematic of enthalpy or volume versus temperature for glass forming materials. See text for discussion.*

Another manifestation of the kinetic features of the glass transition event is a phenomenon referred to as structural recovery [1,2,4–6]. Thus, if one cools the glass at any rate to some temperature below T_g (say temperature A in Fig. 1), then the inherently non-equilibrium state of the glass spontaneously evolves towards equilibrium (shown by the arrow in the Fig. 1). The time evolution of the glassy structure after the temperature is changed from near T_g to below it in a down-jump is shown in Fig. 2 from classic data by Kovacs [2] for volume recovery experiments. In the figure, the volume departure from equilibrium is defined as $\delta = (v - v_\infty)/v_\infty$ where v is the specific volume and v_∞ is its value in equilibrium (infinite time). The family of curves in down-jump conditions to different temperatures is referred to as the family of intrinsic isotherms. As seen in Fig. 2, the rate of volume recovery is a strong function of temperature, i.e., the time to reach equilibrium ($\delta = 0$) increases as the temperature decreases.

Another important definition in studying glasses was initially developed by Tool [7,8] and used frequently in the description of glassy kinetics as discussed subsequently. This is the fictive temperature T_F. Referring again to Fig. 1, T_F is defined as the point of intersection of a line extrapolated from a point (B) in volume (or enthalpy) space to the equilibrium (liquid) line along a line parallel to the glassy line. Hence, along the glass line in Fig. 1, $T_F = T_g$ and along the equilibrium line, $T_F = T$. Obviously T_F and δ are related through the differences in the liquid and glassy slopes and have subtly different meanings when one thinks of the structural state of the glass. However, the formalisms developed to describe the kinetics of glasses using either T_F or δ are mathematically equivalent and give the same results, though the specific parameters used in the relevant equations may have different values that can be readily calculated one set from the other.

The volume and enthalpy responses of glasses are important from a practical view because the changing 'thermodynamic state' of the non-equilibrium glass impacts the mechanical response of the polymer in a process referred to as physical aging [9,10]. In this instance, it is observed that after a down-jump from above T_g to below it, the mechanical properties

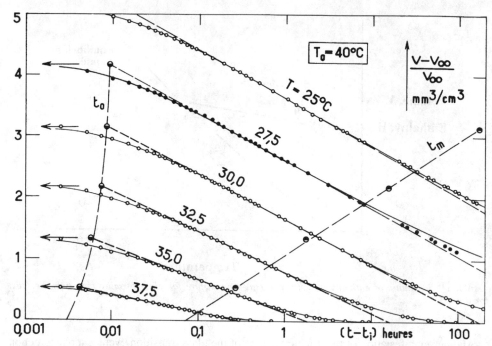

FIG. 2—*Intrinsic isotherms for poly(vinyl acetate) in down-jump experiments from* $T_0 = 40°C$ *to the temperatures indicated in figure. (Data from Ref 2, figure courtesy of A. J. Kovacs). See text for discussion.*

evolve in such a way that, for example, the yield strength, relaxation modulus, and material brittleness increase. In addition, the evolution of the material volume can impact residual stress development. Finally, the thermal properties are affected by the structural recovery and the non-linearities of structural recovery can even have an influence on the way in which one should interpret thermal measurements [11,12]. Hence, an understanding of structural recovery and physical aging phenomena is important [13].

Here we review the phenomenology of structural recovery and the equations used to describe, not only the cooling rate effects on T_g and the intrinsic isotherms, but also the asymmetry of approach, memory (or cross-over) effects. In addition, we show how the equations successfully describe non-isothermal behaviors in thermal measurements typical of a classical differential scanning calorimeter. We then describe the physical aging phenomenon for engineering plastics in both the small and large deformation regimes for the viscoelastic response, its impact on engineering properties, and the effects of structural recovery on residual stresses. Finally, we discuss the impact of structural recovery on temperature modulated differential scanning calorimetry (TMDSC) measurements and interpretation.

Phenomenology of Structural Recovery

As noted above, there is a significant impact of thermal history, even simple thermal histories, on the measured properties of glass forming materials, including polymers. While the rate effects on the glass transition temperature and the observation that the volume evolves after a down-jump in temperature (intrinsic isotherms) are well documented, the behavior of glasses is more complex and richer than evidenced in these experiments. The general richness of behaviors is seen in classic volume dilatometry experiments performed by Kovacs [2] on a poly(vinyl acetate) polymer. His findings are widely recognized as general for both polymers and other glass forming materials [14]. We discuss two sets of his experiments: asymmetry of approach and memory (cross-over) experiments. From these, we define some "essential ingredients" for the description of the general phenomenology of glass forming systems and write the equations that were developed through a series of advances beginning in the 1940's. The strengths and limitations of the equations are discussed and two practical examples of successful analysis using the equations presented: structural recovery in DSC experiments and residual stress build-up in a glass to metal seal.

The Asymmetry of Approach Experiment

The asymmetry of approach experiment is an important one because it demonstrates the highly non-linear response of glass forming materials. While we described it in terms of the volume recovery here, it is important to note that the same sort of response would be seen if one were to measure the enthalpy recovery. The experiment involves performing both down-jump and up-jump experiments to the same final temperature from a condition in which the glass had previously been equilibrated. Then, if the magnitude of the temperature-jump is the same for both conditions, a linear response would show that the up-jump is the mirror image of the down-jump. On the other hand, as shown in Fig. 3, the Kovacs results from experiments for 5 K T-jumps demonstrate that the responses are far from mirror images and,

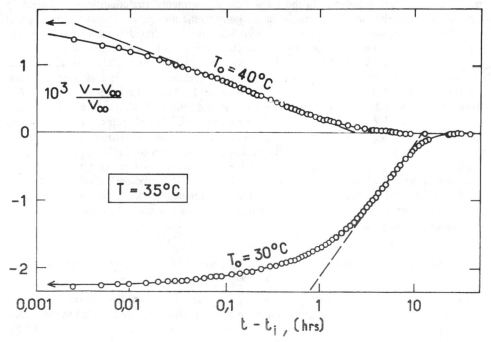

FIG. 3—*Results from an asymmetry of approach experiment for poly(vinyl acetate) at a final temperature of* T = 35°C *for initial temperatures of 30 and 40°C as indicated in figure. (Data from Ref 2. Figure courtesy of A. J. Kovacs). See text for discussion.*

in fact, the up-jump response is observed to begin at very large volume departures δ from equilibrium and exhibits an apparent equilibrium time that appears significantly longer than that obtained in the down-jump experiment. The explanation provided for this was first discussed by Tool [7–8]. The essential argument is that the molecular mobility depends on both the temperature and the thermodynamic state or structure of the glass. Hence, in the down-jump experiment the structure shows a positive departure from equilibrium which decreases with time giving rise to a progressively slowing process—what Kovacs [2] referred to as "autoretarded." On the other hand, in the up-jump experiment, one begins with large negative departures from equilibrium and the mobility increases with increasing time. Hence, one sees an apparently "autocatalytic" response on the logarithmic time scale used in these experiments. As presented subsequently, inclusion of a material time that depends on the departure from equilibrium leads to the strong non-linearity seen in Fig. 3. It is worthwhile to note here that the non-linearity is very strong even though the volume strain (departure from equilibrium δ) is very small, being on the order of 1 to 2×10^{-3}. Hence, our first essential ingredient is a "material clock" or time scale that depends on the instantaneous structure of the glass, as well as on temperature.

The Memory or Cross-Over Experiment

The other experiment that provides important information concerning the nature of the structural recovery response of glass forming materials is the memory or cross-over experiment. The experiment involves a two-step thermal history. First, one does a down-jump to some temperature below the glass transition and allows the structure to recover partially towards equilibrium. The temperature is then increased to T_o after the partial annealing and the volume recovery measured. In the Kovacs [2] experiments, the conditions of annealing were chosen such that upon performing the up-jump to the final test temperature, the volume departure from equilibrium was close to zero (this implies that the partial recovery was performed until the fictive temperature T_F reaches the temperature T_o after the second T-jump.) As shown in Fig. 4, the value of δ starts close to zero (or slightly below) and then increases (crosses over zero) and goes through a maximum before merging with the structural recovery result obtained for a single-step history to T_o. From the time when Tool [7,8] originally observed the cross-over phenomenon and until the 1970's the description of structural recovery was largely considered within the framework of single relaxation time models. Although the asymmetry of approach experiments can be treated qualitatively by a single time that is changing due to the changing structure of the material, the memory experiment demonstrates conclusively that one needs to have multiple (at least two) relaxation times to described structural recovery. Hence, the second "essential ingredient" needed for the description of the structural recovery process is a non-exponential decay function that can be considered either in terms of a sum of exponentials or as a stretched exponential function as discussed subsequently.

The Tool [7,8]-Narayanaswamy [15]-Moynihan [4]-Kovacs [16]-Aklonis-Hutchinson-Ramos (TNM-KAHR) Description of Structural Recovery

The final piece of the puzzle that allows one to very successfully describe the major features of structural recovery is the building of a set of equations that related the stimulus, i.e., temperature history, to the response, i.e., the volume or enthalpy recovery. The TNM-KAHR models do this by postulating that the behavior can be described using a viscoelastic constitutive equation that looks like linear viscoelasticity in reduced time. The equations

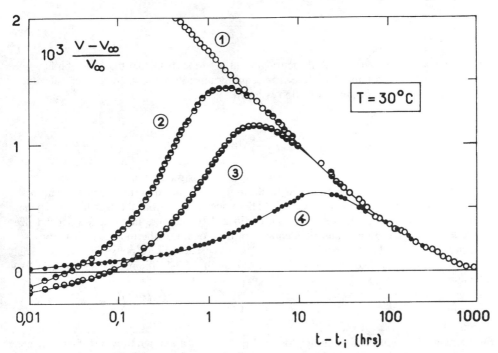

FIG. 4—*Results from Kovacs' memory experiments on a poly(vinyl acetate) glass. The experiments are performed by treating the sample such that there are two temperature steps. First, one quenches the glass from 40°C to an annealing temperature* T_a *and the material is allowed to anneal (recover) partially towards equilibrium for an annealing (aging) time* t_a *before being up-quenched to the final test temperature* T = 30°C. *In the figure, the curve numbers correspond to different thermal treatments: 1)* T_a = 30°C; t_a = 0; 2) T_a = 10°C; t_a = 160 h; 3) T_a = 15°C; t_a = 140 h; 4) T_a = 25°C; t_a = 90 h. (*Data from Ref 2. Figure courtesy of A. J. Kovacs.*).

developed by TNM and KAHR are slightly different in conception, but are mathematically equivalent. Here we follow the development according to KAHR and point out the TNM equivalencies and differences. We also note that, in addition to the "essential ingredients" discussed above, the commonly used forms of the TNM and KAHR models assume that the viscoelastic response follows thermo-rheological simplicity in both temperature and structure, i.e., the molecular mobility depends on both temperature and structure simply by a shifting of the time scale or reduced time. The shape of the response function is unaltered.

The essence of the KAHR model is the following equation

$$\delta(z) = -\Delta\alpha \int_0^z R(z - z') \frac{dT}{dz'} \, dz' \tag{1}$$

where, for volume recovery experiments, $\Delta\alpha$ is the change in coefficient of thermal expansion at the glass transition (the difference between the liquid and glassy coefficients $\alpha_1 - \alpha_g$), T is temperature, $\delta(z)$ is the departure from equilibrium, $R(z)$ is the viscoelastic response function (a retardation function, but common usage has come to use the term relaxation to describe the structural recovery or relaxation response), and z is the reduced time

$$z = \int_0^t \frac{d\xi}{a_T a_\delta} \tag{2}$$

ξ is a dummy variable and the shift factors a_T and a_δ are shift factors that describe how the temperature and the volume departure from equilibrium affect the characteristic relaxation times. The viscoelastic response function is given by

$$R(z) = \sum_{i=1}^N g_{i,r} e^{-z/\tau_{i,r}} \tag{3}$$

Where the N different $\tau_{i,r}$ values indicate a spectrum of retardation times and the $g_{i,r}$ are weighting factors (both are determined in the reference state described below). The non-linearity of the material response in the KAHR model comes from the fact that δ depends on itself in Eq 1 through the reduced time of Eq 2. If there were no structure dependence of z ($a_\delta = 1$) the equation would be identical to those of linear thermo-viscoelasticity.

In the case of the TNM version of the model, the viscoelastic response function is given in terms of the stretched exponential or Kohlrausch [17]-Williams-Watts [18] (KWW) function

$$R(z) = e^{-(z/\tau_{0,r})^\beta} \tag{4}$$

Furthermore, the reduced time is now defined in terms of a temperature shift factor a_T and a structure shift factor a_{T_F} which depends on the fictive temperature T_F rather than on the departure from equilibrium

$$z = \int_0^t \frac{d\xi}{a_T a_{T_F}} \tag{5}$$

The two formalisms are equivalent when the exponential product form for the shift factors given below is used, but there may be subtle differences between the physical interpretations of the parameters. If one speaks of the structure as δ, then the isostructural state is a line approximately parallel to the equilibrium volume or enthalpy (recall Fig. 1). If one assumes that the fictive temperature T_F defines structure, the iso-structural state is defined along lines with a slope equal to the glassy coefficient of thermal expansion. We do not elaborate further on this point.

In the KAHR model the shift factors a_T and a_δ are estimated to be

$$\frac{\tau_i(T,\delta)}{\tau_{i,r}} = a_T a_\delta = e^{-\theta(T-T_r)} e^{-(1-x)\theta\delta/\Delta\alpha} \tag{6}$$

where the first exponential term is a_T and the second is a_δ. $\tau_i(T,\delta)$ is the relaxation time at the relevant values of temperature and structure and the $\tau_{i,r}$ refers to the relaxation time at the reference state, generally taken for $T_r = T_g$ and $\delta = 0$. The parameter x is a partition parameter $0 \leq x \leq 1$ that determines the relative importance of temperature and structure on the relaxation times. The parameter θ is a material constant that characterizes the temperature dependence of the relaxation times in equilibrium. KAHR used $\theta \approx E_a/RT_g^2$ where E_a is an activation energy and R is the gas constant.

In the case of the TNM model the shift factors a_T and a_{T_F} are given as

$$\frac{\tau_0(T,T_F)}{\tau_{0,r}} = a_T a_{T_F} = e^{x\Delta h/R(1/T - 1/T_r)} e^{(1-x)\Delta h/R(1/T_F - 1/T_r)} \tag{7}$$

where now τ_0 is the characteristic time in the KWW expression (Eq 4), the arguments T and T_F are the temperature and fictive temperature respectively. $\tau_{0,r}$ refers to the relaxation time at the reference state, generally taken for $T_r = T_F = T_g$. Δh is an activation energy. Clearly the parameters in Eqs 6 and 7 can be related, yet have subtly different meanings. In addition to the original references, the reader is referred to the review by Hodge [6] for further discussion of the interrelationships between the material parameters for the two approaches.

An important aspect of the models discussed above is that the equations apply to both volume and enthalpy experiments. For enthalpy one substitutes ΔC_p for $\Delta \alpha$ in Eqs 1 and 6. It is worth noting, as well, that the response function $R(z)$ need not be the same for enthalpy and volume. In addition, we note that the solution of the equations in differential form may be easier for many histories. Then, in terms of the fictive temperature based TNM model, we write

$$\frac{dT_F}{dT} = 1 - \exp\left[-\left(\int_0^t \frac{dt}{\tau_0}\right)^\beta\right] \tag{8}$$

and evidently, for any arbitrary thermal history starting in equilibrium at $T = T_r$ we solve Eqs 7 and 8 simultaneously using numerical procedures

$$T_{F,n} = T_r + \sum_{i=1}^n \Delta T_i \left[1 - \exp\left(-\left(\sum_{j=1}^n \frac{\Delta t_j}{\tau_{0,j}}\right)^\beta\right)\right] \tag{9}$$

where $T_{f,n}$ is the fictive temperature after the nth temperature step, Δt_i is the time step, ΔT_i is the temperature step associated with the ith time step and $\tau_{0,j}$ is calculated from Eq 7 using known material parameters.

Strengths and Weaknesses of the Models

The TNM-KAHR frameworks are very powerful and reasonably straight-forward to use. A good example of the success of the models is seen in Figs. 5 and 6 in which the asymmetry of approach and memory experiments are calculated from the KAHR model. In Fig. 5, the asymmetry data for jumps to 140°C for a polycarbonate material are compared with KAHR model calculations. The model parameters were determined [19] from an optimization to data for experiments at 130, 135, and 140°C. It is readily seen that the asymmetry of approach is in reasonable quantitative agreement with the data. Figure 6 compares Kovacs' original memory data with the calculations [20] based on the KAHR model in which the parameters were obtained from data for asymmetry of approach experiments at 30, 35, and 40°C (data not shown here). It can be seen that the model does a reasonable job of reproducing the memory effects. Similarly, if one performs enthalpy recovery experiments followed by a heating scan, as is typically done in DSC measurements, we find that the models can adequately describe the data. A comparison [21] of model predictions with typical DSC data for heating after an aging treatment is shown in Fig. 7. The heating scans themselves are interesting because, for certain thermal treatments, they show both a sub-glass transition peak and a large excess enthalpy peak near to T_g. The models reproduce both features.

Another example of the power of the TNM-KAHR formalisms arises in a specific example of the development of residual stresses in an electronic package. Here, the glass to metal

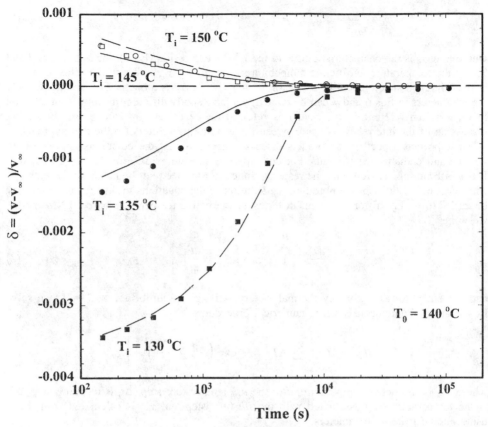

FIG. 5—*Plot of asymmetry of approach experiments in polycarbonate material aged into equilibrium at temperatures* T_i *and then jumped to a final test temperature of* $T_0 = 140°C$. *Points are for experimental data and lines represent KAHR-model description of the data. See text for discussion. (Data and fits from Ref 19).*

seals used in an application for the Department of Energy demanded that the glass that hermetically seals the metallic core be under a residual compressive stress, therefore preventing cracking in long term applications. In the case reported by Chambers [22], tensile cracking was occurring in the seal in spite of the thermo-elastic prediction that the glass should be under a residual compression. At the time, Chambers implemented a finite element analysis that took into account the structural recovery of the glass seal during the cooling stage of the seal formation, which encompassed the glass transition of the seal. For the analysis, the structural recovery was modeled using the TNM formalism discussed above. The nonlinear material response that results upon cooling through the glass transition range was found to cause a large residual tensile stress in the glass at the temperature at which cracking was observed to occur. The residual stress build-up that occurs during cooling, as calculated by Chambers, is shown in Fig. 8. The important point here is that the lack of understanding of the structural recovery that tells us that the volume changes are more complex than those given by a simple thermo-elastic analysis led to a failed part. Similar effects could be expected in any two-component material system. In the case of the Chambers [22] analysis it was possible to predict different processing conditions (cooling conditions)

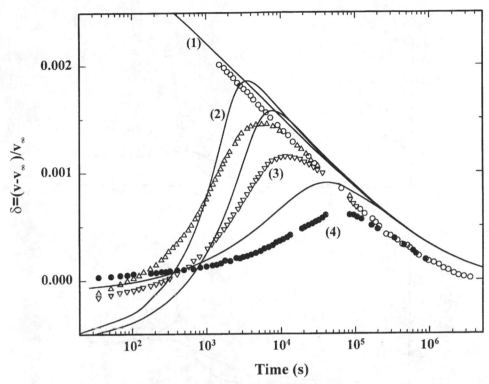

FIG. 6—*Comparison of KAHR-model calculations (lines) with the Kovacs memory experiment data depicted in Fig. 4. Note that the time axis is for time after the beginning of the quench rather than the end as in Fig. 4. Curve numbers have the same meaning as in Fig. 4. (Calculations after C. R. Schultheisz, Ref 20).*

that gave a significant reduction in the magnitude of the tensile residual stress and successful fabrication of the glass-to-metal seal.

In summary, to a very good first approximation, it is clear that the models can reproduce the major phenomena that are observed for the kinetics of structural recovery and discussed in the above paragraphs. However, there is much work that has been performed over the past two decades that shows that the TNM-KAHR models are somewhat deficient and require further refinement. What are the weaknesses? First, it is now reasonably well accepted that the models are not capable of describing experiments outside of a narrow range of temperatures without changing the material parameters [23]. This fact has lead to discussion of the possible problems with the models. One suggestion is that the assumption of thermo-rheological-structural simplicity is incorrect because there seems to be a systematic trend in the value of the KWW β-parameter with changing temperature [23–25]. There has also been considerable effort put into development of somewhat different constitutive equations.

Another weakness of the models is their inability to describe deep quenches or long aging times. Furthermore, the so-called τ-effective paradox [2] and expansion gap [2] are not predicted by the current models [1,26]. These weaknesses need further research, and are often controversial in and of themselves [27,28]. Hence, while the weaknesses of the models are pretty well understood, the successful development of the next generation of models that overcomes these weaknesses is yet to occur. At this point, it can only be said that we have

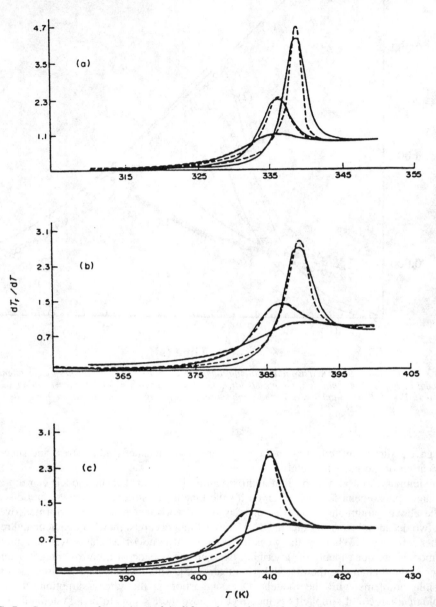

FIG. 7—*Comparison of experimental (solid) and calculated (dashed) enthalpy recovery curves for: a) isotactic PMMA, b) atactic PMMA, and c) syndiotactic PMMA. The annealing was performed approximately 15 K below* T_g*. The curves represent heating experiments after aging of the samples for times of 0, 6000 and 60 000 s, with the peak height increasing as the annealing time increases. (Reprinted with permission of the American Chemical Society©, 1986, Tribone, J. J., O'Reilly, J. M., and Greener, J. "Analysis of Enthalpy Relaxation of Poly(methyl Methacrylate)—Effects of Tacticity, Deuteration, and Thermal History," Macromolecules, Ref 21.) See text for discussion.*

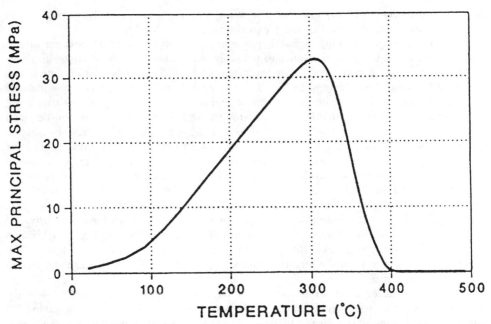

FIG. 8—*Finite element model predictions of stress history upon cooling in a glass-metal seal. Tensile stress arises due to structural recovery effects. Thermo-elastic analysis resulted in a compressive stress of −6.7 MPa. (Reprinted with permission of the American Ceramic Society©, 1989, Chambers, R. S., Gerstle, F. P., and Monroe, S. L. (1989). "Viscoelastic Effects in a Phosphate Glass-metal Seal," Journal of The American Ceramic Society. Ref 22.)*

a good first-order handle on the problem of structural recovery in glass forming systems, including polymers, and the TNM-KAHR approaches have been very successful. Their successors have yet to be developed and accepted.

Physical Aging

Linear Viscoelastic Regime

In the above sections, we have discussed structural recovery and the equations currently used to describe it to a very good approximation. Structural recovery is important to understand, not only because it involves changes in material volume and enthalpy, but also because these changes are accompanied by changes in, e.g., the mechanical properties of the glass forming materials. While some of the changes have been understood since the early 1960's [29] the major impetus given to the study of physical aging came as a result of the work of Struik [9] in the 1970's. At that time, he baptized the changes "Physical Aging" and, in an extensive set of experiments, showed that the viscoelastic response of glassy polymers could be described by a time-aging time superposition principle similar to the time-temperature superposition principle. Hence, much as the structural recovery could be described with a set of equations in which the molecular mobility depends on the structure or fictive temperature of the glass, the viscoelastic properties could be described similarly. It is important to recognize, however, that much of the physical aging work has addressed the aging of materials in down-jump experiments, i.e., as intrinsic isotherms. In more complicated thermal histories, we expect to observe behavior that would vary in a fashion similar to that seen in the asymmetry of approach and memory experiments.

Referring once again to Figs. 1 and 2, physical aging can be expected to occur after a quench from above the glass transition to below it. Associated with the changes in glassy structure, one would expect to find changes in the mechanical properties. Struik introduced a method to probe these changes. In Fig. 9, we show one such method where the material is subjected to a small (in the linear viscoelastic regime) stress or strain probe in a sequence of load-unload events. The important part of the probing is that the perturbation be small and that its duration be short relative to the total aging time after the quench. In general, one asks that the loading time t_i be of the order of 0.10 times the aging time t_e after the quench ($t_i/t_e \lesssim 0.10$). This reduces the magnitude of changes in material response due to structural recovery during the loading and makes the measurement a sort of "snapshot" of the material behavior. Furthermore, it is convenient to carry out the sequence so that the sample is probed at aging times that double with increasing probe number: $t_{e,i} = 2t_{e,i+i}$, e.g., $t_e = 1$ h, 2 h, 4 h, . . .). A typical set of responses is shown in Fig. 10 for a polycarbonate sample in torsional stress relaxation measurements [30]. As seen in the figure, the relaxation response moves to longer times as the aging time increases. The master curve is shown offset an arbitrary amount to show the quality of the superposition. Similar results are obtained in creep or dynamic measurements [9].

Struik [9] introduced an important measure of the aging behavior that is referred to as the aging time shift rate μ which is defined as the slope of a plot of log (a_{te}) versus log (t_e) where a_{te} is the aging time shift factor required to superimpose two modulus or compliance curves obtained at different aging times. A typical plot of log (a_{te}) versus log (t_e) is shown in Fig. 11, again for polycarbonate. There are two things to note from the figure. First, the slope is a function of temperature, decreasing to zero near to the glass transition temperature. Second, at long times the aging rate approaches zero. Given the observed volumetric behavior this is not a surprise. One would expect the volume or enthalpy equilibration to be accompanied by equilibration of the mechanical response. The exact relations among these equilibration times are currently the subject of controversy and the reader is referred to the work by Simon, et al. [31] where this is discussed in detail and for the original references.

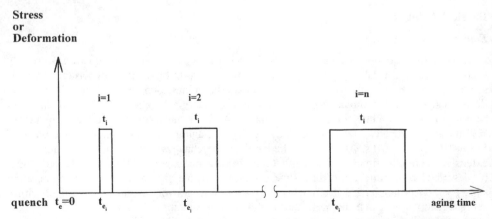

FIG. 9—*Schematic of sequential loading history applied to samples after quench. Note that the* t_{ei} *represent increasing aging times where* $t_{ei+1} \approx 2t_{ei}$. *The* t_i *are the loading times and* $t_i/t_{ei} < 0.10$.

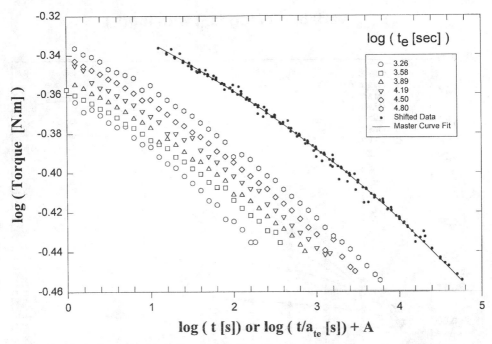

FIG. 10—*Typical aging behavior for a polycarbonate quenched from above the glass transition to a temperature of 70°C where aging begins. The tests were conducted at a torsional strain of 4.5%. The logarithms of the aging times t_e are depicted in the legend. The reduced or master curve is shifted one decade for clarity. (After Ref 30.)*

The complications of relating the material response to the volume or enthalpy departures from equilibrium alluded to in the previous paragraph make it difficult to give a firm model for the viscoelastic response: does it depend on the volume, the enthalpy or some other internal ordering parameter? If the volume departure from equilibrium is the determining factor, then the viscoelastic response would depend on the reduced time through the structure shift factor a_δ and a_{te} should depend on a_δ in a straight-forward manner. Struik [32,33] makes strong arguments that this is indeed the case. Yet there is other work by McKenna and co-workers [34–37], Delin and co-workers [38] that dispute the findings of Struik. Here we refer the reader to the original references, but accept that, to a first approximation, one could follow Struik and the equations relating the viscoelastic response to the thermal and mechanical history would be

$$\sigma(z) = \int_0^z E(z - z') \frac{d\epsilon}{dz'}\, dz' \tag{10}$$

and the reduced time z is as defined in Eq 2 above, $\sigma(z)$ is the stress, $\epsilon(z)$ is the strain and $E(z)$ is the relaxation modulus. When one only has aging time information, Eq 2 can be modified by substituting a_{te} for a_δ. Obviously, Eqs 1 and 2 must be used for complex thermal histories since δ can even vary non-monotonically, as in the instance of the memory experiment.

variation of log (a$_{te}$) with log(t$_e$)

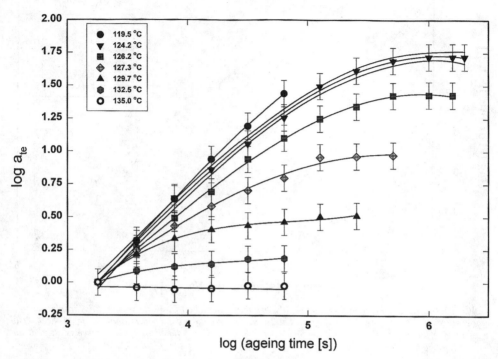

FIG. 11—*Typical behavior for the aging time shift factor versus aging time for a polycarbonate tested at 2% torsional strain and for different temperatures. See text for discussion. (Unpublished data).*

Nonlinear Viscoelastic Regime

Although the small strain behavior in physical aging is reasonably well understood, the larger strain behavior is less well so. However, there are certain features of the phenomenology upon which there is general agreement. It is in the interpretation of the results that one finds controversy. Here, we outline the experimental observations and briefly discuss the possible interpretations providing original references for the reader interested in a more in-depth look at the subject.

If one performs physical aging experiments using the sequential loading procedure discussed above and shown in Fig. 9, one observes that the value of μ decreases as the magnitude of the applied stress or deformation of the probe increases. Such behavior is shown in Fig. 12. Importantly, similar behavior is seen when individual tests, rather than sequential tests, are run for each aging time, although the response may be different in detail. This is because at large deformations, the time between individual probe steps becomes an important parameter as a large probe can affect the subsequent probe for very long times. In any event, it is clear that the apparent rate of aging is affected by the magnitude of the probe stress or strain. Struik [9] originally attributed this response to an erasure of aging by the large mechanical deformations. However, subsequent work by McKenna and co-workers [34–37] using the NIST torsional dilatometer [39] demonstrated that the volumetric response was only temporarily affected by the large deformations, hence suggesting a decoupling of the

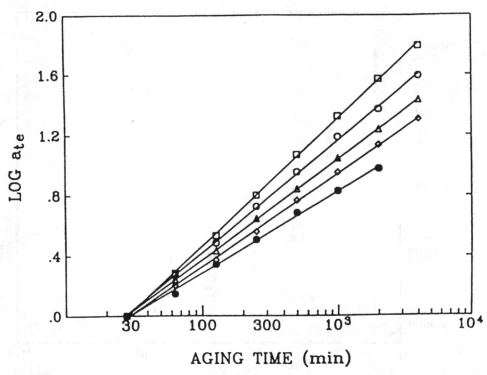

FIG. 12—*Double logarithmic representation of the aging time shift factor* a_{te} *versus aging time* t_e *for an epoxy glass aged at* $T_g - 13.2$ *K. Symbols represent results from tests in which aging response was "probed" at different levels of applied stress:* (□) *1 MPa;* (○) *5 MPa;* (△) *10 MPa;* (◇) *15 MPa;* (●) *20 MPa. (After Ref 44.) See text for discussion.*

volumetric response from the mechanical stimuli. This is shown in Fig. 13. As an additional possibility Cama, Myers, and Sternstein [40] argued that the mechanical stimuli actually accelerate the aging.

Without going into details of the responses here, suffice it to say that the impact of aging on the nonlinear viscoelastic response of a polymer glass appears to be less than the impact on the linear response. At the same time, the reader is referred to works by Struik [9,32,33], Waldron, Santore, and McKenna [41], McKenna and Zapas [42,43], Lee and McKenna [44], Yee, et al. [45], Boyce and co-workers [46,47], Oleinik [48], Ricco and Smith [49] for some insight into the range of issues surrounding the problem of interpreting the various experiments.

Engineering Properties

There is relatively little work performed to examine the impact of structural recovery on engineering properties. Several such properties that have been studied are yield behavior, static fatigue (creep rupture), craze initiation, and fracture toughness. The first engineering property that we consider here is yield. It is well known [9,50] that the yield stress increases with increasing aging time. An example [51] of how dramatic the increase can be is provided in Fig. 14 where the compressive stress-strain curves for an epoxy are shown for aging times

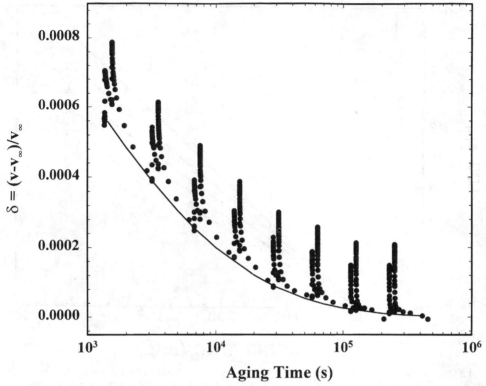

FIG. 13—*Volume departure from equilibrium versus aging time for an epoxy glass at* $T_g - 8.9°C$ *after a down-jump from above* T_g. *Points are data obtained for a torsional "probe" strain of 3% and solid line represents data obtained for a nearly undeformed sample (strain = 0.25%). (Data from Ref 41.) See text for discussion.*

ranging from a quarter hour to over 1100 h. Clearly, the yield stress can be seen to increase dramatically as aging progresses. This is true near to the glass transition. However, further from the glass transition the impact of aging on the yield stress is less important, though still present. The reason for this may arise because the yield stress, in a simple approximation depends on the modulus $E(t)$ times the strain at yield. Near to the glass transition $E(t)$ varies in magnitude in addition to shifting along the time scale. As one moves far below the glass transition, even though the time scale is still shifting by some amount a_{te}, the magnitude of the modulus $E(t)$ is not changing nearly as much. Hence, the aging response of the yield behavior will be somewhat different from that of the viscoelastic properties.

In addition to the creep deformation that occurs, when a polymer is subjected to a constant load, it is also found that the material will eventually rupture. This behavior is referred to as creep rupture or static fatigue (to differentiate from dynamic, oscillatory fatigue). Crissman and McKenna [52,53] showed that the lifetime in creep rupture experiments for a poly(methyl methacrylate) was affected by structural recovery. As seen in Fig. 15, the creep rupture lifetime is greater for the material that has been aged at room temperature for 5 years than it is for the same material that was quenched and aged for a week prior to testing. In other work, Arnold [54], found that polystyrene actually decreased in lifetime upon physical aging.

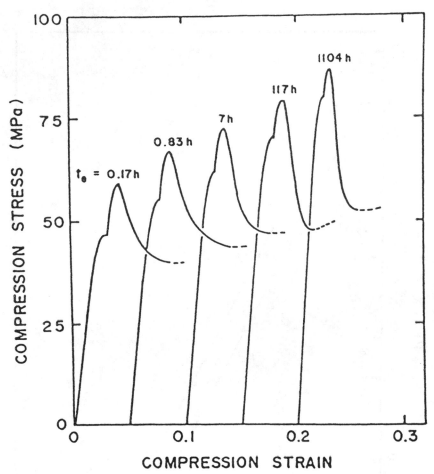

FIG. 14—*Typical compression stress-strain curves for an epoxy glass aged for different amounts of time, as indicated. Test and aging temperatures were* $T_g - 10$ *K.* (*After Ref* 51.) *"Double-yield" is due to increasing strain rate at top of first yield.*

This suggests that the mechanism that determines failure may determine how aging impacts the failure behavior.

In related work, Gusler and McKenna [55] found that craze initiation could be affected by physical aging. However, these results were both material dependent (polystyrene versus a styrene acrylonitrile (SAN) copolymer) and loading condition dependent (uniaxial stress relaxation versus biaxial creep). Interestingly, Delin and McKenna [56] have found that the craze growth undergoes a dramatic change in rate upon physical aging in a SAN copolymer. The existence of the transition is strongly dependent on temperature.

Finally, it is found that physical aging can impact the fracture toughness [57,58]. Normally one finds an embrittlement, however, it is also possible to find no effect. Perhaps because fracture toughness can depend on the yield strength, the choice of temperature conditions will have a strong effect on the fracture toughness, much as it does on the yield strength itself. Also, because the fracture event is controlled by the large deformation process zone

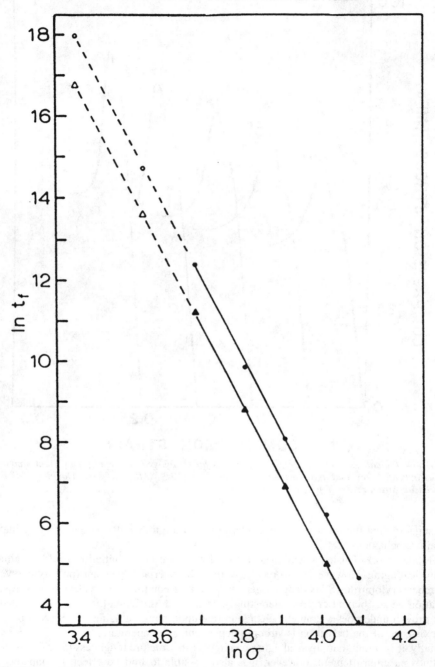

FIG. 15—*Double logarithmic representation of the creep rupture time* t_f *versus applied stress for poly(methyl methacrylate) tested at 22.5°C. Triangles represent data obtained for freshly quenched samples. Circles are for samples aged for 5 years at room temperature. (After Ref 52.)*

at the crack tip, physical aging may occur at a slower rate than it does for small strain processes as discussed above.

Impact of Structural Recovery on Temperature Modulated DSC: Measurements and Interpretation

It is clear from the above that structural recovery and physical aging are important events that can impact the performance of a material and that can lead to dimensional changes or residual stress development that are undesirable. On the other hand, it is less well appreciated that the structural recovery can also affect measurements and their interpretation. A specific example of this has to do with a new thermal analysis technique referred to as modulated or temperature-modulated DSC (TMDSC). Here we take the TNM version of the structural recovery model and show that structural recovery of the enthalpy can cause apparent distortions of the thermal signal in the TMDSC and, correspondingly, potentially spurious results.

Description of the TMDSC Experiment

Briefly, TMDSC experiments are run by subjecting a sample to a sinusoidal temperature perturbation over the normal temperature scan as shown in Fig. 16. The purported advantages of TMDSC include the ability to separate overlapping phenomena, as well as improved resolution and sensitivity [59]. However, the proper analysis of the TMDSC data is currently unclear. One approach commonly used is to treat the thermal signal using linear response

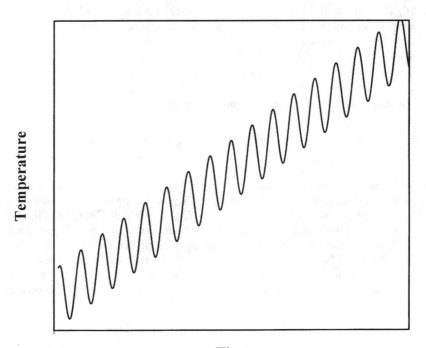

Time

FIG. 16—*Schematic of temperature-time profile in a temperature modulated DSC experiment.*

theory [60,61]. If the response is linear, one expects to obtain a sinusoidal output signal out of phase with the thermally oscillating input. The in-phase and out-of-phase components of the response are then labeled as dynamic heat capacity components $C_p'(\omega)$ and $C_p''(\omega)$. If the signal is distorted due to some non-linearity in the material (or equipment) response, such an analysis becomes very problematic, particularly, if the distortions are large and a harmonic analysis loses its validity. Consequently, the use of TMDSC and a linear response theory analysis to measure nonlinear processes such as melting and reaction has been explicitly questioned [12,62,63]. Here, we show that the structural recovery events occurring near the glass transition can also lead to nonlinear response, with a resultant need to interpret measurements with care.

TNM Analysis of TMDSC

The experimental observable in the TMDSC is the heat flow as a function of time (or temperature). We calculate the heat flow noting that it is the time derivative of the enthalpy, which is a function only of temperature (and pressure) in the equilibrium state. As noted above, however, in the glassy state, the enthalpy depends on the temperature and the structure of the glass, the latter being defined by the fictive temperature T_F. When we assume that the enthalpy of the equilibrium liquid at zero Kelvin is zero, then the enthalpy H of the amorphous material can be written

$$H = \int_T^{T_F} \Delta C_p dT + \int_0^T C_{pl} dT \tag{11}$$

where ΔC_p is the difference between the heat capacity in the liquid and glassy states ($C_{pl} - C_{pg}$). Further assuming that ΔC_p and C_{pl} are independent of temperature. Then the equation for the enthalpy becomes

$$H = \Delta C_p(T_F - T) + C_{pl}T = \Delta C_p T_F + C_{pg}T \tag{12}$$

and, as mentioned above, heat flow Q is simply the time derivative of H

$$Q = \frac{dH}{dt} = \Delta C_p \frac{dT_F}{dt} + C_{pg} \frac{dT}{dt} \tag{13}$$

Here, we treat the ideal experiment in which there is no thermal lag in the sample. Then the problem of modeling the heat flow during a TMDSC temperature ramp is one of modeling the structural evolution of the material, i.e., $dT_F/dt = (dT_F/dT)/(dT/dt)$. To do this we use Eq 8 above with the temperature and fictive temperature shift factors being given by Eq 7. The evolution of the fictive temperature during a given thermal history is obtained by numerical solution of Eq 9 coupled with Eq 7. The thermal history used for modeling a TMDSC experiment includes the cooling leg from above T_g to a point below T_g and the subsequent heating leg, with modulation, to above T_g. Assuming the ideal experiment, the thermal history is

$$T = T_0 - qt \qquad t \le t_1 \text{ (cooling leg)} \tag{14}$$

$$T = T_0 - qt_1 + m(t - t_1) + A \sin[\omega(t - t_1)] \qquad t > t_1 \text{ (heating leg)} \tag{15}$$

where t_1 is the time at which the cooling leg is completed, q is the cooling rate, m is the heating rate, A is the amplitude of the temperature modulation and ω is the radian frequency of the modulation.

Lissajous Loop Analysis

In the results presented here, we use the Lissajous [64] plot analysis. We have previously given the appropriate equations [65] and summarize the important results here. In the case of a linear response between the heat flow Q and the rate of change of temperature dT/dt, one would expect an elliptical Lissajous loop. If the loop is non-elliptical then the response is non-linear. If the loop is extremely distorted, one might then suspect that even a harmonic analysis of the response would become invalid [66] for the purposes of estimating C_p' and C_p''.

Figure 17 depicts the Lissajous loop response for an ideal dynamic heat capacity, i.e., the heat flow is a cosine that is out of phase by an amount ϕ with the sinusoidally oscillating temperature. Since Q is dependent on dT/dt, we have plotted the Lissajous loop accordingly. If the linear response theory is appropriate to analyze such signals the values of the heat capacity C_p' and C_p'' are found from

$$C_p' = C_p \cos \phi \tag{16}$$

$$C_p'' = C_p \sin \phi \tag{17}$$

where $C_p = B/A$ is the modulus of the complex heat capacity, B is the amplitude of the

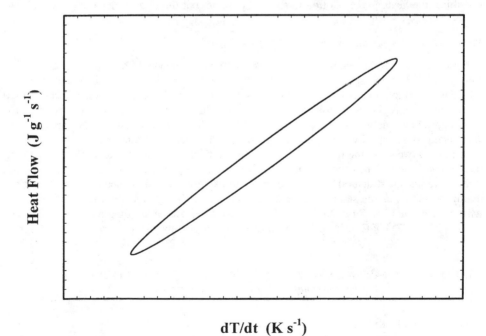

FIG. 17—*Schematic of Lissajous representation for a dynamic thermal experiment. See text for discussion.*

oscillatory component of the heat flow and A is the amplitude of the oscillating heating rate dT/dt. The phase lag ϕ is determined from the area of the Lissajous loop

$$\sin \phi = \frac{\text{area}}{\omega^2 \pi BA} \tag{18}$$

and

$$\text{area} = -\int_{t=0}^{t=2\pi} Qd\left(\frac{dT}{dt}\right) \tag{19}$$

Description of the Simulated Experiments

We performed simulations of the impact of thermal history on the TMDSC response for the ideal TMDSC experiment (e.g., no thermal lag) for three experimental conditions using the TNM model parameters for a poly(vinyl acetate) material. The parameters are presented in Table 1. Note that the definition of the reference relaxation (retardation) time is

$$\ln \tau_0 = \ln A + \frac{x\Delta h}{RT} + \frac{(1-x)\Delta h}{RT_F} \tag{20}$$

Three separate experiments are considered. First, one in which the cooling rate and heating rate are only slightly different; second, the cooling rate is much slower than the heating rate, and third, the heating and cooling rates are the same, but the period of modulation is long.

Results of TNM Modeling of the TMDSC Experiments

Figure 18 depicts the results for the TNM calculations for which the heating rate is 0.5°C/min after cooling the sample from T_g + 50°C at 1°C/min. The period of oscillation was 60 s and the amplitude of the oscillation was 1°C. As seen in the figure, the Lissajous loop in the transition region is slightly distorted from an ellipse. In this case, it would be expected that a harmonic analysis might give reasonable results. The insert shows the apparent instantaneous heat capacity during the experiment as a function of temperature.

In Fig. 19, we show the results of simulating a similar experiment, except that the cooling rate is much slower than the heating rate. In such an experiment, the slow cooling would lead to significant structural recovery during the experiment and consequent significant enthalpy overshoot as the sample temperature traverses the glass transition regime, similar to what is seen in Figs. 7a-c. In Fig. 19 we depict three Lissajous diagrams from different regions of the TMDSC scan, offset arbitrarily by an amount C for clarity. In the diagram

TABLE 1—*TNM model parameters used* [65] *in calculations of simulated TMDSC experiments on poly(vinyl acetate). (Note that in ln(A) is evaluated at* T_g.)

$T_g(K)$	$\Delta h/R$ (K)	x	β	$\ln(A/s)$	C_{pg} $(J/g/K)$	C_{pl} $(J/g/K)$
310	39 900	0.35	0.57	−124.8	1.18	1.86

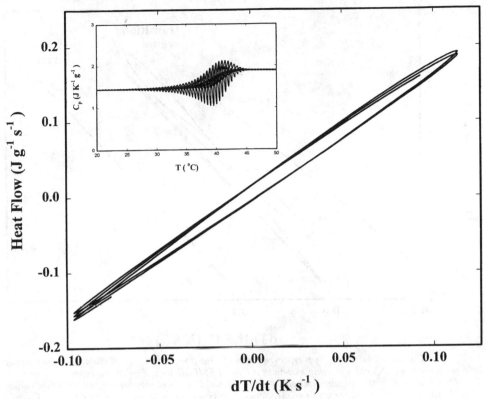

FIG. 18—*Lissajous loop representation of heat flow versus rate of change of temperature in simulated TMDSC experiment on a poly(vinyl acetate) glass. Loops are in transition regime and show slight lack of ellipticity. Cooling rate = −1°C/min; Heating rate = 0.5°C/min; Period of oscillation is 60 s; Amplitude of oscillation is 1°C. Insert shows apparent instantaneous heat capacity (Q/dT/dt) versus temperature during scan. See text for discussion.*

labeled "glass" the Lissajous loop is a straight line whose slope would be the heat capacity in the glassy state. Similarly, that labeled melt is from calculations well above the glass transition regime and the slope would give the melt or equilibrium heat capacity. In the transition regime, however, we see that the structural recovery is causing a significant distortion and rotation of the loops. Here, one begins to question the meaning of such data when interpreted in terms of C_p' and C_p''. Similar results were found previously by us [11,12] for polystyrene and poly(vinyl chloride) material parameters.

In Fig. 20, we show an additional potential problem with the TMDSC when very low modulation frequencies are used—the period of modulation interacts with the changing properties, even though a higher frequency modulation might not show a huge distortion of the Lissajous loop, as in Fig. 18. In Fig. 20, we show the calculation for the TNM model for a case in which the heating and cooling rates are equal at 1°C/min, the amplitude of oscillation is 1°C and the period is 600 s. As in Fig. 19, the glass and melt regimes result in straight lines, the slopes of which would be the heat capacity in the glassy and melt states. In the transition regime, on the other hand, one sees a significant distortion of the Lissajous loop, suggesting that harmonic or linear analysis may be problematic for such experimental conditions. In fact, a recent analysis by Snyder and Mopsik [67] suggests that temperature

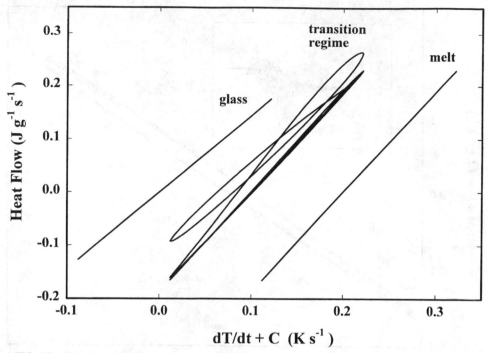

FIG. 19—*Calculated Lissajous loop representation of heat flow versus rate of change of temperature for a poly(vinyl acetate) glass forming system in a typical TMDSC scan. Figure shows response in glass (C = 0) and melt (C = 0.2) regimes as nearly straight lines. Structural recovery in the transition regime (C = 0.1) leads to large distortions of loops. Simulation is for a Cooling rate of 0.01°C/min; Heating rate is 1°C/min; Oscillation period is 60 s; Oscillation amplitude is 1°C.*

scanning methods in which an oscillatory response, such as dielectric or mechanical, is measured requires care for all conditions, unless the property is changing very slowly relative to the oscillatory probe itself. The nature of the glass transition may make such measurements difficult in an oscillatory mode.

Summary

The kinetic nature of the glass transition event has been reviewed. The richness of the glass transition phenomenology has been described using the asymmetry of approach and memory experiments as illustrations of this complexity. The essential ingredients that the Tool-Narayanaswamy-Moynihan-Kovacs-Aklonis-Hutchinson-Ramos (TNM-KAHR) models were presented: a retardation spectrum that exhibits "non-exponential" behavior (this implies that the response cannot be described by a single exponential response); a characteristic time that depends on the current glassy structure (or departure from equilibrium) and temperature through a structure-temperature shift factor; additivity of responses in reduced time (rather than laboratory time) that looks like linear viscoelasticity through an appropriate convolution integral.

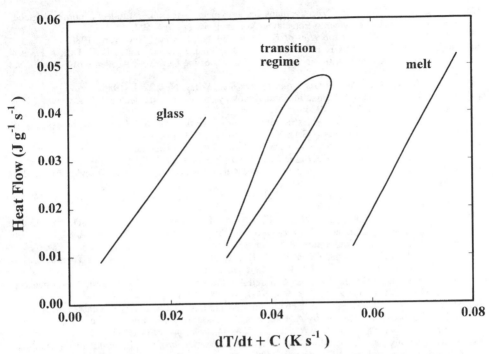

FIG. 20—*Calculated Lissajous loop representation of heat flow versus rate of change of temperature for a poly(vinyl acetate) glass forming system in a typical TMDSC scan. Figure shows response in glass (C = 0) and melt (C = 0.05) regimes as nearly straight lines. Structural recovery and interaction between period of oscillation and rapidly changing heat capacity in the transition regime lead to large distortions of loops. Simulations for a Cooling rate of 1°C/min; Heating rate is 1°C/min; Oscillation period is 600 s; Oscillation amplitude is 1°C.*

"Physical Aging" was defined as the impact of structural recovery on mechanical properties and the physical aging response in the linear and nonlinear viscoelastic regimes was reviewed. In addition, physical aging effects on engineering properties such as yield, creep rupture, and crazing were briefly discussed. An example of the impact of structural recovery on residual stress development in a composite seal was described.

Finally, the impact of structural recovery on the measurement of heat capacity in a temperature modulated differential scanning calorimetry experiment was simulated. Results were given in a Lissajous loop representation and it was shown that structural recovery can distort the loops dramatically, leading to possible difficulties in linear or harmonic analyses. Finally, one simulation shows that, when the period of temperature oscillation is long relative to the heating rate, additional loop distortions or non-linearities can arise.

References

[1] McKenna, G. B., "Glass Formation and Glassy Behavior," *Comprehensive Polymer Science. Vol. 2, Polymer Properties,* by C. Booth and C. Price, Eds. Pergamon, Oxford, 1989, pp. 311–363.

[2] Kovacs, A. J., "Transition Vitreuse dans les Polymères Amorphes. Etude Phénoménologique," Fortschritte der Hochpolymeren-Forschung, Vol. 3, 1964, pp. 394–507.

[3] Ferry, J. D., Viscoelastic Properties of Polymers, 3rd, ed., J. Wiley, New York, 1980.

[4] Moynihan, C. T., Macedo, P. B., Montrose, C. J., Gupta, P. K., DeBolt, M. A., Dill, J. F., et al. "Structural Relaxation in Vitreous Materials," Annals of the New York Academy of Sciences, Vol. 279, 1976, pp. 15–35.

[5] Scherer, G. W., Relaxation in Glass and Composites, Wiley, New York, 1986.

[6] Hodge, I. M., "Enthalpy Relaxation and Recovery in Amorphous Materials," Journal of Non-Crystalline Solids, Vol. 169, 1984, pp. 211–266.

[7] Tool, A. Q., "Viscosity and the Extraordinary Heat Effects in Glass," Journal of Research of the National Bureau of Standards (USA), Vol. 37, 1946, pp. 73–90.

[8] Tool, A. Q., "Relation Between Inelastic Deformability and Thermal Expansion of Glass in Its Annealing Range," Journal of the American Ceramic Society, Vol. 29, 1946, pp. 240–253.

[9] Struik, L. C. E., Physical Aging in Polymers and Other Amorphous Materials, Elsevier, Amsterdam, 1978.

[10] Hutchinson, J. M., "Physical Aging of Polymers," Progress in Polymer Science, Vol. 20, 1995, p. 703–760.

[11] Simon, S. L. and McKenna, G. G., "The Effects of Structural Recovery and Thermal Lag in Modulated DSC Measurements," Proceedings of the North American Thermal Analysis Society, (NATAS) 25th Annual Conference, 1997, pp. 358–365.

[12] Simon, S. L. and McKenna, G. B., "The Effects of Structural Recovery and Thermal Lag in MDSC," Thermochimica Acta, Vol. 307, 1997, pp. 1–10.

[13] McKenna, G. B., "On the Physics Required for the Prediction of Long Term Performance of Polymers and Their Composites," J. Res. NIST, Vol. 99, 1994, pp. 169–189.

[14] McKenna, G. B., "Comments on 'Isobaric Volume and Enthalpy Recovery of Glasses. II. A Transparent Multiparameter Theory,' by A. J. Kovacs, J. J. Aklonis, J. M. Hutchinson and A. R. Ramos," Journal of Polymer Science, Part B: Physics, Vol. 34, 1996, pp. 2463–2465.

[15] Narayanaswamy, O. S., "A Model of Structural Relaxation in Glass," Journal of the American Ceramic Society, Vol. 54, 1971, pp. 491–498.

[16] Kovacs, A. J., Aklonis, J. J., Hutchinson, J. M., and Ramos, A. R., "Isobaric Volume and Enthalpy Recovery of Glasses. II. A Transparent Multiparameter Theory," Journal of Polymer Science, Polymer Physics Edition, Vol. 17, 1979, pp. 1097–1162.

[17] Kohlrausch, R., "Theorie des Elektrischen Rückstandes in der Leidener Flasche," Annalen der Physik und Chemie von J. C. Poggendorff, Vol. 91, 1854, pp. 179–214.

[18] Williams, G. and Watts, D. C., "Nonsymmetrical Dielectric Relaxation Behavior Arising from a Simple Empirical Decay Function," Transactions of the Faraday Society, Vol. 66, 1970, pp. 80–85.

[19] Schultheisz, C. R. and McKenna, G. B., "Volume Recovery, Physical Aging and the Tau-Effective Paradox in Glassy Polycarbonate Following Temperature Jumps," Proc. NATAS 25th Annual Conference, September 1997, pp. 366–373.

[20] Schultheisz, C. R., NIST, Gaithersburg, MD 20899, (unpublished data).

[21] Tribone, J. J., O'Reilly, J. M., and Greener, J., "Analysis of Enthalpy Relaxation in Poly(methyl Methacrylate)-Effects of Tacticity, Deuteration, and Thermal History," Macromolecules, Vol. 19, 1986, pp. 1732–1739.

[22] Chambers, R. S., Gerstle, F. P., and Monroe, S. L., "Viscoelastic Effects in a Phosphate Glass-metal Seal," Journal of The American Ceramic Society, Vol. 72, 1989, pp. 929–932.

[23] McKenna, G. B., Angell, C. A., Rendell, R. W., Moynihan, C. T., Kovacs, A. J., McKenna, G. B., Hutchinson, J. M., Oguni, M., Oreilly, J., Struik, L., Hodge, I. M., Bauwens, J. C., Oleynick, E., Rekhson, S., Williams, G., and Matsuoka, S., "The Phenomenology and Models of The Kinetics of Volume and Enthalpy in The Glass-transition Range," Journal of Non-crystalline Solids, Vol. 131, 1991, pp. 528–536.

[24] O'Reilly, J. M., "Review of Structure and Mobility in Amorphous Polymers," CRC Critical Reviews in Solid State and Materials Sciences, Vol. 13, 1987, pp. 259–277.

[25] S. L. Simon, "Enthalpy Recovery of Poly(ether Imide): Experiment and Model Calculations Incorporating Thermal Gradients," Macromolecules, Vol. 30, 1997, pp. 4056–4063.

[26] Kovacs, A. J., Hutchinson, J. M., and Aklonis, J. J., "Isobaric Volume and Enthalpy Recovery of Glasses (I) A Critical Survey of Recent Phenomenological Approaches" in The Structure of Non-Crystalline Materials, ed by P. H. Gaskell, Taylor and Francis, New York, Vol. 153, 1977.

[27] Struik, L. C. E., "Volume-Recovery Theory: 1. Kovacs' τ_{eff} Paradox," Polymer, Vol. 38, 1997, pp. 4677–4685.

[28] McKenna, G. B., Vangel, M. R., Rukhin, A. L., Leigh, S. D., Straupe, C., and Lotz, B., "The τ-Effective Paradox Revisited: An Extended Analysis of Kovacs' Volume Recovery Data on Poly(vinyl Acetate)," *Polymer,* Vol. 40, 1999, pp. 5183–5205.

[29] Kovacs, A. J., Stratton, and Ferry, J. D., "Dynamic Mechanical Properties of Polyvinyl Acetate in Shear in the Glass Transition Temperature Range," *Journal of Physical Chemistry,* Vol. 67, 1963, 152–161.

[30] O'Connell, P. A. and McKenna, G. B., "Large Deformation Response of Polycarbonate: Time-Temperature, Time-Aging Time and Time-Strain Superposition," *Polymer Engineering and Science,* Vol. 37, 1997, pp. 1485–1495.

[31] Simon, S. L., Plazek, D. J., Sobieski, J. W., and McGregor, E. T., "Physical Aging of a Polyetherimide: Volume Recovery and its Comparison to Creep and Enthalpy Measurements," *Journal of Polymer Science, Part B: Polymer Physics,* Vol. 35, 1997, pp. 929–936.

[32] Struik, L. C. E., "On the Rejuvenation of Physically Aged Polymers by Mechanical Deformation," *Polymer,* Vol. 38, 1997, pp. 4053–4057.

[33] Struik, L. C. E., "Dependence of Relaxation Times of Glassy Polymers on Their Specific Volume," *Polymer,* Vol. 29, 1988, pp. 1347–1353.

[34] McKenna, G. B., "Dilatometric Evidence for the Decoupling of Glassy Structure from the Mechanical Stress Field," *Journal of Non-Crystalline Solids,* Vol. 172–174, 1994, pp. 756–764.

[35] Santore, M. M., Duran, R. S., and McKenna, G. B., "Volume Recovery in Epoxy Glasses Subjected to Torsional Deformations: The Question of Rejuvenation," *Polymer,* Vol. 32, 1991, pp. 2377–2381.

[36] McKenna, G. B., Leterrier, Y., and Schultheisz, C. R., "The Evolution of Material Properties During Physical Aging," *Polymer Engineering and Science,* Vol. 35, 1995, pp. 403–410.

[37] McKenna, G. B., Santore, M. M., Lee, A., and Duran, R. S., "Aging in Glasses Subjected to Large Stresses and Deformations," *Journal of Non-Crystalline Solids,* Vol. 131, 1991, pp. 497–504.

[38] Delin, M., Rychwalski, R. W., Kubat, J., Klason, C., and Hutchinson, J. M., "Physical Aging Time Scales and Rates for Poly(vinyl Acetate) Stimulated Mechanically in the T_g-region," *Polymer Engineering and Science,* Vol. 36, 1996, pp. 2955–2967.

[39] Duran, R. S. and McKenna, G. B., "A Torsional Dilatometer for Volume Change Measurements on Deformed Glasses: Instrument Description and Measurements on Equilibrated Glasses," *Journal of Rheology,* Vol. 34, 1990, pp. 813–839.

[40] Myers, F. A., Cama, F. C., and Sternstein, S. S., "Mechanically Enhanced Aging of Glassy Polymers," Annals of the New York Academy of Sciences, Vol. 279, 1976, pp. 94–99.

[41] Waldron, W. K., McKenna, G. B., and Santore, M. M., "The Nonlinear Viscoelastic Response and Apparent Rejuvenation of an Epoxy Glass," *Journal of Rheology,* Vol. 39, 1995, pp. 471–497.

[42] McKenna, G. B. and Zapas, L. J., "Superposition of Small Strains on Large Deformations as a Probe of Nonlinear Response in Polymers," *Polymer Engineering and Science,* Vol. 26, 1986, pp. 725–729.

[43] McKenna, G. B. and Zapas, L. J., "The Superposition of Small Deformations on Large Deformations: Measurements of the Incremental Modulus for a Polyisobutylene Solution," *Journal of Polymer Science, Polymer Physics. Edition.,* Vol. 23, 1985, pp. 1637–1656.

[44] Lee, A. and McKenna, G. B., "The Physical Aging Response of an Epoxy Glass Subjected to Large Stresses," *Polymer,* Vol. 31, 1990, pp. 423–430.

[45] Yee, A. F., Bankert, R. J., Ngai, K. L., and Rendell, R. W., "Strain and Temperature Accelerated Relaxation in Polycarbonate," *Journal of Polymer Science,* Part B: Polymer Physics, Vol. 26, 1988, pp. 2463–2483.

[46] Hasan, O. A., Boyce, M. C., and Berko, S., "An Investigation of the Yield and Postyield Behavior and Corresponding Structure of Poly(methyl Methacrylate)," *Journal of Polymer Science, Part B: Polymer Physics,* Vol. 31, 1993, pp. 185–197.

[47] Hasan, O. A. and Boyce, M. C., "Energy Storage During Inelastic Deformation of Glassy Polymers," *Polymer,* Vol. 34, 1993, pp. 5085–5092.

[48] Oleinik, E. F., "Epoxy-aromatic Amine Networks in The Glassy State Structure and Properties," *Advances in Polymer Science,* Vol. 80, 1986, pp. 49–99.

[49] Ricco, T. and Smith, T. L., "Rejuvenation and Physical Aging of a Polycarbonate Film Subjected to Finite Tensile Strains," *Polymer,* Vol. 26, 1985, pp. 1979–1984.

[50] Chow, T. S., "Stress-strain Behavior of Physically Aging Polymers," *Polymer,* Vol. 34, 1993, pp. 541–545.

[51] G'Sell, C. and McKenna, G. B., "Influence of Physical Aging on the Yield Behavior of Model DGEBA/Poly(propylene oxide) Epoxy Glasses," *Polymer,* Vol. 33, 1992, pp. 2103–2113.

[52] Crissman, J. M. and McKenna, G. B., "Relating Creep and Creep Rupture in PMMA Using a Reduced Variables Approach," *Journal of Polymer Science, Polymer Physics Edition,* Vol. 25, 1987, pp. 1667–1677.

[53] Crissman, J. M. and McKenna, G. B., "Physical and Chemical Aging in PMMA and Their Effects on Creep and Creep Rupture Behavior," *Journal of Polymer Science, Polymer Physics Edition,* Vol. 28, 1990, pp. 1463–1473.

[54] Arnold, J. C., "The Effects of Physical Aging on the Brittle Fracture Behavior of Polymers," *Polymer Engineering and Science,* Vol. 35, 1995, pp. 165–169.

[55] Gusler, G. M. and McKenna, G. B., "The Craze Initiation Response of a Polystyrene and a Styrene-Acrylonitrile Copolymer During Physical Aging," *Polymer Engineering and Science,* Vol. 37, 1997, pp. 1442–1448.

[56] Delin, M. and McKenna, G. B., "Craze Growth in Stress Relaxation Conditions: Effects of Physical Aging," ANTEC '98, II. *Materials,* 1998, pp. 1669–1671.

[57] Chang, T. D. and Brittain, J. O., "Studies of Epoxy Resin Systems. Iv. Fracture Toughness of an Epoxy Resin: a Study of the Effect of Crosslinking and Sub-T_g Aging," *Polymer Engineering and Science,* Vol. 22, 1982, pp. 1228–1237.

[58] Hill, A. J., Heater, K. J., and Agrawal, C. M., "The Effects of Physical Aging in Polycarbonate," *Journal of Polymer Science, Part B: Polymer Physics,* Vol. 28, 1990, pp. 387–405.

[59] See the special issue of Thermochimica Acta, Vol. 304/305, 1997 and articles therein.

[60] Schawe, J. E. K. and Hone, G. W. H., "The Analysis of Temperature Modulated DSC Measurements by Means of the Linear Response Theory," *Thermochimica Acta,* Vol. 287, 1996, pp. 213–223.

[61] Weyer, S., Hensel, A., Korus, J., Donth, E., and Schick, C., "Broad Band Heat Capacity Spectroscopy in the Glass-transition Region of Polystyrene," *Thermochimica Acta,* Vol. 305, 1997, pp. 251–255.

[62] Wunderlich, B., Boller, A., Okazaki, I., and Ishikiriyama, K., "Heat-capacity Determination by Temperature-modulated DSC and its Separation from Transition Effects," *Thermochimica Acta,* Vol. 305, 1997, pp. 125–136.

[63] Gill, P. S., Sauerbrunn, S. R., and Reading, M., "Modulated Differential Scanning Calorimetry," *Journal of Thermal Analysis,* Vol. 40, 1993, pp. 931–939.

[64] Brodt, M., Cook, L. S., and Lakes, R. S., "Apparatus for Measuring Viscoelastic Properties over Ten Decades: Refinements," *Review of Scientific Instruments,* Vol. 66, 1995, pp. 5292–5297.

[65] Simon, S. L. and McKenna, G. B., "Interpretation of the Dynamic Heat Capacity Observed in Glass-Forming Liquids," *Journal of Chemical Physics,* Vol. 107, 1997, pp. 8678–8685.

[66] Ganeriwala, S. N. and Rotz, C. A., "Fourier Transform Mechanical Analysis for Determining the Nonlinear Viscoelastic Properties of Polymers," *Polymer Engineering and Science,* Vol. 27, 1987, pp. 165–178.

[67] Snyder, C. R. and Mopsik, F. I., "Dynamically Induced Loss and Its Implications on Temperature Scans of Relaxation Processes," *Journal of Chemical Physics,* Vol. 110, 1999, pp. 1106–1111.

Maria L. Cerrada[1] *and Gregory B. McKenna*[2]

Creep Behavior in Amorphous and Semicrystalline PEN

REFERENCE: Cerrada, M. L., McKenna, G. B., **"Creep Behavior in Amorphous and Semicrystalline PEN,"** *Time Dependent and Nonlinear Effects in Polymers and Composites, ASTM STP 1357,* R. A. Schapery and C. T. Sun, Eds., American Society for Testing and Materials, West Conshohocken, PA, 2000, pp. 47–69.

ABSTRACT: Poly(ethylene-2,6-naphthalate), PEN, is a suitable candidate to replace poly (ethylene terephthalate), PET, in some applications. PEN can be produced in either the amorphous or semi-crystalline states, depending on the processing conditions. Here, we report on results from uniaxial tension experiments in creep conditions in which we probed the viscoelastic and physical aging responses for both amorphous and semi-crystalline forms of PEN. The data show the existence of overlapping β and α relaxations in the experimental creep time range studied. The β process is stronger in the amorphous than in the semi-crystalline material and, in both cases, shows different aging time and temperature dependencies. A model in which the α process is treated as a stretched exponential process and the β process as a Cole-Cole process is developed and its validity examined for both the amorphous and semi-crystalline PEN materials.

KEYWORDS: PEN, creep, viscoelasticity, stress relaxation, time-aging time superposition, glasses, semi-crystalline, polymer

Poly(ethylene-2,6-naphthalate) or PEN is a thermoplastic material found in numerous commercial applications in the form of films of different thickness, crystallinity, and orientation. The primary modes of PEN processing are blow molding, vacuum forming, drawing, and biaxial stretching. During these processing modes, the polymer can take either an oriented amorphous or semicrystalline structural state. Understanding of the structure and orientation changes that occur in a polymer during drawing are important from both fundamental and applied views. Furthermore, once different structures are produced, they can have a pronounced impact on the final properties of the polymer product. The occurrence of crystallization during processing can provide increased stiffness and hardness as well as better dimensional stability and creep resistance. For such reasons, PEN is being increasingly used as a substitute for poly(ethylene terephthalate) (PET) in commercial applications. It displays some properties in common with PET, such as ability to crystallize under strain or stress or exhibition of very desirable stress-strain characteristics (high modulus of elasticity, high elongation or strain, high yield strength, and high breaking strength). Moreover, it has a higher modulus, higher glass transition temperature, and higher melting temperature than

[1] Research Scientist, Instituto de Ciencia y Tecnologia de Polimeros (CSIC), Juan de la Cierva 3, 28006-Madrid, Spain; e-mail: ictg26@Fresno.csic.es, previously Visiting Scientist, Structure and Mechanics Group, Polymers Division, NIST, Gaithersburg, MD 20899.

[2] Professor, Department of Chemical Engineering, Texas Tech University, Lubbock, TX 79409-3121; e-mail: greg.mckenna@coe.ttu.edu; previously Group Leader, Structure and Mechanics Group, Polymers Division, NIST, Gaithersburg, MD 20899.

PET. Here, we report on a study of the viscoelastic (creep) and physical aging properties of PEN films obtained in the semi-crystalline and amorphous states.

In general, it is well understood that cooling of a polymer through its glass transition results in a non-equilibrium glass whose thermodynamic properties (structure) spontaneously evolve towards equilibrium in a process known as structural recovery [1–5]. Clearly, as the glassy structure evolves towards equilibrium, other properties, such as the viscoelastic response, yield or failure, also change and this observation was baptized physical aging by Struik [6] in his landmark work. For the linear viscoelastic response of materials, it is recognized that physical aging can often be described using a time-aging time or time-structure superposition principle where the material characteristic time depends on structure or aging time analogous to the temperature dependence observed in time-temperature superposition [7].

The success of the time-aging time superposition approach has been primarily obtained with amorphous polymers in which there is no significant sub-glass transition peak. In fact, early on, Struik [6] commented that the aging regime should be considered to lie between the sub-glass transition β process and the glass transition α process. Read, et al. [8–11] have shown some success in working with materials having widely separated β and α processes by considering superposition of the individual mechanisms. Furthermore, studies of aging in semi-crystalline polymers have been reported to be complicated due to possible broadening of the glass transition due to crystalline constraints on the amorphous regions [12–15], although in some instances, the sub-T_g aging is found to be qualitatively similar to that seen in amorphous polymers [16,17].

Poly(ethylene naphthalate) exhibits [18] a very strong and broad β relaxation with a peak at approximately 65°C that overlaps with the α relaxation. In addition, as mentioned above, PEN can be obtained in both the amorphous and semi-crystalline states. Semi-crystalline PEN also exhibits strongly overlapping β and α processes, though the β process is somewhat weaker than in the amorphous material. The purpose of the current study is to investigate and compare the physical aging responses of amorphous and semi-crystalline PEN. The responses are fit to a two-mechanism model in which each mechanism has a separate temperature and structure dependence. Our results are reported below.

Method of Analysis

Viscoelastic Data

The analysis of the physical aging response of polymeric materials depends on defining a viscoelastic response function and of measuring how the response function depends on the time elapsed in the aging experiment. Here, we deal with the creep compliance response in uniaxial extension $D(t)$ and the aging time t_e. A very widely used response function is the so-called Kohlrausch [19]-Williams-Watts [20] stretched exponential function, which has often [6,11,16,21,22] been written for creep as

$$D(t) = D_0 \exp[t/\tau_0)^\beta]$$ (1)

where D_0 is a "zero-time" compliance, t is the experimental time, τ_0 is a characteristic retardation time, and β is a parameter that is related to the width of the viscoelastic dispersion described by the function. Compliances calculated using Eq 1 will tend to infinity at long times, therefore, this function can describe the time dependent compliance only over a limited timescale.

An alternative function that has been successfully used to describe the creep of predominantly semi-crystalline polymers [23] is based on the Williams-Watts model for describing charge decay in dielectrics [20] and is consistent with a series representation for the creep recovery function (cf. Ferry [7]). This creep function is of the form

$$D(t) = D_0 + \Delta D_\alpha [1 - \exp(-(t/\tau_0)^\beta)] \tag{2}$$

The parameters D_0, τ_0 and β have a similar significance as in Eq 1 although it should be noted that the values of the parameters required to describe a given creep curve will be different. Equation 2 reaches an upper compliance limit at long times defined by ΔD_α, which represents the magnitude of the relaxation process.

Equations 1 and 2 are applicable only when a single relaxation mechanism is involved in the creep response. However, as noted above, it is common for polymers to exhibit several relaxation processes. In a European round robin comparison of aging models, Tomlins and Read [24] report on a European Technical Committee evaluation of several approaches to model the aging response of a semi-crystalline polypropylene. Each dealt with the aspect of two mechanisms in a different way and we leave the reader to examine the original reference for a discussion of the approaches used. It is interesting that, Tomlins and Read did not include in their own evaluation a successful approach from the National Physical Laboratory (NPL) reported previously [23,25] for aging in poly(vinyl chloride), poly(methyl methacrylate), and other polymers. One possible reason for the omission is the closeness of the β relaxation to the α relaxation and how this affects the choice of material representation in a limited time and temperature window. In any event, we have chosen to follow the NPL approach. Then, for a given aging time t_e and temperature T, the creep compliance may be written [23,25]

$$D(t) = D_{U\beta} + D_\beta(t) + D_\alpha(t) \tag{3}$$

Where $D_{U\beta}$ is the unrelaxed compliance for the β retardation $D_\beta(t)$ and $D_\alpha(t)$ are the compliance contributions from the respective β and α processes. Due to the symmetry of the dielectric β relaxation [23] for several polymers, the function $D_\beta(t)$ was chosen to follow a Cole-Cole equation, transformed into the time-domain. On the other hand, $D_\alpha(t)$ was taken to be a KWW retardation function (Eq 2). Consequently, the tensile creep compliance can be expressed as

$$D(t) = D_{U\beta} + \Delta D_\beta \left(\frac{(t/\tau_\beta)^n [(t/\tau_\beta)^n + \cos(n\pi/2)]}{1 + 2(t/\tau_\beta)^n \cos(n\pi/2) + (t/\tau_\beta)^{2n}} \right) + \Delta D_\alpha [1 - \exp(-(t/\tau_\alpha)^m)] \tag{4}$$

where $\Delta D_\beta(=D_{R\beta} - D_{U\beta})$ and $\Delta D_\alpha(=D_{R\alpha} - D_{R\beta})$ are retardation strengths and τ_β and τ_α are mean retardation times. The constants n $(0) < n \leq 1)$ and m $(0 < m \leq 1)$ are distribution parameters which decrease in magnitude as the widths of the respective β and α retardation spectra increase. A value for these parameters can be obtained by using a least-squares optimization routine of all the parameters in the function to a series of short-term creep curves or some of them can be estimated from an analysis of dynamic mechanical data [26]. The others are obtained by least-squares optimization. Here, we use least squares optimization for the data analysis.

Time-Temperature and Time-Aging Time Superposition

Once one has a viscoelastic response function that describes a material's behavior at a single temperature and aging time, the problem of describing the behavior at another temperature and aging time becomes one of considerable importance as the model determines one's ability to make longer or shorter time predictions than those obtainable in the limited laboratory experimental window. For a material with a single viscoelastic mechanism, the approach is straight-forward when the principles of time-temperature and time-structure (or time-aging time) superposition are valid. In such an instance, imagine that we have a mechanism as in Eq 2 with a characteristic time τ_0. Then, the response at another temperature or aging time is obtained by performing both vertical and horizontal shifts of the data on the double-logarithmic representation of compliance versus time curves

$$\tau_0(T,t_e) = a_T a_{te} \tau_0(T_r,t_{e,r}) \tag{5}$$

or

$$a_T = \frac{\tau_0(T,t_e)}{\tau_0(T_r,t_e)}; \qquad a_{te} = \frac{\tau_0(T,t_e)}{\tau_0(T,t_{e,r})} \tag{6}$$

These are the horizontal shifts, where a_T is the temperature shift factor, a_{te} is the aging time shift factor, T is temperature, t_e is aging time and the subscripts r refer to the chosen reference condition. For the vertical shifts one writes in a similar fashion

$$\Delta D_\alpha(T,t_e) = b_T b_{te} \Delta D_\alpha(T_r,t_{e,r}) \tag{7}$$

and

$$D_0(T,t_e) = b'_T b'_{te} D_0(T_r,t_{e,r}) \tag{8}$$

where the b_T and b_{te} are the vertical shifts due to temperature and aging respectively. Depending on the complexity of material behavior, b'_T and b'_{te} may or may not equal b_T and b_{te}. What is clear is that upon addition of a second mechanism, one needs to consider whether or not the shift factors for the different viscoelastic process have the same temperature or aging time dependencies. When the dependencies are the same, then classical thermo-rheological simplicity obtains. However, when the dependencies are different, as we found with the material under investigation here, the material appears thermo-rheologically complex and temperature and aging time dependence of each mechanism needs to be considered separately.

In this paper, we examine the influence of physical aging on the short-term tensile creep behavior of commercial poly(ethylene 2,6-naphthalate) (PEN) at different temperatures. The differences in the semi-crystalline and amorphous materials are evaluated. As seen in the subsequent paragraphs, two overlapping retardation regions were evident in the creep behavior of both the semi-crystalline and amorphous materials. Therefore, we found that the creep responses of both materials could be modeled by means of Eq 4. Each mechanism required both vertical and horizontal shifting to fit the data.

Experimental[3]

Material

Amorphous and semi-crystalline PEN film specimens from Eastman Kodak Company were used in this study. The amorphous films had a nominal thickness of 140 μm. The semi-crystalline films had a nominal thickness of 89 μm. Samples for creep testing were prepared by cutting strips of material 43.0 mm long, 9.6 mm wide. The sample to sample variability of geometry in width and length was measured to be less than 1% based on the range of measurements. Each sample was measured with a micrometer which could be read to within 2.5 μm. This leads to an uncertainty in the thickness measurement of less than 2%, which is greater than the ability to measure the film thickness variability. The glass transition temperature for the amorphous material was obtained using differential scanning calorimetry (DSC) at a heating rate of 10°C/min. The value of the glass transition was obtained from the mid-point of the heat capacity change and it was found to be 120°C for the amorphous PEN. The DSC measurement could not detect the T_g of the semi-crystalline material, although Gillmor and Greener [18] report a value of approximately 140°C based on dynamic mechanical measurements. The reader should see Ref 26 for a discussion of the measurement uncertainties typical of DSC measurements as well as for a discussion of the differences in assignment of T_g values based on different measurement techniques.

Mechanical Testing

All creep tests were performed on a Dynastat Mark II dynamic testing machine in a tensile mode. Because of the difficulty of testing thin films, new clamps were specially designed to accommodate the thin films used in this study. All experiments were performed at a stress of 11 MPa which is close to the linear viscoelastic response regime.

For the aging of amorphous specimens, the samples were first heated to 125°C, a temperature above the nominal glass transition of amorphous PEN, for 30 min to thermally erase prior aging in the material. The samples were then removed and placed into the testing chamber that had been preheated to the aging temperature where the physical aging experiments were carried out. On the other hand, for the semi-crystalline PEN, the as-received samples were placed directly into the testing chamber at the aging temperature in order to avoid possible changes in crystallinity and crystal morphology which could occur if the material were heated to above the glass transition to erase any prior aging.

Subsequent to placing the sample in the testing chamber, sequential creep tests were performed to probe the aging response. The test sequence followed the protocol originally proposed by Struik [6]. Aging times considered were 0.5 to 128 h. At a given aging temperature, the same specimen was used for the entire test. Loads of duration t_1 were applied at increasing aging times, t_e such that the ratio t_1/t_e, was maintained constant at 0.1. The applied loads are essentially probes into the material structure and are of a sufficiently short duration that the structural recovery (aging) does not significantly influence the measurements. By allowing the sample to recover for a time, $9 t_e$, the material essentially forgets the effect of the previous loading cycle. Experiments were performed at temperatures of

---

[3] Certain commercial materials and equipment are identified in this paper to specify adequately the experimental procedure. In no case does such identification imply recommendation or endorsement by the National Institute of Standards and Technology, nor does it imply necessarily that the product is the best available for the purpose.

30°C, 50°C, 80°C, and 100°C. The temperature of the apparatus was measured to fluctuate about the mean temperature over a range of ± 0.1°C.

The creep compliance $D(t) = \varepsilon(t)/\sigma$ was determined from knowing the applied stress σ and the grip displacement δ measured using the instrument displacement transducer on the ± 0.5 mm range. Hence, $\varepsilon(t) = \delta(t)/l_0$ where l_0 is the initial sample length (grip separation).

Measurement Uncertainty

The experimental precision in the creep data, based on two standard deviations from multiple measurements, on different samples was less than 10% of the reported value. As mentioned above, the same specimen was used throughout each aging experiment. This minimizes the variation associated with material variability and sample misalignment. In order to estimate the precision of the multiple measurements in a single aging test, we measured the magnitudes of the vertical shifts (see above and subsequently) required to superimpose the measurements at different aging times. The trends of the shifts with aging time were plotted against the logarithm of aging time and a least squares line fit to the data. The deviations of the vertical shifts from the regression line were less than 0.5% of the total measured compliance. This is typical of such experiments as performed in this laboratory for other materials [16,28,29]. An extensive analysis of the errors involved in shifting data such as those presented here is given by Bradshaw and Brinson [30].

Results

Creep Data for Amorphous PEN

Results of the short-term creep experiments on the amorphous PEN are shown in Fig. 1 as double logarithmic plots of the creep compliance versus time for each of the four aging temperatures and for different aging times. In all of the plots in Fig. 1, it is seen that, as expected, the compliance decreases as the aging time increases. For many amorphous polymers, e.g., polycarbonate, the curves primarily shift along the time axis and time-aging time superposition is obeyed. However, the behavior of amorphous PEN is significantly different. Beginning with the data at 30°C, the logarithm of the compliance changes quite linearly with respect to the logarithm of time for short aging times. As aging time increases, the compliance undergoes a transition in shape. Two different processes seem to appear: "power law" behavior at short times and, at longer times, an "exponential-like" behavior in which the expected upward curvature is exhibited. At 50°C, the PEN behavior was qualitatively similar to that exhibited at 30°C. At the shortest aging times, the PEN response seems to follow a power law which simply shifts along the time axis, while at the long aging times, the two different processes are observed. The aging effects at 50°C were greater than at 30°C.

The compliance curves for 80°C at different aging times exhibit the same two processes as observed at 30 and 50°C. In this case, however, the second exponential-like process is evident throughout the entire aging time range investigated. The power law response is now observed for only a very narrow experimental time window (at short times) for each aging time investigated. The response of amorphous PEN at 100°C is similar to that observed at 80°C. The exponential-like mechanism is observed over the whole aging time range and most of the creep time range. Although the 100°C is very close to the glass transition, the power-law process is still occurring at the shortest creep times, i.e., over a very small part of the experimentally accessible time window.

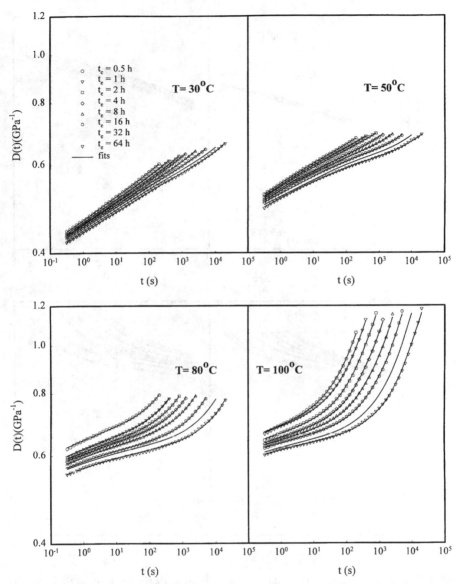

FIG. 1—*Double logarithmic representation of amorphous PEN compliance at the indicated temperatures and aging times and the corresponding fits to Eq 4.*

Creep Data for Semi-crystalline PEN

Results of the short-term creep experiments on the semi-crystalline PEN are shown in Fig. 2 as double logarithmic plots of the creep compliance versus time for each of the four aging temperatures and for different aging times. In all of these plots, the creep compliance is considerably lower than for the amorphous material at each aging temperature. Apparently,

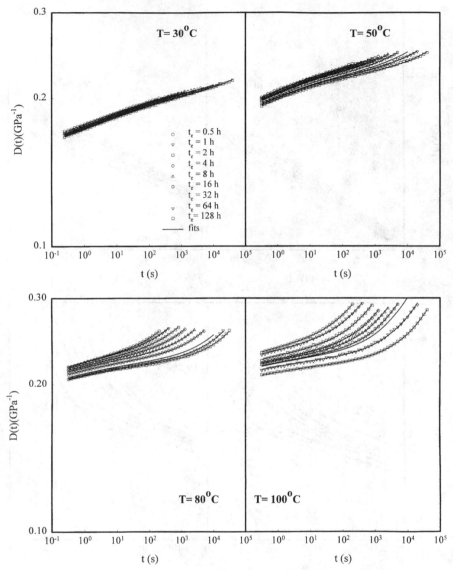

FIG. 2—*Double logarithmic representation of semi-crystalline PEN compliance at the indicated temperatures and aging times and the corresponding fits to Eq 4.*

the crystallites behave as stiff fillers in an amorphous matrix. At 30°C, the compliance changes nearly linearly on the double logarithmic scale at each aging time. Moreover, the physical aging effect is smaller at this temperature than in the amorphous material. (We note that this may be associated with the difference in thermal treatments between the materials because 30°C is close to the ambient temperature of storage for the materials prior to testing.) At 50°C, the PEN behavior at short aging times was similar to that exhibited at 30°C. For longer aging times, the behavior is no longer linear in the logarithmic representation and begins to display evidence of a second viscoelastic mechanism, as was observed for the amorphous PEN at 30°C. At 80°C, the two processes observed at 50°C are more obvious over the entire aging time range. As in the case of the amorphous PEN, the first process is observed only at very small creep times in the results from the 100°C tests.

Time-Aging Time and Time-Temperature Reduction

As discussed above, it is common in the analysis of physical aging experiments to super-impose the data obtained at different aging times or temperatures by shifting them along the time axis using time-aging time or temperature superposition [3,6,28–30]. In the case of the PEN data depicted in Figs. 1 and 2, it is clear that time-aging time superposition does not work. The presence of the large β mechanism that interferes with the main α or glass transition in PEN [18] apparently makes ready superposition of the data difficult. In addition, physical aging is still occurring at temperatures lower than T_β as measured by means of dynamic mechanical analysis [18]. This is somewhat at variance with the proposal originally put forth by Struik [6] that the temperature range in which the aging occurs in a polymer is between the glass transition temperature and its first secondary transition (the α and β re-laxations in amorphous polymers, respectively). A similar conclusion was obtained by Mc-Kenna and Kovacs [31], who observed aging in PMMA in the region in which the β transition becomes strong, and by Lee and McGarry [32], who observed volume recovery at temper-atures below the β relaxation in polystyrene.

Because of the two mechanisms evident in the creep response of PEN, a simple time-aging time superposition would not be expected unless both processes had the same time-aging time dependencies. Manual shifting of the curves demonstrates this as shown in Fig. 3 for the amorphous PEN. In the figure, the shifting is performed relative to the short time process at 30°C and 50°C and relative to the long time process at 80 and 100°C. At 30 and 50°C, the compliance displays some deviations from the "master curve" at the longest times while at 80 and 100°C the deviations occur at the short times. Similar results (Fig. 4) were obtained for the semi-crystalline PEN, even though the β process is somewhat weaker for this material than for the amorphous PEN.

Similarly, one would anticipate that the two viscoelastic processes might have different temperature dependencies and that time-temperature superposition would also be violated. This is indeed the case, and it is demonstrated clearly in Fig. 5 for the amorphous PEN and in Fig. 6 for the semi-crystalline PEN.

Because of the manifestation of two viscoelastic mechanisms for both amorphous and semi-crystalline PEN, we turned to Eq 4 in our attempts to reconcile the aging and temper-ature data. A least-squares optimization routine was used to obtain the different parameters in the Eq 4. In this equation the two relaxation processes that contribute to the creep response are separated into the β and α contributions. The Cole-Cole function characterizes the β contribution and KWW function characterizes the α process. Here, it needs to be remarked that a pure least-squares approach was not used. Rather, the data were first fitted allowing the parameters to float. Then, "representative" values were chosen for the exponent n of the Cole-Cole contribution to the response and for the exponent m of the KWW contribution to the response. This was done to assure that the time-aging time and time-temperature super-positions hold for each process. The lines in Fig. 1 represent the fits to the data for the amorphous PEN and it is seen that the fit of Eq 4 to these data is very good. Similar quality representations were obtained for the semi-crystalline PEN and are represented by the lines in Fig. 2. The parameters found from the fits to the data are presented in Table 1 for the amorphous PEN and in Table 2 for the semi-crystalline PEN.

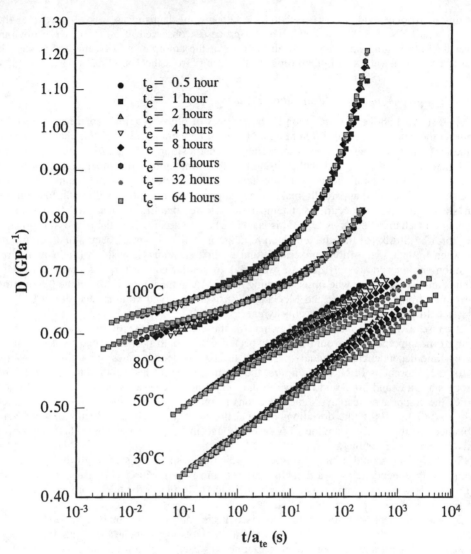

FIG. 3—*Manual time-aging time reduction at the different temperatures in amorphous PEN. Figure demonstrates lack of time-aging time superposition.*

Discussion

The above observations indicate that both amorphous and semi-crystalline PEN do not follow simple time-temperature and time-aging time superposition. On the other hand, we do find that the representation of the data for each material by an equation that includes two viscoelastic mechanisms reconciles all of the data presented here. We note that in recent work comparing dielectric with mechanical responses, Zorn, et al. [33] and McKenna, et al. [34] discuss the difficulty of using narrow time or frequency windows of observation to validate time-temperature superposition. In this study, the degree of success obtained and the possibility that we have superposition for each mechanism separately seems to stem from

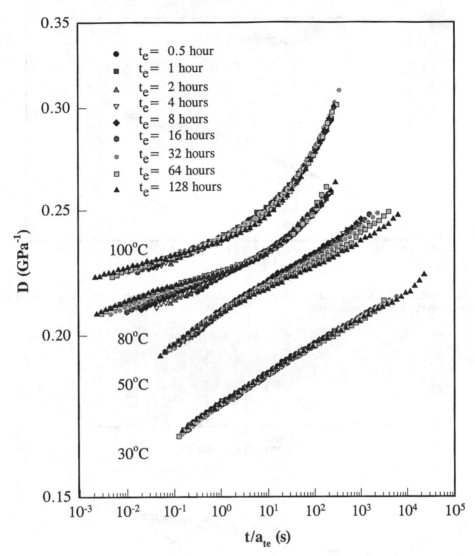

FIG. 4—*Manual time-aging time reduction at the different temperatures in semi-crystalline PEN. Figure demonstrates lack of time-aging time superposition.*

the fact that both mechanisms are observable over most of the range of the measurements, i.e., they are very close together. Therefore, it is of some interest to further investigate the aging time and temperature dependencies of the two processes along with the changes in the relevant parameters from Eq 4. Furthermore, the data allow us to make comparisons of the behavior of the amorphous material with that of the semi-crystalline material.

One of the first points to be considered arises from the observation that, while all of the data for each material could be fit with Eq 4 and with constant values of n and m, the values of n and m are very different between the two materials. The semi-crystalline PEN exhibits a higher value of n and lower value of m than does the amorphous material. This is consistent

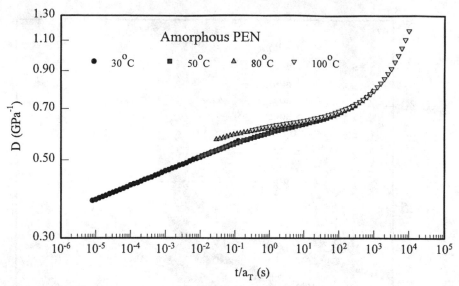

FIG. 5—*Reduced compliance curves for amorphous PEN at different temperatures. Aging time is 32 h and reference temperature is 100°C.*

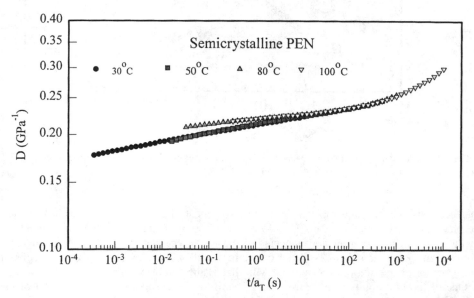

FIG. 6—*Reduced compliance curves for semi-crystalline PEN at different temperatures. Aging time is 32 h and reference temperature is 100°C.*

TABLE 1—*Creep parameters and standard deviations obtained by least-squares optimization of Eq 4 in amorphous PEN.*

Aging Time (h)	$D_{u\beta}$ (GPa^{-1})	ΔD_β (GPa^{-1})	τ_β (s)	n	ΔD_α (GPa^{-1})	τ_α (s)	m
30°C							
0.5	0.305 ± 0.007	0.48 ± 0.01	32 ± 2.0	0.19	0.87 ± 0.03	7.9 10^4 ± 2.6 10^3	0.65
2	0.305 ± 0.004	0.451 ± 0.006	31.5 ± 0.9	0.19	0.87 ± 0.03	1.6 10^5 ± 2.3 10^3	0.65
4	0.305 ± 0.004	0.445 ± 0.004	31.5 ± 0.8	0.19	0.87 ± 0.02	2.4 10^5 ± 3.1 10^3	0.65
8	0.305 ± 0.004	0.434 ± 0.003	31.5 ± 0.8	0.19	0.87 ± 0.02	4.3 10^5 ± 5.5 10^3	0.65
16	0.305 ± 0.004	0.423 ± 0.002	31.5 ± 0.7	0.19	0.87 ± 0.02	7.8 10^5 ± 9.2 10^3	0.65
32	0.305 ± 0.002	0.412 ± 0.001	31.5 ± 0.4	0.19	0.87 ± 0.01	1.4 10^6 ± 7.9 10^4	0.65
64	0.305 ± 0.003	0.401 ± 0.001	31.5 ± 0.5	0.19	0.87 ± 0.01	1.7 10^6 ± 1.1 10^4	0.65
50°C							
0.5	0.322 ± 0.025	0.427 ± 0.018	0.49 ± 0.04	0.19	0.90 ± 0.04	3.4 10^4 ± 1.3 10^3	0.65
1	0.322 ± 0.021	0.419 ± 0.015	0.49 ± 0.03	0.19	0.90 ± 0.02	5.7 10^4 ± 1.9 10^3	0.65
2	0.322 ± 0.019	0.414 ± 0.010	0.49 ± 0.03	0.19	0.90 ± 0.02	9.4 10^4 ± 2.1 10^3	0.65
4	0.322 ± 0.017	0.100 ± 0.008	0.49 ± 0.02	0.19	0.90 ± 0.01	1.6 10^5 ± 3.1 10^3	0.65
8	0.322 ± 0.015	0.392 ± 0.004	0.49 ± 0.02	0.19	0.90 ± 0.01	3.1 10^5 ± 4.1 10^3	0.65
16	0.322 ± 0.012	0.382 ± 0.002	0.49 ± 0.02	0.19	0.90 ± 0.01	6.5 10^5 ± 6.4 10^3	0.65
32	0.322 ± 0.005	0.370 ± 0.004	0.49 ± 0.01	0.19	0.90 ± 0.01	1.1 10^6 ± 4.3 10^4	0.65
64	0.322 ± 0.008	0.358 ± 0.005	0.49 ± 0.01	0.19	0.90 ± 0.01	1.7 10^6 ± 7.5 10^4	0.65
80°C							
0.5	0.33 ± 0.05	0.42 ± 0.01	4 10^{-3} ± 4 10^{-4}	0.19	0.95 ± 0.04	6.4 10^3 ± 5.3 10^2	0.65
1	0.33 + 0.04	0.395 + 0.007	5 10^{-3} ± 3 10^{-4}	0.19	0.95 ± 0.04	1.1 10^4 ± 5.2 10^2	0.65
2	0.33 ± 0.04	0.387 ± 0.005	6 10^{-3} ± 3 10^{-4}	0.19	0.95 ± 0.03	1.8 10^4 ± 6.1 10^2	0.65
4	0.33 ± 0.03	0.380 ± 0.003	6 10^{-3} ± 2 10^{-4}	0.19	0.95 ± 0.03	3.0 10^4 ± 5.9 10^2	0.65
8	0.33 ± 0.03	0.371 ± 0.004	7 10^{-3} ± 2 10^{-3}	0.19	0.95 ± 0.03	5.0 10^4 ± 1.6 10^3	0.65
16	0.33 ± 0.03	0.355 ± 0.004	7 10^{-3} ± 1.10^{-4}	0.19	0.95 ± 0.01	9.8 10^4 ± 3.4 10^3	0.65
32	0.33 ± 0.03	0.349 ± 0.004	7 10^{-3} ± 1 10^{-4}	0.19	0.95 ± 0.01	2.0 10^5 ± 8.9 10^3	0.65
64	0.33 ± 0.02	0.332 ± 0.003	8 10^{-3} ± 1 10^{-4}	0.19	0.95 ± 0.01	3.3 10^5 ± 7.1 10^3	0.65
100°C							
0.5	0.36 ± 0.04	0.40 ± 0.02	7 10^{-4} ± 5 10^{-5}	0.19	1.20 ± 0.04	1.1 10^3 ± 4.8 10^1	0.65
1	0.36 ± 0.04	0.39 ± 0.01	7 10^{-4} ± 5 10^{-5}	0.19	1.20 ± 0.05	1.6 10^3 ± 5.9 10^1	0.65
2	0.36 ± 0.05	0.37 ± 0.01	7 10^{-4} ± 6 10^{-5}	0.19	1.20 ⊥ 0.03	2.6 10^3 ⊥ 1.1 10^2	0.65
4	0.36 ± 0.03	0.36 ± 0.01	7 10^{-4} ± 5 10^{-5}	0.19	1.20 ± 0.03	4.6 10^3 ± 1.5 10^2	0.65
8	0.36 ± 0.03	0.35 ± 0.01	7 10^{-4} ± 5 10^{-5}	0.19	1.20 ± 0.03	8.1 10^3 ± 2.5 10^2	0.65
16	0.36 ± 0.03	0.342 ± 0.009	8 10^{-4} ± 5 10^{-5}	0.19	1.20 ± 0.02	1.5 10^4 ± 3.8 10^2	0.65
32	0.36 ± 0.03	0.327 ± 0.009	8 10^{-3} ± 5 10^{-5}	0.19	1.20 ± 0.01	2.7 10^4 ± 7.8 10^2	0.65
64	0.36 ± 0.03	0.319 ± 0.007	8 10^{-3} ± 5 10^{-5}	0.19	1.20 ± 0.01	5.2 10^4 ± 8.0 10^2	0.65

with an interpretation that the β process in PEN is slightly broader in the amorphous material than in the semi-crystalline one. On the other hand, the fact that m is higher for the amorphous material suggests that the effect on the α mechanism is just the opposite. The crystallites are expected to reduce the mobility in some parts of the amorphous regions. Struik [38] defined two different amorphous segments: the amorphous segments near the crystals are highly constrained and those further away from the crystal surface are unconstrained, hence, more mobile. The effect is to broaden the distribution of retardation times. This is consistent with the decreased value for m for the semi-crystalline PEN.

Referring to Figs. 7 and 8, we now examine the shifting behaviors for the characteristic times that describe the individual mechanisms. Recalling the definition of the time-aging time shift factor a_{te} the characteristic times in Tables 1 and 2 can be used to determine the shift factors. The double logarithmic shift rate μ is defined as [6].

TABLE 2—*Creep parameters and standard deviations obtained by least-squares optimization of Eq 4 in semi-crystalline PEN.*

Aging Time (h)	$D_{u\beta}$ (GPa^{-1})	ΔD_β (GPa^{-1})	τ_β (s)	n	ΔD_α (GPa^{-1})	τ_α (s)	m
30°C							
0.5	0.130 ± 0.004	0.089 ± 0.011	2.2 ± 0.2	0.20	0.15 ± 0.02	6.0 10^6 ± 1.3 10^4	0.51
1	0.130 ± 0.003	0.089 ± 0.001	2.6 ± 0.2	0.20	0.15 ± 0.01	6.2 10^6 ± 1.4 10^4	0.51
2	0.130 ± 0.003	0.089 ± 0.001	2.6 ± 0.1	0.20	0.150 ± 0.009	6.5 10^6 ± 1.5 10^4	0.51
4	0.130 ± 0.003	0.088 ± 0.001	2.9 ± 0.1	0.20	0.150 ± 0.008	7.0 10^6 ± 1.8 10^4	0.51
8	0.130 ± 0.002	0.088 ± 0.001	3.03 ± 0.09	0.20	0.150 ± 0.005	7.5 10^6 ± 1.8 10^4	0.51
16	0.030 ± 0.002	0.087 ± 0.001	3.08 ± 0.08	0.20	0.150 ± 0.004	7.9 10^6 ± 2.2 10^4	0.51
32	0.130 ± 0.001	0.087 ± 0.001	3.10 ± 0.07	0.20	0.150 ± 0.002	8.4 10^6 ± 1.5 10^4	0.51
64	0.030 ± 0.001	0.085 ± 0.001	3.11 ± 0.05	0.20	0.150 ± 0.001	9.5 10^6 ± 1.5 10^4	0.51
128	0.130 ± 0.001	0.085 ± 0.001	3.29 ± 0.04	0.20	0.150 ± 0.001	1.0 10^7 ± 1.5 10^4	0.51
50°C							
0.5	0.13 ± 0.01	0.119 ± 0.002	4 10^{-2} ± 5 10^{-3}	0.20	0.29 ± 0.02	5.6 10^5 ± 3.2 10^4	0.51
1	0.13 ± 0.01	0.119 ± 0.002	4 10^{-2} ± 4 10^{-3}	0.20	0.29 ± 0.02	7.8 10^5 ± 1.7 10^4	0.51
2	0.130 ± 0.008	0.118 ± 0.002	5 10^{-2} ± 3 10^{-3}	0.20	0.29 ± 0.02	8.0 10^5 ± 1.6 10^4	0.51
4	0.130 ± 0.007	0.117 ± 0.001	5 10^{-2} ± 3 10^{-3}	0.20	0.29 ± 0.01	9.3 10^5 ± 2.5 10^4	0.51
8	0.130 ± 0.006	0.117 ± 0.001	5 10^{-2} ± 3 10^{-3}	0.20	0.29 ± 0.01	1.2 10^6 ± 1.7 10^4	0.51
16	0.130 ± 0.006	0.115 ± 0.001	6 10^{-2} ± 3 10^{-3}	0.20	0.287 ± 0.009	1.6 10^6 ± 1.9 10^4	0.51
32	0.130 ± 0.005	0.113 ± 0.001	6 10^{-2} ± 2 10^{-3}	0.20	0.287 ± 0.009	2.6 10^6 ± 1.7 10^4	0.51
64	0.130 ± 0.005	0.111 ± 0.001	6 10^{-2} ± 3 10^{-3}	0.20	0.287 ± 0.009	4.4 10^6 ± 2.2 10^4	0.51
128	0.130 ± 0.005	0.110 ± 0.001	6 10^{-2} ± 3 10^{-3}	0.20	0.287 ± 0.009	9.5 10^6 ± 1.3 10^4	0.51
80°C							
0.5	0.13 ± 0.01	0.112 ± 0.009	7 10^{-4} ± 5 10^{-5}	0.20	0.37 ± 0.01	4.3 10^4 ± 4.8 10^3	0.51
10^4	0.13 ± 0.01	0.112 ± 0.005	7 10^{-4} ± 4 10^{-5}	0.20	0.37 ± 0.01	6.7 10^4 ± 4.3 10^3	0.51
2	0.13 ± 0.01	0.110 ± 0.003	7 10^{-4} ± 4 10^{-5}	0.20	0.373 ± 0.008	1.2 10^5 ± 5.1 10^3	0.51
4	0.130 ± 0.006	0.108 ± 0.003	7 10^{-4} ± 3 10^{-5}	0.20	0.373 ± 0.007	2.0 10^5 ± 8.1 10^3	0.51
8	0.130 ± 0.006	0.106 ± 0.001	7 10^{-4} ± 3 10^{-5}	0.20	0.373 ± 0.005	4.0 10^5 ± 1.1 10^4	0.51
16	0.130 ± 0.005	0.103 ± 0.001	7 10^{-4} ± 3 10^{-5}	0.20	0.373 ± 0.005	7.1 10^5 ± 1.3 10^4	0.51
32	0.130 ± 0.005	0.999 ± 0.001	7 10^{-4} ± 2 10^{-5}	0.20	0.373 ± 0.007	1.6 10^6 ± 3.4 10^4	0.51
64	0.130 ± 0.004	0.097 ± 0.001	7 10^{-4} ± 2 10^{-5}	0.20	0.373 ± 0.005	2.2 10^6 ± 3.5 10^4	0.51
128	0.130 ± 0.004	0.096 ± 0.001	7 10^{-4} ± 2 10^{-5}	0.20	0.373 ± 0.004	3.3 10^6 ± 1.9 10^4	0.51
100°C							
0.5	0.13 ± 0.01	0.124 ± 0.007	9 10^{-5} ± 8 10^{-6}	0.20	0.38 ± 0.01	1.1 10^4 ± 9.1 10^2	0.51
1	0.13 ± 0.01	0.122 ± 0.006	9 10^{-5} ± 7 10^{-6}	0.20	0.380 ± 0.009	1.8 10^4 ± 1.7 10^3	0.51
2	0.130 ± 0.009	0.116 ± 0.004	9 10^{-5} ± 5 10^{-6}	0.20	0.380 ± 0.008	3.3 10^4 ± 2.3 10^3	0.51
4	0.130 ± 0.008	0.112 ± 0.003	9 10^{-5} ± 4 10^{-6}	0.20	0.380 ± 0.008	6.3 10^4 ± 1.3 10^3	0.51
8	0.130 ± 0.008	0.111 ± 0.003	9 10^{-5} ± 2 10^{-6}	0.20	0.380 ± 0.006	1.0 10^5 ± 1.9 10^3	0.51
16	0.130 ± 0.007	0.109 ± 0.002	9 10^{-5} ± 2 10^{-6}	0.20	0.380 ± 0.006	1.6 10^5 ± 1.5 10^3	0.51
32	0.130 ± 0.006	0.108 ± 0.002	9 10^{-5} ± 2 10^{-6}	0.20	0.380 ± 0.006	2.3 10^5 ± 1.8 10^3	0.51
64	0.130 ± 0.004	0.104 ± 0.002	9 10^{-5} ± 1 10^{-6}	0.20	0.380 ± 0.005	5.4 10^5 ± 5.6 10^3	0.51
128	0.130 ± 0.004	0.098 ± 0.001	9 10^{-5} ± 1 10^{-6}	0.20	0.380 ± 0.03	1.2 10^6 ± 4.1 10^4	0.51

$$\mu = \frac{d \log(a_{te})}{d \log(t_e)} \qquad (9)$$

and we determine μ from double logarithmic representations of a_{te} versus t_e.

Figure 9 shows such a representation for both the semi-crystalline and amorphous materials for the four temperatures studied and for the alpha process. The values of μ_α are also given in the plot. It can be seen that, for the amorphous PEN, μ_α increases from 30 to 50°C and remains approximately constant between 50, and 80, and 100°C. This result is con-

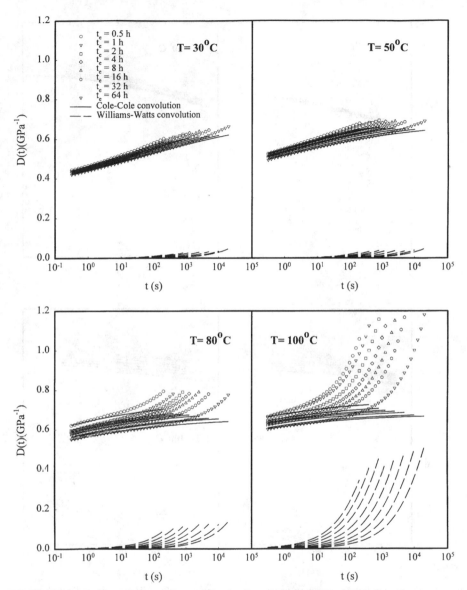

FIG. 7—*Representation of the creep compliance of amorphous PEN at the indicated aging temperatures and aging times and the corresponding contributions from the β and the α terms in Eq 4.*

sistent with those observed for the shifting of the α process in other amorphous materials [6,16,29].

In the semi-crystalline PEN, the aging behavior is different from that of the amorphous material. At 80°C and 100°C, the shift factor for the α process follows normal behavior and is approximately the same as for the amorphous material at the same temperatures. However, at 30°C and 50°C, the shifting behavior appears to be anomalous. At 50°C, there are two

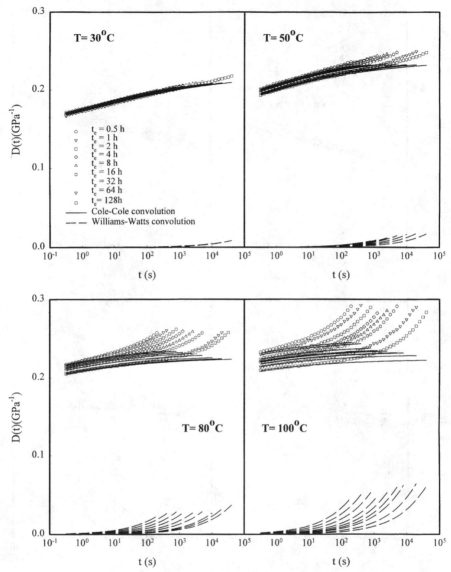

FIG. 8—*Representation of the creep compliance of semi-crystalline PEN at the indicated aging temperatures and aging times and the corresponding contributions from the β and α terms in Eq 4.*

regimes of behavior. At short aging times, the shift rate is very low ($\mu_1 \approx 0.28$) and increases towards a value typical of the shift rates at 80°C and 100°C ($\mu \approx 0.86$). At 30°C, the shift rate remains very low throughout the full range of aging times investigated ($\mu \approx 0.1$). While it is expected [6] that the aging rate might decrease as one gets further below the glass transition or approaches the β "temperature," to our knowledge, the observation of the two aging rates reported here (the long time shift rate is higher than that obtained at short aging times) for the data obtained at 50°C are novel. One possible explanation for the apparently

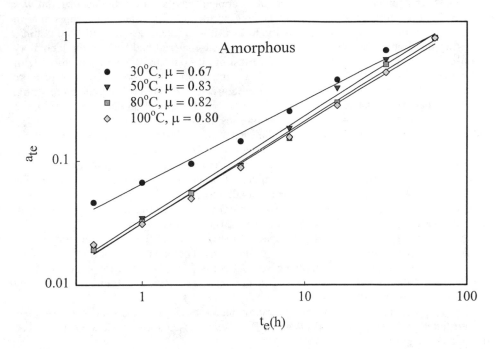

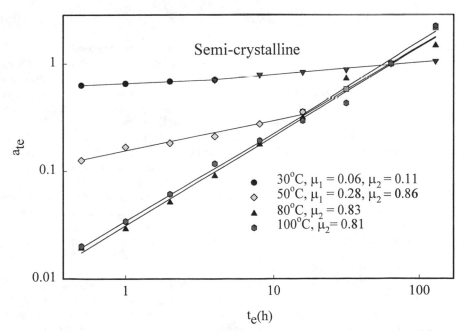

FIG. 9—*Aging time shift factors for the α process as a function of aging time at the different temperatures in amorphous and semi-crystalline PEN. The inverted solid triangles and the solid squares in the lower figure refer to the shift factors at 30°C and 50°C aftr the slope changes to the second regime. See text for discussion.*

anomalous behavior arises from the thermal treatment. The semi-crystalline material was tested without a "normalization" step of heating above the glass transition temperature in order not to perturb the state of crystallinity. Rather, it was taken "as-received" and put into the testing machine at the aging temperature. Since the material had been in the laboratory and in transit for up to several months before testing, it could have aged at the room temperature of approximately 24°C for this time. Testing at 30°C is close to room temperature and the reported aging times would need to be corrected for the room temperature aging. One would have a similar result for the 50°C tests, but the correction due to aging at room temperature would be less, apparently becoming insignificant after about 20 h of aging.

While the plots in Figs. 7 and 8 show that the β process is affected by the aging of the material, it is clear from Tables 1 and 2 that there is little, if any, significant effect of aging on τ_β and, therefore, on the aging time shift factor. As a result, to good approximation, $0 < \mu_\beta \leq 0.15$ for all of the temperatures examined. This is true for both the semi-crystalline and amorphous materials. These changes are small enough that it is unclear if they are real. If the characteristic time for the β process is unaffected by aging, this is consistent with what has been observed by Read [35]. This is discussed further below.

The above discussion focused on the impact of the physical aging process and temperature on the characteristic times for the β and α mechanisms. The other aspect of thermorheological simplicity is the practice of applying vertical shifts [7,33,34,39] to the data as defined by Eq 7. Here, the factors that are related to the vertical shifts are ΔD_β and ΔD_α and they reflect the strength of the dispersion of interest.

For the β process ΔD_β decreases as aging time increases. Such a result for other polymers was attributed by Read [35] to a decrease in $D_{R\beta}$. He proposed a linear relationship between ΔD_β and the aging time

$$\Delta D_\beta = Bt_e^{-k} \tag{10}$$

In addition, at the very short creep times $D_\alpha(t)$ was assumed to be negligible and τ_β was assumed to be essentially independent of aging time [35]. As discussed above, we found τ_β to be nearly independent of the aging time. Also, as seen in Fig. 10, log ΔD_β follows a linear dependence on log t_e, as suggested by Eq 10 for both amorphous and semi-crystalline PEN materials. ΔD_β decreases with increasing aging time and this has been interpreted as a reduction in the number of active groups that participate in the β process [8,36,37].

Further examination of Fig. 10 suggests that the values of k (Eq 10) for both the amorphous and semi-crystalline PEN increase slightly as temperature increases. This would imply that the number of active groups is less affected by the aging process at low temperature. We also note that the magnitudes of ΔD_β for amorphous PEN are larger than those in the semi-crystalline material. Also, we note that the value of ΔD_β decreases with increasing temperature for the amorphous material, while it seems to increase with increasing temperature for the semi-crystalline material. Because the magnitudes of ΔD_β for the semi-crystalline material are much smaller than those of the amorphous material, it is unclear whether or not the uncertainties in the measurements allow this conclusion to be sustained, and further work would be needed to substantiate this result.

On the other hand, ΔD_α, referring to Tables 1 and 2, is independent of the aging time at each temperature for both amorphous and semi-crystalline materials. However, at the higher temperatures, it is still a significant proportion of the total response, even at short times. This is a somewhat different expectation from that of Read [35] in the materials he studied where the β and α dispersions were more widely separated.

Finally, we can see the impact of temperature on the individual mechanisms for the amorphous and semi-crystalline PEN by examination of Figs. 11 and 12. Here we take data

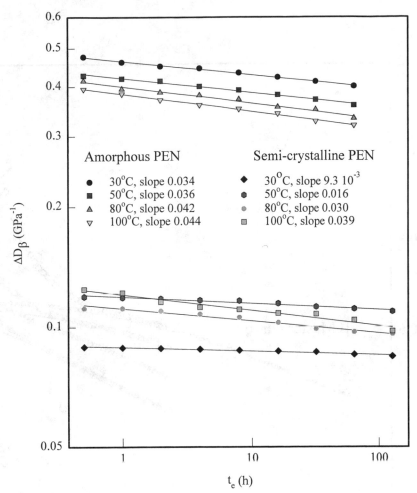

FIG. 10—*Double logarithmic representation of retardation strength in the β process with aging time at different temperatures in amorphous and semi-crystalline PEN.*

obtained at 32 h of aging and compare them at the four different temperatures tested. As can be seen, the model permits formation of different "master curves" for each mechanism because it is constructed such that thermo-rheological simplicity holds for the separate mechanisms. Also seen is the fact that the data themselves are not thermo-rheologically simple, i.e., they would not superimpose to form a master curve.

Summary

The physical aging response of amorphous and semi-crystalline PEN films has been investigated using sequential creep experiments for temperatures from 30 to 100°C and aging times from 0.5 to 128 h. The results show two distinct viscoelastic mechanisms which are changed by both aging time and temperature. For both the amorphous and semi-crystalline PEN the body of experimental observations could be reconciled by using a creep compliance

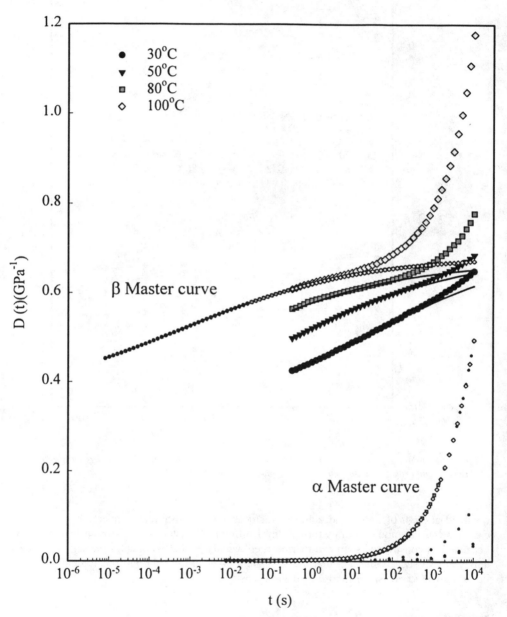

FIG. 11—*Compliance curves for amorphous PEN at different temperatures at an aging time of 32 h. Corresponding β (-----) and α (.......) contributions to the creep response and corresponding β and α master curves at 100°C as reference temperature.*

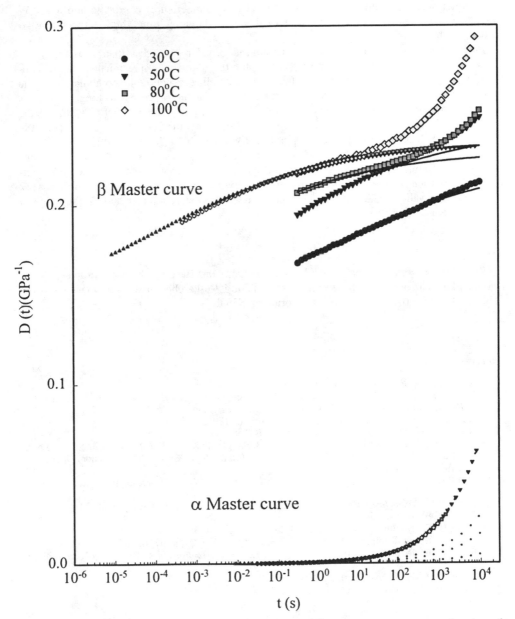

FIG. 12—*Compliance curves for semi-crystalline PEN at different temperatures at an aging time of 32 h. Corresponding β (------) and α (.........) contributions to the creep response and corresponding β and α master curves at 100°C as reference temperature.*

function that was thermo-rheologically complex for the total response, but for which each mechanism was thermo-rheologically simple. The function chosen (Eq 4) includes a Cole-Cole function (time domain) for the short time (low temperature) β process and a KWW stretched exponential—type of function to describe the long time (high temperature) α process.

It is found that the amorphous PEN exhibits a broader β process than does the semi-crystalline material, while the opposite is true for the α process—presumably due to constraining effects of the crystallites in the semi-crystalline material. In both materials, the aging process shifts the characteristic retardation time for the α process to longer times, as is commonly observed in aging experiments. However, there is little effect of aging on the characteristic times for the β process. On the other hand, the strength of the viscoelastic retardation ΔD_β of the β dispersion is significantly changed by the physical aging while the strength of the α dispersion ΔD_α is virtually unaffected. The results show that strict thermo-rheological simplicity need not obtain in order to describe the viscoelastic behavior of polymer glasses and semi-crystalline polymers that exhibit multiple relaxations. Further work is required to validate the material description in a quantitative manner outside the range of measurements.

Acknowledgments

Part of this work was supported by a grant from the Eastman Kodak Company. MLC would like to thank NIST for support as a Visiting Researcher in the Polymers Division. The work was carried out in the laboratories of NIST.

References

[1] Kovacs, A. J., *Fortschr. Hochpolym.-Forsch.*, Vol. 3, 1964, p. 394.
[2] McKenna, G. B., "Glass Formation and Glassy Behavior," *Comprehensive Polymer Science, Vol. 2, Polymer Properties,* C. Booth and C. Price, Eds., Pergamon, Oxford, England, 1989.
[3] Hutchinson, J. M., *Prog. Polym. of Science,* Vol. 20, 1995, p. 703.
[4] Hodge, I. M., *Journal of Non-Crystalline Solids* Vol. 169, 1994, p. 211.
[5] Moynihan, C. T., Macedo, P. B., Montrose, C. J., Gupta, P. K., DeBolt, M. A., Dill, J. F., Dom, B. E., Drake, P. W., Easteal, A. J., Elterman, P. B., Moeller, R. P., Sasabe, H. and Wilder, J. A., *Ann. N.Y. Acad. Sci.,* Vol. 279, 1976, p. 15.
[6] Struik, L. C. E., *Physical Aging in Amorphous Polymers and Other Materials,* Elsevier, Amsterdam, 1978.
[7] Ferry, J. D., *Viscoelastic Properties of Polymers,* 3rd. Edition, Wiley, New York, 1980.
[8] Read, B. E., Tomlins, P. E., and Dean, G. D., *Polymer,* Vol. 31, 1990, p. 1204.
[9] Read, B. E., Tomlins, P. E., and Dean G. D., *Polymer,* Vol. 35, 1994, p. 4376.
[10] Read, B. E. and Tomlins, P. E., *Polymer Eng. and Sci.,* Vol. 37, 1997, p. 1572.
[11] Read, B. E., Dean, G. D., Tomlins, P. E., and Lesniarek-Hamid, J. L., *Polymer,* Vol. 33, 1992, p. 2689.
[12] Struik, L. C. E., *Polymer,* Vol. 28, 1987, p. 1521.
[13] Struik, L. C. E., *Polymer,* Vol. 28, 1987, p. 1534.
[14] Struik, L. C. E., *Polymer,* Vol. 30, 1989, p. 799.
[15] Struik, L. C. E., *Polymer,* Vol. 30, 1989, p. 815.
[16] Beckmann, J., McKenna, G. B., Landes, B. G., Bank, D. H., and Bubeck, R. A., *Polymer Eng. and Sci.,* Vol. 37, 1997, p. 1459.
[17] Spinu, I. and McKenna, G. B., *Journal of Plastic Film and Sheeting,* Vol. 13, 1997, p. 311.
[18] Gillmor, J. R. and Greener, J., *Proceedings ANTEC' 97,* SPE, Vol. II, 1997, p. 1582.
[19] Kohlrausch, F., *Pogg. Ann. Phys.,* Vol. 12, 1847, p. 393.
[20] Williams, G. and Watts, D. C., *Trans. Faraday Soc.,* Vol. 66, 1970, p. 80.

[21] Lee, A. and McKenna, G. B., *Polymer*, Vol. 31, 1990, p. 423.
[22] Tomlins, P. E., *Polymer*, Vol. 37, 1996, p. 3907.
[23] Read, B. E., Dean, G. D., and Tomlins, P. E., *Polymer*, Vol. 29, 1988, p. 2159.
[24] Tomlins, P. E. and Read, B. E., *Polymer*, Vol. 39, 1998, p. 355.
[25] Dean, G. D., Read, B. E., and Small, G. D., *Plast. Rubber Process. Appl.*, Vol. 9, 1988, p. 173.
[26] McCrum, N. G., Read, B. E., and Williams, G., *Anelastic and Dielectric Effects in Polymeric Solids*, Wiley, New York, 1967.
[27] Seyler, R. J., editor, *Assignment of the Glass Transition, ASTM STP 1249*, American Society for Testing and Materials, Philadelphia, PA, 1994.
[28] Lee, A. and McKenna, G. B., *Polymer*, Vol. 29, 1988, p. 1812.
[29] O'Connell, P. A. and McKenna, G. B., *Polymer Eng. and Sci.*, Vol. 37, 1997, p. 1485.
[30] Bradshaw R. D. and Brinson, L. C., *Polymer Eng. and Sci.*, Vol. 37, 1997, p. 31.
[31] McKenna, G. B. and Kovacs, A. J., *Polymer Eng. and Scie.*, Vol. 24, 1984, p. 1131.
[32] Lee, H. H.-D. and McGarry, F. J., *Polymer*, Vol. 34, 1993, p. 4267.
[33] Zorn, R., Mopsik, F. I., McKenna, G. B., Willner, L., and Richter, D., *J. Chem. Phys.*, Vol. 107, 1997, p. 3645.
[34] McKenna, G. B., Mopsik, F. I., Zorn, R., Willner, L., and Richter, D., *Proceedings ANTEC' 97*, SPE, Vol. II, 1997, p. 1027.
[35] Read. B. E., *Journal of Non-Crystalline Solids*, Vol. 131–133, 1991, p. 408.
[36] Diaz-Calleja, R., Ribes-Greus, A., and Gomez-Ribelles, J. L., *Polymer*, Vol. 30, 1989, p. 1433.
[37] Guerdoux, L. and Marchal, E., *Polymer*, Vol. 22, 1981, p. 1199.
[38] Struik, L. C. E., *Polymer*, Vol. 28, 1987, p. 1521.
[39] Markovitz, H., *Journal of Polymer Science*, Symposia, Vol. 50, 1975, p. 431.

Ihor D. Skrypnyk,[1] Jan L. Spoormaker,[2] and Prabhu Kandachar[2]

A Constitutive Model for Long-Term Behavior of Polymers

REFERENCE: Skrypnyk, I. D., Spoormaker, J. L., and Kandachar, P., **"A Constitutive Model for Long-Term Behavior of Polymers,"** *Time Dependent and Nonlinear Effects in Polymers and Composites, ASTM STP 1357,* R. A. Schapery and C. T. Sun, Eds., American Society for Testing and Materials, West Conshohocken, PA, 2000, pp. 70–82.

ABSTRACT: This paper describes a constitutive model for long-term behavior of plastics. The Struik theory of physical aging is adapted for use with the earlier proposed generalization of the Schapery model. Experimental data on creep and recovery behavior for different elapsed times are necessary to build the model. The parameter identification procedure is based on a minimization of the total error between experimental data and model prediction.

The model has been verified using extensive experimental data on long-term creep and recovery of polypropylene, reported by the National Physical Laboratory (NPL, Teddington, U.K.). The model has shown the ability to work with the independent data; it sufficiently describes the variation of strains in time for loading history, which contains several loading steps. The maximum deviation of the model prediction from the experimental data in a control set is less than 11% to 12%. The average deviation is less than 4%.

KEYWORDS: nonlinear viscoelasticity, short-term and long-term creep, recovery, physical aging of polymers

Nomenclature

σ, $\sigma(t)$	Stress and stress history (scalar notation)
$\varepsilon[\sigma,t]$	Strain as a function of stress and time (one-dimensional formulation)
$\hat{\varepsilon}[\sigma_k,t]$	Experimental data from creep tests with stress level σ_k
t, t_e	Time under load and elapsed time (prior to loading), respectively
$t_{1,2}$	Time moments, when the stress levels change in multiple-steps-loading history
$\xi(t)$	Reduced time
ξ	Dummy variable in the hereditary integral $\zeta < t$
$\log(a_t)$	Horizontal shifting factor for time-elapsed time superposition principle
$\log(b_t)$	Vertical shifting factor for time-elapsed time superposition principle
μ	Horizontal shifting rate
η, η_i	Vertical shifting rate (rates)
T_g	Glass transition temperature
T_g^L, T_g^U	Lower and upper limit for glass transition for semicrystalline polymers

[1] Senior researcher, Delft University of Technology, on leave from Karpenko Physico-Mechanical Institute, National Academy of Sciences of Ukraine, 5 Naukova st., 290601, Lviv, Ukraine.

[2] Professor and associate professor, respectively, Laboratory of Mechanical Reliability, Faculty of Industrial Design Engineering, Delft University of Technology, Jaffalaan 9, 2628 BX, Delft, The Netherlands.

$J_0[\sigma]$ Instantaneous (time-independent) strain as a function of applied stresses
$J(t)$ Time-dependent creep compliance function
$F_i(t)$ Time-dependent part of creep function
τ_i Retardation times (for the Kelvin-Voigt type models)
$\phi_i(\sigma)$ Preintegral stress function in Schapery type models
$g_i(\sigma)$ Nonlinear stress function in hereditary integral in Schapery type models
D_i, λ_i, γ_i, α_i,   Constants in functions $\phi_i(\sigma)$, $g_i(\sigma)$, $F_i(t)$
A, β, θ, ϑ Constants in function $\phi(\sigma,t,t_e)$
$H(x)$ Heavyside function

According to the present-day requirements, plastic products should last from several months to several years. Hence, for designing sustainable products, modeling of the long-term behavior of polymers is necessary. Since simulation of the long-term behavior of plastic products is usually performed using finite-element analysis packages, the model for long-term behavior need not be too simple, but should be easily understood and handled by design engineers to enable them to apply these tools. The model also should be effective and hence the accuracy of the prediction must be sufficient ($<10\%$ of error).

There is another important question—material characterization for long-term behavior of plastics. This point precedes (and is related to) the question of modeling of materials behavior. A set of tests lasting several months or years would be extremely expensive. In addition, it would delay introduction of new polymers and products on the market. Thus, there is an important technical challenge: the development of a method to predict the long-term behavior based on short-term tests. It is generally accepted that for this purpose the physical aging phenomenon has to be taken into account.

This paper aims at the development of a constitutive model for long-term viscoelastic behavior of plastics. For this purpose the Struik theory of physical aging [1] is adapted for use with the earlier proposed generalization of the Schapery model on viscoelasticity. The model developed has been verified on extensive experimental data on long-term creep and recovery of polypropylene (PP), reported by the National Physical Laboratory (NPL, Teddington, U.K.) [2,3].

Some Physical Basics of Polymer Aging

After a quick transition from a temperature above T_g to a temperature below T_g, the amorphous phase of polymers is out of thermodynamic equilibrium. In order to reach equilibrium, a continuous structural transformation proceeds in the material. This transformation relates to the steady reduction of the free volume and to decrease in molecular mobility. The latter leads to an increase of retardation times τ_i. Because of this transformation, the material becomes stiffer and more brittle, while its damping decreases. These phenomena are referred to as the physical aging of polymers. This is a thermodynamically reversible process: if the polymer is heated above T_g, it reaches thermodynamic equilibrium and all the previous aging history will be erased. On the other hand, below the β-transition temperature the mobility of molecular chains is restricted to such an extent that no aging occurs [4].

The majority of the creep curves for different elapsed times prior to loading (i.e., aging periods) t_e can be shifted over the logarithmic time scale (Fig. 1) to a single "master-curve." Let us consider two different elapsed times t_e and $t_e^{(1)}$. Then the relation between creep curves, obtained for t_e and $t_e^{(1)}$, can be also given in mathematical form:

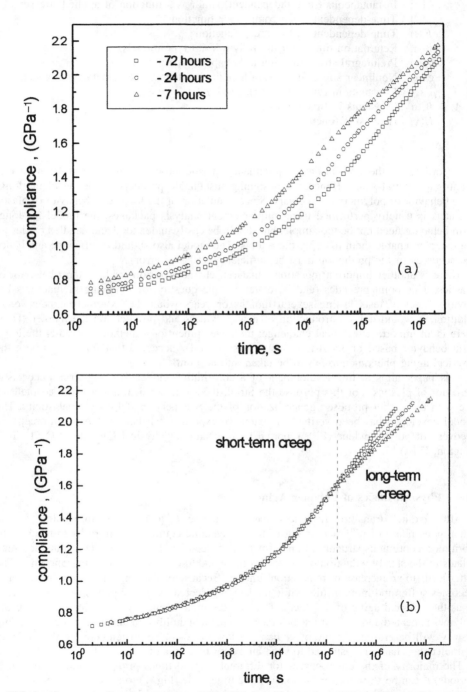

FIG. 1—*Creep compliance curves (σ = 2.96 MPa) for polypropylene (PP) at different elapsed times* t_e *(a), [2,3] and "single" curve, created by shifting (b).*

$$J(t,t_e^{(1)}) = J\left(\frac{t}{a_t}, t_e\right) \tag{1}$$

where $a_t = a_t(t_e^{(1)},t_e)$ is the aging shift factor. This relation is often referred to as the "time-elapsed time" superposition principle. It also can be described in a more specific way: an increase in the elapsed time t_e causes shifting of the retardation (or relaxation) times spectrum to longer times because a_t increases with $t_e^{(1)}$.

Usually, it is impossible to merge the creep compliance curves completely (Fig. 1b). This serves for introduction of the terms "short-term" and "long-term" creep [1]. The elapsed times t_e for all the three curves presented in Fig. 1 are less than 2.5×10^5 s. For the time interval $t < 10^5$ s, the structural effects, caused within this material by physical aging, are nearly constant and these parts of the creep curves coincide. The effects of creep within this time range can be considered separately from the physical aging effects. The creep behavior within this time period is the "short-term creep": $t < t_e$. The physical aging process proceeds also during the creep tests. Therefore, for the tests times considerably larger than the shortest t_e (in this case $\sim 2.0 \times 10^4$ s), the effects caused by physical aging also will grow. In these cases the creep effects can not be superimposed on the aging effects. This creep behavior is called the "long-term creep": $t > t_e$.

Struik found [1] that much aging data obeyed the equation

$$\mu = \frac{d \log a_t}{d \log(t_e)} \tag{2}$$

where μ (the "shift rate") is a positive constant. Introducing normalization

$$a_t(t_e^{(1)},t_e) = 1 \tag{3}$$

Equation 2 can be rewritten in the form

$$a_t = \left(\frac{t_e^{(1)}}{t_e}\right)^{\mu} \tag{4}$$

For the loading that starts at t_e, long-term creep may be accounted for by using $t_e^{(t)} = t_e + t$ in Eq 4 for the shift factor, giving the age at time-under-load, t, and then using the differential form for effective time [1]

$$\partial\zeta = \frac{\partial t}{a_t} \tag{5}$$

If this agrees with the experimental data obtained for a certain material (i.e., the dependencies Eqs 1–5 describe properly the behavior of material investigated), it also means that modeling of long-term behavior of this material is possible based on the data from short-term creep tests for different elapsed time t_e [4,5].

Equation 5 can be rewritten in integral form, to calculate the integral effective time through the complete loading history

$$\zeta(t) = \int_0^t \frac{dt}{a_t(t_e + t, t_e)} \tag{6}$$

For semicrystalline polymers the aging effects are even more complicated due to the complex molecular structure of these materials. Struik has proposed to consider the semicrystalline polymers as two-phase composites [1]. The time-dependent behavior (viscoelasticity) is related to the amorphous phase, while the crystallites are considered as fillers. The amorphous phase in turn can be divided into two regions: the region close to the crystals and the region remote from the crystalline phase. The chain mobility in the first region is restricted by the crystallites [6]. As a result, the creep rate in this region is also restricted and T_g is increased compared with the second region.

Therefore, the amorphous phase does not have a single T_g, but a range of temperatures, where the glass transition occurs. The lower limit T_g^L is related to the glass transition of the amorphous material remote from the crystals, while the upper limit T_g^U corresponds to the material closest to the crystallite phase.

The upper limit of the aging effect occurrence is related to T_g. Hence, for the semicrystalline polymers, this limit cannot be strictly defined. For the temperature $T_g^L < T < T_g^U$, the amorphous region, which is closest to the crystals, undergoes aging, while the material remote from the crystals is already above its glass transition point and is, therefore, not effected by aging.

Struik has studied thoroughly the aging effects in semicrystalline polymers [7–10] and, in particular, the question of applicability of "time-elapsed time" and "time-temperature" superposition principles. For this purpose he has split the range of temperatures investigated into four regions: $T < T_g^L$ (region 1), $T \sim T_g^L$ (region 2), $T_g^L < T < T_g^U$ (region 3), and $T_g^U < T$ (region 4). It has been shown, that in general (i.e., for all regions studied) it is possible to combine the creep curves for different elapsed times t_e using simultaneously horizontal and vertical shifting. Some times only horizontal shifting is necessary.

For the case when both horizontal and vertical shifting are necessary, the relation between creep compliance curves for two different elapsed times t_e and $t_e^{(1)}$ can be written in the form

$$J(t, t_e^{(1)}) = \frac{1}{b_t} J\left(\frac{t}{a_t}, t_e\right) \tag{7}$$

where $t_e^{(1)} = t_e + t$ and the functions

$$a_t = \left(\frac{t_e + t}{t_e}\right)^\mu, \quad b_t = \left(\frac{t_e + t}{t_e}\right)^\eta \tag{8}$$

are the horizontal and vertical shifting factors.

As mentioned before, the above dependencies have been proposed to describe the long-term behavior of plastics based on the short-term test data for different t_e. It also means that for further progress with constitutive model development, the elements of short-term creep modeling [11–13] can be used.

Generalized Form of the Schapery Model

At a temperature just below a transition point, the stresses applied to a polymer can interact with temperature to induce the transition. For instance in PMMA the β-transition occurs at ambient temperature and stresses ~22 to 27 MPa [14]. In this case the shape of creep curves can be different below and above the transition point and quite nonregular at the conditions of transitions. As a result, the time-stress superposition principle cannot provide the model for a whole range of loading: from low stresses to the yield point [15,16]. At the same time, constitutive equations should describe the viscoelastic behavior for the whole range of loading in order to be suitable for the engineering calculations (FE analysis).

The following equation, which is similar to expressions known from harmonic analysis, has shown its ability to describe the experimental creep curves with different, often nonregular shape [12,13]

$$\hat{\epsilon}[\sigma_k, t] = J_0[\sigma_k] + \sum_i F_i(t)g_i(\sigma_k) \tag{9}$$

To a certain extent [12] this representation can be considered as an extension of generalized Kelvin-Voigt model for the nonlinear region. There are also other models [4,5,11], which can be mentioned in this respect. However, only the model, proposed by Lai [4], has qualitatively the same features. Except for use of a stress-dependent reduced time [11], traditional models [5,11] assume a linear difference between stress functions $g_i(\sigma) = D_i * g(\sigma)$ for different terms in series, while the proposed approach allows nonlinear variation.

Similar to [4,5,11], Eq 9 will be rewritten in the hereditary integral form to account for a complex loading history

$$\epsilon[\sigma(t), t] = J_0[\sigma(t)] + \int_0^t \sum_i F_i(t - \xi) \frac{\partial g_i[\sigma(\xi)]}{\partial \xi} d\xi \tag{10}$$

The functions $J_0[\upsilon(t)]$, $F_i(t)$ and $g_i[\upsilon(t)]$ are material functions, which should be estimated from the experimental results $\hat{\epsilon}[\sigma_k, t]$ for the creep tests (Eq 9). The power law form for the functions $J_0[\sigma(t)]$ and $g_i[\sigma(t)]$

$$J_0[\sigma] = A[\sigma + A_1\sigma^\beta]; \quad g_i[\sigma] = D_i\sigma^{\alpha_i} \tag{11}$$

is sufficiently accurate for many materials.

Concerning the time functions $F_i(t)$, the form of Prony series term

$$F_i(t) = (1 - \exp(-\lambda_i t)) \tag{12}$$

seems to be preferable, because it enables efficient implementation of hereditary integral model into FEA [17]. However, while choosing this form of time functions, one should be aware of the peculiarities and restrictions, which come along with this choice. The terms in the Prony series (Eq 12) change their values from 0 to 1. The main growth occurs within two time decades (two orders of change of t variable). Therefore, if behavior of material for certain period of time has to be described using terms of Prony series, the number of terms in Eq 9 should be equivalent to the number of decades [18] in order to reach suitable accuracy. This situation is similar to the representation of geometrical solids by elements and functions by shape functions in FEA. Consequently, the number of model parameters in the Eqs 9, 11, and 12 grows with number of time decades to be described. Another disadvantage is that the model with a Prony series cannot be used to predict the material behavior outside the time region of the curve fit.

On the other hand the localization of the area of the function growth makes a procedure of parameter identification effective.

Equation 10 implies that the strain recovery behavior of material can be described, based on the information gained from creep tests. For instance, this model assumes that the elastic (or instantaneous) part of strains is constant during the loading history. However, for plastics with nonlinear viscoelastic response it is not always the case. Following Schapery [11], the preintegral functions $\phi_i[\sigma(t)]$ can be introduced in order to account for the difference between the creep and recovery behavior [12,13]

$$\varepsilon[\sigma(t),t] = J_0[\sigma(t)] + \sum_i^n \phi_i[\sigma(t)] \int_0^t F_i(t - \xi) \frac{\partial g_i[\sigma(\xi)]}{\partial \xi} d\xi \tag{13}$$

where functions $\phi_i[\sigma(t)]$ is modeled by the exponential form

$$\phi_i[\sigma] = \exp(\gamma_i \sigma) \tag{14}$$

The preintegral function should be not equal to zero when stress is zero. Otherwise the viscous strain would "disappear" with stresses going to zero and the description of strain recovery will be impossible. From all the functions which meet this requirement, the exponential function has minimum variables (only one).

Combining the Eqs 6–8, known from the theory of physical aging of polymers, and Eqs 9–13 for short-term viscoelastic behavior, the hereditary integral model for the long-term behavior of polymers can be formulated. However, based on the previous analysis, some additional assumptions will be introduced. First of all, due to physical aging the polymer becomes stiffer. Assuming this aging applies to the initial compliance J_0, we will use the form

$$J_0[\sigma,t,t_e] = A[\sigma + A_1\sigma^\beta] \left(\frac{t_e + t}{t_e}\right)^\theta \left(\frac{t_e}{t_e^*}\right)^\vartheta \tag{15}$$

The last term in Eq 15 is introduced for normalization. The value of elapsed time $t_e^* = 86\,400$ s has been chosen as a basic level, since most of the data available for the study are obtained for this particular elapsed time. The material parameters A, A_1, β, θ, ϑ were established to fit the data.

Secondly, it has been mentioned above, that the physical aging of polymers affects the retardation times spectrum of material by shifting it as a whole to the longer times. At the same time, it is reported [19] that the relaxation time spectrum of PP also changes the shape during physical aging. In order to account for this peculiarity, different vertical shifting rates η_i for every time decade are introduced, similar to the assumption about the different stress functions $g_i(\sigma)$ for every term of Prony series (Eq 9). As a result, a hereditary integral model for the long-term creep will be written in the following form

$$\varepsilon[\sigma(t),t,t_e] = J_0[\sigma(t),t,t_e] + \sum_i^n \left(\frac{t_e + t}{t_e}\right)^{\eta_i} \phi_i[\sigma(t)] \int_0^t F_i(\zeta(t) - \zeta(\xi)) \frac{\partial g_i[\sigma(\xi)]}{\partial \xi} d\xi \tag{16}$$

The following notation has been used for the reduced time

$$\zeta(t) = \int_0^t \frac{dt}{a_t(t_e + t, t_e)}, \quad \zeta(\xi) = \int_0^\xi \frac{d\xi}{a_t(t_e + \xi, t_e)} \tag{17}$$

In order for these equations to be used, the model parameters have to be estimated, based on the data from the creep and recovery tests for material of interest. The searching procedure is based on the minimization of the function of total error between the experimental data and model prediction. For this, the Powell method [20] for multidimensional minimization has been used. Representation of the strain response $\varepsilon[\sigma,t]$ in the form of locally separated stress and time functions (Eq 9) enables fast and robust parameter estimation. An additional advantage gives the choice of the time functions $F_i(t)$ in the form of Prony series terms. Because the growth of these functions is localized in two time decades, it restricts the number of model parameters, which can actually influence the predicted value of strains for certain periods of time.

Model Verification and Discussion

The result of searching for parameters is presented in the Table 1. This set of parameters gives a fair description for both test type: creep (Fig. 2) and recovery (Fig. 3). The total error reached is less than 2.7%. Certain deviation of the model prediction from the experimental data is observed only for the latter stages of the tests and is caused by peculiarity of the Prony series terms.

At the same time, the prediction of creep for stress level of $\sigma = 2.96$ MPa and different t_e levels (Fig. 4) is not as good as previous results (Figs. 2 and 3) obtained for elapsed time $t_e = 24$ h. The peak of error encountered for the data for $t_e = 72$ h is about 11%. This is caused by a comparatively small amount of data for the different elapsed times used in the parameter identification procedure. (The weight of these data during the error evaluation was low.)

Further, the constitutive Eqs 16–17 have been partly verified, using data, which have not been used for parameter estimation.

The creep under stress $\sigma = 8.97$ MPa and 11.8 MPa for the elapsed time $t_e = 7$ h is described rather well (Fig. 5). The integral error is not more than 3.3%. The maximum deviation between the experimental data and model prediction has been found for the end of the creep test for $\sigma = 8.97$ MPa and amounts to 10%.

TABLE 1—*Set of model parameters for modeling creep-recovery behavior of PP.*

Term, i	A	β	μ	θ	ϑ
0	3.811×10^{-4}	0.9560	0.9897	9.114×10^{-2}	-4.6536×10^{-2}
	D_i	γ_i	α_i	η_i	λ_i
1	1.947×10^{-6}	1.694×10^{-1}	2.056	3.505×10^{-1}	10^{-0}
2	1.646×10^{-5}	-1.407×10^{-2}	2.067	2.280×10^{-1}	10^{-1}
3	5.693×10^{-4}	1.570×10^{-2}	1.5829	5.254×10^{-2}	10^{-2}
4	9.563×10^{-5}	4.147×10^{-2}	1.307	1.071×10^{-1}	10^{-3}
5	3.929×10^{-4}	5.323×10^{-2}	0.9212	-1.302×10^{-1}	10^{-4}
6	1.487×10^{-4}	-3.312×10^{-1}	2.134	9.905×10^{-2}	10^{-5}
7	5.746×10^{-6}	-1.383×10^{-2}	4.252	-2.444×10^{-1}	10^{-6}

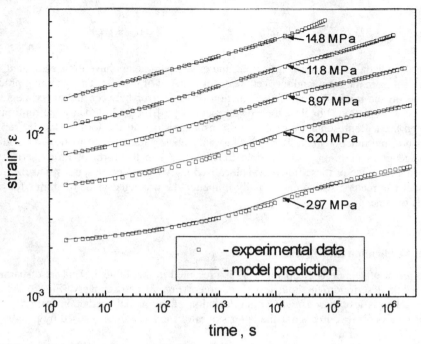

FIG. 2—*Experimental data and model prediction for the long-term creep behavior of PP.*

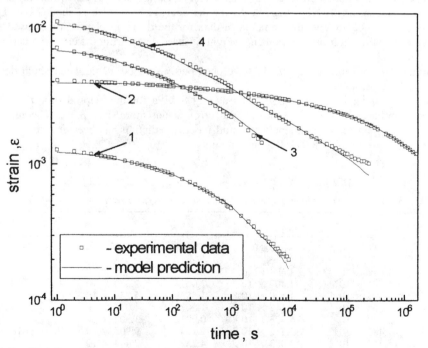

FIG. 3—*Modeling of strain recovery in PP after creep during: 1 h, under σ = 2.96 MPa − (1); 481 h, under σ = 2.96 MPa − (2); 1.09 h, under σ = 9.02 MPa − (3); 1 h, under σ = 11.8 MPa − (4).*

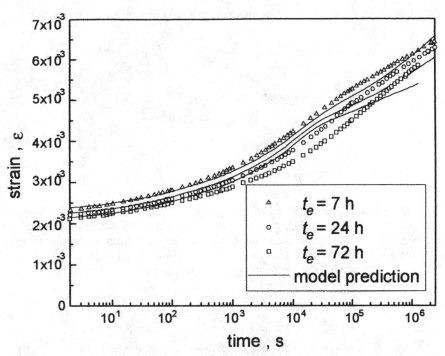

FIG. 4—*Results of the modeling of creep behavior of PP under loading* σ = 2.96 *MPa for different elapsed time (symbols denote experimental data).*

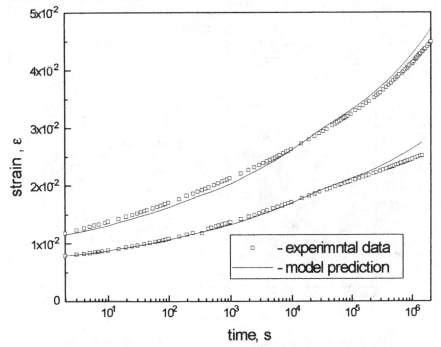

FIG. 5—*Prediction of long-term creep behavior of PP* (t_e = 7 *h*) *for loading level* σ = 8.97 *MPa and* σ = 11.8 *MPa.*

The creep strain for the second step of loading history can be described as follows:

$$\varepsilon[\sigma(t),t,t_e] = A[\sigma_2 + A_1\sigma_2^\beta]\left(\frac{t_e + t}{t_e}\right)^\theta \left(\frac{t_e}{t_e^*}\right)^\vartheta$$

$$+ \sum_i^n \left(\frac{t_e + t}{t_e}\right)^{\eta_i} D_i \exp(\gamma_i\sigma_2) \times [\sigma_1^{\alpha_i}(\exp(-\lambda_i(\zeta - \zeta_1)) - \exp(-\lambda_i\zeta)) \tag{18}$$

$$+ \sigma_2^{\alpha_i}(1 - \exp(-\lambda_i(\zeta - \zeta_1)))]$$

$$\zeta = \frac{t_e}{1 - \mu}\left[\left(\frac{t_e + t}{t_e}\right)^{1-\mu} - 1\right] \quad \text{and} \quad \zeta_1 = \frac{t_e}{1 - \mu}\left[\left(\frac{t_e + t_1}{t_e}\right)^{1-\mu} - 1\right] \tag{19}$$

The model prediction, Eqs 18 and 19, for the second step of loading history is compared with the experimental result (Fig. 6). Also the theoretical prediction gives too high values for the beginning of the second step (about 12% of deviation form experimental data), the data for the end of test do not differ a lot (less than 2% of error).

The strains for the third stage of test can be estimated as:

$$\varepsilon[\sigma(t),t,t_e] = A[\sigma_1 + A_1\sigma_1^\beta]\left(\frac{t_e + t}{t_e}\right)^\theta \left(\frac{t_e}{t_e^*}\right)^\vartheta + \sum_i^n \left(\frac{t_e + t}{t_e}\right)^{\eta_i} D_i \exp(\gamma_i\sigma_1)$$

$$\times [\sigma_1^{\alpha_i}(1 - \exp(-\lambda_i\zeta) + \exp(-\lambda_i(\zeta - \zeta_1)) - \exp(-\lambda_i(\zeta - \zeta_2)))] \tag{20}$$

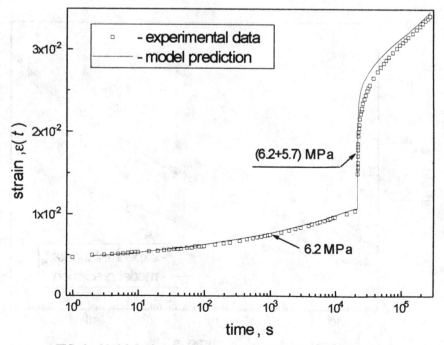

FIG. 6—*Model verification on the creep of PP under two-step loading.*

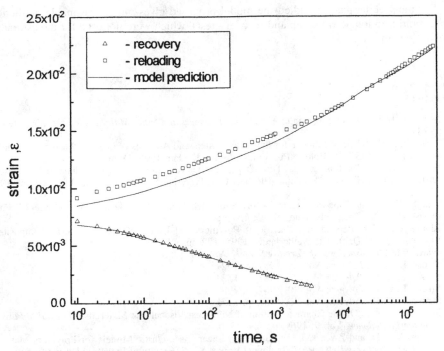

FIG. 7—*Model verification for two consequent steps of three-step loading history. Lower curve corresponds to strain recovery curve in PP after 1.1 h loading at 9 MPa. Upper curve describes creep of PP at resumed loading (9 MPa) after 1 h of recovery ($t_e = 7$ h).*

$$\zeta = \frac{t_e}{1-\mu}\left[\left(\frac{t_e+t}{t_e}\right)^{1-\mu}-1\right], \; \zeta_i = \frac{t_e}{1-\mu}\left[\left(\frac{t_e+t_1}{t_e}\right)^{1-\mu}-1\right],$$

$$\xi_2 = \frac{t_e}{1-\mu}\left[\left(\frac{t_e+t_2}{t_e}\right)^{1-\mu}-1\right] \quad (21)$$

The results of prediction using the representation, Eqs 20 and 21, are compared with the experimental results. The test was performed in following way: the specimen was loaded by stress 9.0 MPa and crept for 1.1 h. Afterwards it was unloaded completely and was allowed to recover for 1 hour. Next it was loaded again by stress 9.0 MPa and crept for 75 h. The second and third steps of the experiment (recovery and reloading) are presented in Fig. 7 together with the model prediction. The horizontal coordinate corresponds to the time calculated from the beginning of loading step (recovery or reloading respectively). Again, after the perturbation period the deviation decreases to less than 1%.

Conclusions

1. The earlier proposed generalization of the Schapery model for nonlinear viscoelasticity is enhanced to describe the time-dependent strain-stress relations in polymers that feature physical aging.

2. For the specific plastic studied the model proposed gives rather reasonable predictions for the multi-step long-term creep and recovery tests, which were not used for the estimation of model parameters.

References

[1] Struik, L. C. E., *Physical Ageing in Amorphous Polymers and Other Materials,* Elsevier, Amsterdam 1978.

[2] Tomlins, P. E., "Code of Practice for the Measurement and Analysis of Creep in Plastics," Technical Report NLP MMS 002: 1996, NPL, Teddington, U.K., Jan. 1996, 18 pp.

[3] Read, B. E. and Tomlins, P. E., "Time-Dependent Deformation of Polypropylene in Response to Different Stress Histories," Technical Report CMMT(A)16, NPL, Teddington, U.K., March 1996, 21 pp.

[4] Lai, J., Non-linear Time-dependent Deformation Behavior of High Density Polyethylene," Ph.D. thesis, TU Delft, The Netherlands, 1995, 157 pp.

[5] Zhang, L., "Time-dependent Behavior of Polymers and Unidirectional Polymeric Composites," Ph.D. thesis, TU Delft, The Netherlands, 1995, 172 pp.

[6] Ward, I. M., *Transactions of Faraday Society,* Vol. 56, 1960, p. 648.

[7] Struik, L. C. E., *Polymer,* Vol. 28, 1987, p. 1521.

[8] Struik, L. C. E., *Polymer,* Vol. 28, 1987, p. 1534.

[9] Struik, L. C. E., *Polymer,* Vol. 30, 1987, p. 799.

[10] Struik, L. C. E., *Polymer,* Vol. 30, 1987, p. 815.

[11] Schapery, R. A., "On the Characterization of Nonlinear Viscoelastic Materials," *Journal of Polymer Engineering Science,* Vol. 9, 1969, pp. 295–310.

[12] Skrypnyk, I. D. and Zweers, E. W. G., "Non-linear Visco-elastic Models for Polymers Materials (Creep and Recovery Behavior)," Technical Report K345, Faculty of Industrial Design Engineering, Delft University of Technology, Delft, The Netherlands, 1996, 40 pp.

[13] Skrypnyk, I. D. and Spoormaker, J. L., "Modelling of Non-linear Visco-elastic Behaviour of Plastic Materials," *Proceedings,* 5th European Conference on Advanced Materials/Euromat 97/, *Materials, Functionality and Design,* Vol. 4, pp. 4/491–4/495.

[14] Read, B. E. and Dean, G. D., *Polymer,* Vol. 25, 1984, p. 1679.

[15] Crissman, J. M. and McKenna, G. B. J., "Relating Creep and Creep Rupture in PMMA Using a Reduced Variable Approach," *Journal of Polymer Science, Part B—Polymers Physics,* Vol. 25, 1987, p. 1667.

[16] Crissman, J. M. and McKenna, G. B. J., "Physical and Chemical Aging in PMMA and Their Affect on Creep and Creep Rupture Behavior," *Journal of Polymer Science, Part B—Polymers Physics,* Vol. 28, 1990, p. 1463.

[17] Henriksen, M., "Nonlinear Viscoelastic Stress Analysis—A Finite Element Approach," *Computers & Structures,* Vol. 18, 1984, p. 133.

[18] Schapery, R. A., "Stress Analysis of Viscoelastic Composite Materials," *Journal of Composite Materials,* Vol. 1, 1967, p. 228.

[19] Chai, C. K. and McCrum, N. G., *Polymer,* Vol. 21, 1980, p. 706.

[20] *Numerical Recipes (FORTRAN version),* by W. H. Press, B. P. Flannery, S. A. Teukolsky, W. T. Vetterling, Cambridge University Press, U.K., 1989.

Ihor D. Skrypnyk,[1] *Jan L. Spoormaker,*[2] *and Willem Smit*[2]

Implementation of Constitutive Model in FEA for Nonlinear Behavior of Plastics

REFERENCE: Skrypnyk, I. D., Spoormaker, J. L., and Smit, W., **"Implementation of Constitutive Model in FEA for Nonlinear Behavior of Plastics,"** *Time Dependent and Nonlinear Effects in Polymers and Composites, ASTM STP 1357,* R. A. Schapery and C. T. Sun, Eds., American Society for Testing and Materials, West Conshohocken, PA, 2000, pp. 83–97.

ABSTRACT: The main elements of implementation into a finite-element analysis (FEA) package of the earlier developed model for nonlinear viscoelastic behavior of plastics are described.

The Henriksen scheme of discretization of the hereditary integral has been chosen for implementation. This scheme enables development of a fast procedure for modeling of viscoelastic behavior. As a result, the time necessary for calculation of problems of viscoelasticity is not much larger than the calculation time required for simulation of elasto-plastic behavior.

Several discretization schemes have been analyzed, implemented in FEA software MARC and verified. The numerical algorithm, which is chosen as a result of comparison, allows us to reach a total deviation of less than 8% to 10% for the modeling of creep and recovery of PMMA and HDPE for the broad range of loading levels.

The case study of a thick plate under distributed transversal loading is examined to compare the results achieved using the Schapery model with the newly proposed approach.

KEYWORDS: nonlinear viscoelasticity, prediction of creep and recovery, implementation into FEA

Notations and Symbols

σ, $\sigma(t)$	Stress and stress history (one-dimensional formulation)
$\hat{\sigma}$	Effective stress
$\varepsilon[\sigma,t]$	Strain as a function of stress and time (one-dimensional formulation)
$\hat{\varepsilon}[\sigma_k,t]$	Experimental data from creep-recovery tests with stress level σ_k during creep
$\bar{\sigma}$, $\bar{\varepsilon}$	Pseudo vector of stress and strain tensor components (see Eq 19)
t	Time variable
$t_{1,2}$	Time moments, when stress level changes in multiple-steps-loading history
$\tilde{t}$	Effective time (elapsed from the moment, when loading was changed last time in multiple-steps-loading history)
$a_\sigma(\sigma)$	Exponent of shifting factor for "time-stress" superposition principle
$\psi(t)$	Stress-reduced time
ξ	Dummy variable in the hereditary integral $\xi < t$
$J_0[\sigma]$	Instantaneous (time-independent) strain as a function of stresses applied

[1] Senior researcher, Delft University of Technology, on leave from Karpenko Physico-Mechanical Institute, National Academy of Sciences of Ukraine, 5 Naukova st., 290601, Lviv, Ukraine.

[2] Professor and associate professor, respectively, Laboratory of Mechanical Reliability, Faculty of Industrial Design Engineering, Delft University of Technology, Jaffalaan 9, 2628 BX, Delft, The Netherlands.

$\tilde{\varphi}(\hat{\sigma})$ Reduced form of instantaneous strain function J_0 for 3-D representation (see Eq 17)

$J(t)$ Time-dependent creep compliance function

$F(t)$, $F_i(t)$ Time-dependent part of creep function

$\phi_i(\sigma)$ Preintegral stress function in Schapery type models

$g(\sigma)$, $g_i(\sigma)$ Nonlinear stress function within the hereditary integral in Schapery or Leaderman type models

$\tilde{g}_i(\hat{\sigma})$ Reduced form of the g_i-functions for the 3-D representation (see Eq 17)

D_i, γ_i, a_i Constants in functions $\phi_i(\sigma)$, $g_i(\sigma)$, $F_i(t)$

λ_i Inverse retardation times in Prony series

v_0, v_1 Poisson's ratio (for instantaneous and long-time deformation of viscoelastic bodies)

$\mathbf{M_D}$, $\mathbf{M_H}$ Matrixes that are related to deviatoric and hydrostatic parts in stress-strain relations (see Eq 18)

θ_i Internal parameters, related to hereditary integrals

Because of improvement in their properties, engineering plastics can be used in many load-bearing applications. While loaded, however, plastics feature the nonlinear time-dependent behavior (viscoelasticity, physical aging, etc.) to a much larger degree than metals. Although a number of nonlinear models are available to describe these phenomena, they have not yet been implemented into commercially available FEA packages. Therefore, engineers rarely use the modern models of nonlinear viscoelasticity for design purposes. Instead, the linear viscoelasticity theory is being used to predict the time-dependent behavior of plastic products.

In this paper the main elements of implementation into an FEA package of the earlier developed nonlinear viscoelasticity model are presented. The scope is limited to modeling of isothermal creep and recovery behavior of nonaging plastic materials.

Generalized Form of the Schapery Model

The first nonlinear model of hereditary integral type, proposed by Leaderman [1]

$$\varepsilon(t) = J_0[\sigma(t)] + \int_0^t F(t - \xi) \frac{\partial[g(\sigma)]}{\partial \xi} \, d\xi \tag{1}$$

simply assumes that the principle of loading (stress) superposition is valid also for plastics, which display nonlinear response to loading (Fig. 1). In other words, it was a generalization of the Boltzmann linear superposition principle in the case of nonlinear response. The material functions J_0, $F(t)$ and $g(\sigma)$ in Eq 1 can be estimated, based on the results of creep tests, using the relation which follows from Eq 1 for $\sigma = \text{const}$

$$\hat{\varepsilon}[\sigma_k, t] = J_0[\sigma_k] + F(t)g(\sigma_k) \tag{2}$$

This model, however, is not able to describe the recovery of viscous strains in plastics properly. Schapery [4] has proposed another variant of a nonlinear hereditary integral model

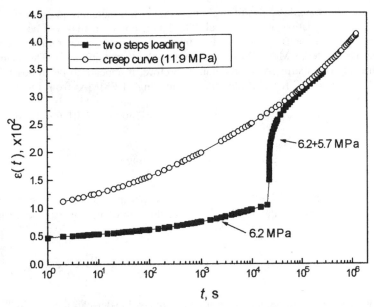

FIG. 1—*Nonlinear response of PP to two-step loading. The principle of the loading step superposition is still valid. (Experimental data from the report of NPL [2,3].)*

$$\varepsilon(t) = J_0 g_0(\sigma)\sigma + g_1(\sigma) \int_0^t F(\psi(t) - \psi(\xi)) \frac{\partial[g_2(\sigma)\sigma]}{\partial\xi} \, d\xi, \qquad \psi(t) = \int_0^t \frac{d\xi}{a_\sigma[\sigma(\xi)]} \qquad (3)$$

Here, function $g_1(\sigma)$, multiplied by the hereditary integral, should account for the difference between the creep and recovery behavior. In addition, this model incorporates the idea about the "time-stress" superposition principle (in the form of the stress-reduced time $\psi(t)$). The physical meaning of this principle is that stress level σ induces the same amount of creep during the time increment $\partial\psi$, as a certain reference stress level σ_0 (which usually should be chosen relatively small, so that $a_\sigma(\sigma_0) \approx 1$) within the time period ∂t.

The material functions in Eq 3 should be estimated, based on data from creep and recovery tests

$$\text{for creep} \qquad \hat{\varepsilon}[\sigma_k, t] = J_0[\sigma_k] + F\left(\frac{t}{a_\sigma(\sigma_k)}\right) g_1(\sigma_k) g_2(\sigma_k)\sigma_k \qquad (4)$$

$$\text{for recovery} \qquad \hat{\varepsilon}[\sigma_k, t] = [F(\tilde{t} + t_1/a_\sigma(\sigma_k)) - F(\tilde{t})]g_2(\sigma_k)\sigma_k \quad \tilde{t} = t - t_1 \qquad (5)$$

Equation 5 implies that functions $g_1(\sigma)$ and $a_\sigma(\sigma)$ are chosen in such a way that $g_1(0) = 1$ and $a_\sigma(0) = 1$. The latter makes this model correspond to the Leaderman Eq 1 for small stress levels.

Often the "time-stress" superposition principle is referred to as the shifting rule. It means that if experimental creep data correspond to the above principle, the creep compliance curves for different stress levels can be shifted over the logarithmic time scale to coincide and to form a so-called "master-curve." On the other hand, the creep curves should be regular enough in order to coincide after shifting. They should not have any jog, but should change flowingly with the variation of stress level. The latter, however, may not be the case when material undergoes α- or β-transitions. For instance, at room temperature and within the stress range σ = 22 to 28 MPa, PMMA encounters the β-transition [5]. The shape of creep curves in this region is strongly nonregular. Therefore, it appears to be impossible to build the single "master-curve" for a whole range of loading of PMMA: from low stresses up to yield point [6,7].

Recently, the authors have proposed another representation [8,9] for the description of creep data

$$\hat{\varepsilon}[\sigma_k,t] = J_0[\sigma_k] + \sum_i F_i(t)g_i(\sigma_k) \tag{6}$$

This expression has shown its ability to describe the experimental creep curves with different, even nonregular shapes. Similar to Eqs 2 and 4, this representation can be written in the hereditary integral form to account for a complex loading history

$$\varepsilon[\sigma(t),t] = J_0[\sigma(t)] + \int_0^t \sum_i F_i(t - \xi) \frac{\partial g_i[\sigma(\xi)]}{\partial \xi} d\xi \tag{7}$$

It is obvious that Eq 7 represents a generalization of the Leaderman model, Eq 1. Similarly, a generalization of the Schapery model, Eq 3, can be written as follows

$$\varepsilon[\sigma(t),t] = J_0[\sigma(t)] + \sum_i^n \phi_i[\sigma(t)] \int_0^t F_i(t - \xi) \frac{\partial g_i[\sigma(\xi)]}{\partial \xi} d\xi \tag{8}$$

The conception of stress-reduced time is missing in Eq 8, since the representation of Eq 6 enables proper fit to the creep data without any "shifting."

Although $J_0[\sigma(t)]$, $F_i(t)$, $\phi_i[\sigma(t)]$, and $g_i[\sigma(t)]$ might be chosen among the functions of different types (polynomial, exponential, etc.), the following forms seem to be preferable [8,10] for description of experimental data for many plastics

$$J_0[\sigma] - A\sigma \quad \text{or} \quad A[\sigma + A_1\sigma^\beta] \tag{9}$$

$$g_i[\sigma] - D_i\sigma^{\alpha_i} \quad \text{or} \quad D_{i,1}\sigma^{\alpha_{i,1}} + D_{i,2}\sigma^{\alpha_{i,2}} \tag{10}$$

$$\phi_i[\sigma] - \exp(\gamma_i\sigma) \quad \text{or} \quad 1 + B_{i,1}\sigma^{\gamma_{i,1}} + B_{i,2}\sigma^{\gamma_{i,2}} \tag{11}$$

$$F_i(t) = (1 - \exp(-\lambda_i t)) \tag{12}$$

As always, the simpler form should be preferable. The more sophisticated form is chosen in order to reach a better fit of the experimental data; the longer calculation time is necessary for FE simulation and, more likely, that numerical instability might occur. Taking into account that, on the one hand, the creep tests on plastics cannot be reproduced with less than 3 to 5% of deviation [2,8,11,12] and, on the other hand, the error of the FE analysis usually also is about 5 to 10%, there is no reason to strive for a fit better than 3 to 5% of total deviation between data and model prediction.

There are convincing advantages in choosing the time functions $F_i(t)$ in the form of Prony series terms. First of all, it allows efficient numerical calculation [13] of hereditary integrals.

Second, together with the representation of the strain response $\hat{\varepsilon}[\sigma_k,t]$ in the form of locally separated variables (Eq 6) it gives better possibilities for estimation of the parameters.

The procedure of parameter identification for the material functions, Eqs 9–11, is based on the idea of minimization of the relative deviation between experimental data and model prediction. This relative deviation can be considered as a function that depends on the model parameters $I(D_i,\alpha_i,\lambda_i,\gamma_i) = I_1 + I_2$, where I_1 denotes the function of errors for the creep stage

$$I_1(D_i,\alpha_i,\lambda_i,\gamma_i) = \frac{1}{Nt_{end}} \sum_{k=1}^{N} \int_0^{t_1} \left| \frac{\hat{\varepsilon}[\sigma_k,t] - \sum_{i=1}^{n} D_i \exp(\gamma_i\sigma_k)\sigma_k^{\alpha_i}(1 - \exp(-\lambda_i t))}{\hat{\varepsilon}[\sigma_k,t]} \right| dt \quad (13)$$

and I_2 represents the function for the strain recovery

$$I_2(D_i,\alpha_i,\lambda_i) = \frac{1}{Nt_{end}} \sum_{k=1}^{N} \int_{t_1}^{t_{end}} \left| \frac{\hat{\varepsilon}[\sigma_k,t] - \sum_{i=1}^{n} D_i\sigma_k^{\alpha_i}[\exp(-\lambda_i(t - t_1)) - \exp(-\lambda_i t)]}{\hat{\varepsilon}[\sigma_k,t]} \right| dt$$

$$(14)$$

N is number of loading levels σ_k.

From this point of view the parameter identification procedure involves a multidimensional minimization. The algorithm used is based on Powell's multi-dimensional minimization procedure [14]. For the evaluation of the integrals (Eqs 13 and 14) the Romberg integration procedure [14] has been chosen.

The prediction of creep and recovery behavior for HDPE [15] is given in Fig. 2. In order to evaluate the adequacy of the newly proposed representation, Eq 8, the "curve fitting" has been performed for two models: The Schapery model, Eq 3, and the generalized representation, Eq 8. The set of parameters for material functions, Eqs 9–12, for Eq 8 is given in Table 1.

The following material functions were used in Schapery model, Eq 3, to describe the time-dependent behavior of HDPE

$$J_0 = 1 \cdot 10^{-3}; \; g_1[\sigma] = \exp(0.041 \cdot \sigma); \; g_2[\sigma] = \sigma^{0.396}; \; a[\sigma] = \exp[-0.15 \cdot \sigma]$$

$$F(t) = \sum_{n=1,6} D_n(1 - \exp(-t/10^n))$$

$$(15)$$

$$D_n = \{0.281 \cdot 10^{-5}, 0.224 \cdot 10^{-3}, 0.283 \cdot 10^{-3},$$

$$0.720 \cdot 10^{-3}, 0.161 \cdot 10^{-8}, 0.824 \cdot 10^{-3}\}$$

The total error reached for the Schapery model is less than 6.7%, while Eq 8 enabled us to reach 2.2%. Both approaches model the creep behavior well, but generalized representation gives a better fit for the recovery behavior.

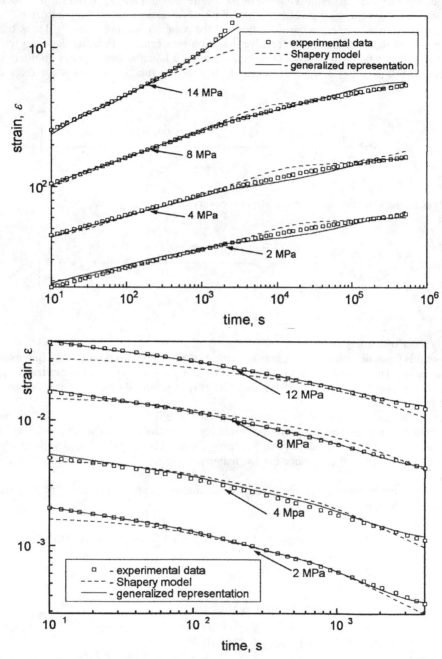

FIG. 2—*Experimental data [15] and model prediction for the creep and recovery behavior of HDPE.*

TABLE 1—*The set of model parameters for description of HDPE.*

Term, i	A			β		
0	7.852×10^{-4}			0.0		
	$D_{i,1}$	$\alpha_{i,1}$	$D_{i,2}$	$\alpha_{i,2}$	γ_i	λ_i
1	0.1278×10^{-4}	2.759	0.2592×10^{-3}	1.059	0.2404×10^{-1}	10^{-1}
2	0.3364×10^{-3}	1.075	0.5910×10^{-5}	2.872	0.1473×10^{-3}	10^{-2}
3	0.3729×10^{-3}	1.118	0.9672×10^{-5}	2.740	0.6727×10^{-1}	10^{-3}
4	0.5814×10^{-4}	2.193	0.2842×10^{-9}	6.254	0.4689×10^{-1}	10^{-4}
5	0.6187×10^{-3}	1.547	0.6753×10^{-11}	9.779	0.9244×10^{-3}	10^{-5}
6	0.4855×10^{-1}	1.637	0.2791×10^{-2}	2.786	-3.377	10^{-6}

Extension of the Model to 3-D Formulation

There is no clear understanding yet for establishing a three-dimensional formulation of the constitutive equations for nonlinear viscoelasticity. Straightforward multiaxial tests are complicated and require special experimental equipment. On the other hand, to establish a 3-D model extended tests program would be necessary. Therefore, sometimes the experimental program is restricted to simple tensile and torsion creep tests [16] with a further attempt to combine the phenomenological dependencies obtained.

Another way to develop the 3-D formulation is to assume that the behavior of viscoelastic material is, to a certain extent, similar to the elastic behavior. Since this is a very general assumption, it leaves room for some additional phenomenological hypotheses. As in Refs 11 and 12, the following assumptions have been used for deriving the model [8]:

- material is compressible and initially isotropic;
- strains are small enough to apply stress and strain tensors definitions, conventional for small deformations;
- rate of viscous flow is proportional to the effective stress $\hat{\sigma}$;
- deviatoric and hydrostatic parts of the deformation process are completely uncoupled.

As a result, a three-dimensional formulation of the viscoelastic model has been proposed

$$\bar{\varepsilon} = [(1 + v_0)\mathbf{M}_D + (1 - 2v_0)\mathbf{M}_H]\tilde{\varphi}(\hat{\sigma})\bar{\sigma}$$

$$+ [(1 + v_1)\mathbf{M}_D + (1 - 2v_1)\mathbf{M}_H] \sum_{i=1}^{n} \phi_i(\hat{\sigma}) \int_0^t F_i(t - \xi) \frac{\partial[\tilde{g}_i(\hat{\sigma})\bar{\sigma}]}{\partial \xi} d\xi \quad (16)$$

Here, for the case of uniaxial loading

$$J_0[\sigma] = \tilde{\varphi}(\sigma)\sigma; \quad g_1[\sigma] = \tilde{g}_1(\sigma)\sigma \quad (17)$$

The following matrix and vector notations have been used above

$$\mathbf{M}_D = \begin{cases} \dfrac{2}{3}, & \text{if } i = j \text{ and } i,j \leq 3; \\[2mm] -\dfrac{1}{3}, & \text{if } i \neq j \text{ and } i,j \leq 3; \\[2mm] 2, & \text{if } i = j \text{ and } i,j > 3; \\[2mm] 0, & \text{if } i \neq j \text{ and } i,j > 3. \end{cases} \qquad \mathbf{M}_H^{i,j}\bigg|_{i,j=1\ldots3} = \dfrac{1}{3} \qquad (18)$$

$$\bar{\sigma} = \{\sigma_{xx}\ \sigma_{yy}\ \sigma_{zz}\ \tau_{xy}\ \tau_{yz}\ \tau_{zx}\}^T; \qquad \bar{\varepsilon} = \{\varepsilon_{xx}\ \varepsilon_{yy}\ \varepsilon_{zz}\ \gamma_{xy}\ \gamma_{yz}\ \gamma_{zx}\}^T \qquad (19)$$

Main Elements of Numerical Algorithm

The MARC FEA package has been chosen for implementation of the viscoelasticity model, Eq 8. This package has an open structure that enables easy access to the variables (such as stresses, strains, time increment). Moreover, it has extended facilities to handle the geometrical and physical nonlinear problems that are essential for simulation of the visco-elastic behavior of plastics. The nonlinear problems can be solved by FEA software only incrementally. Since MARC is based on the displacement method, it requires the stress-strain relation to be represented in terms of increments as follows

$$\Delta\bar{\sigma} = \mathbf{L}(\Delta\bar{\varepsilon},\Delta t \ldots) \qquad (20)$$

Therefore, Eq 16 has to be discretized and inverted to be used with MARC. For enough regular functions $g_i(\sigma)$, such that $\partial^2[g_i(\sigma)]/\partial t^2 \ll 1$, the hereditary integral with the kernel function in the form of Prony series term can be discretized, following the Henriksen scheme [13], (Fig. 3). For easier analysis, further derivations have been assembled into a diagram (Fig. 4).

To implement the general representation, Eq 16, in an FEA package, the total differential has to be derived. In Ref 12, similar derivations have been presented for the case of the Schapery model, Eq 3. As a result, a quite unmanageable expression has been obtained which shows low convergence ability.

Therefore, it has been proposed [10] first of all to discretize the representation, Eq 8

$$\Delta\varepsilon = \varphi'(\sigma)\Delta\sigma + \sum_{i=1}^{n} [(\phi_i'(\sigma)g_i(\sigma) + \phi_i(\sigma)g_i'(\sigma))(1 - \Gamma_i(\Delta t))$$

$$+ \phi_i'(\sigma)g_i(\sigma(t - \Delta t))\Gamma_i(\Delta t) + \phi_i'(\sigma)\exp(-\lambda_i\Delta t)\theta_i(t - \Delta t)]\Delta\sigma \qquad (21)$$

$$+ \sum_{i=1}^{n} \phi_i(\sigma)(\exp(-\lambda_i\Delta t) - 1)\theta_i(t - \Delta t)$$

While deriving the above relation it has been implied that the loading history begins always from zero $\sigma(0) = 0$. Further, Eq 21 has to be expanded to 3-D form similarly to Eq 16

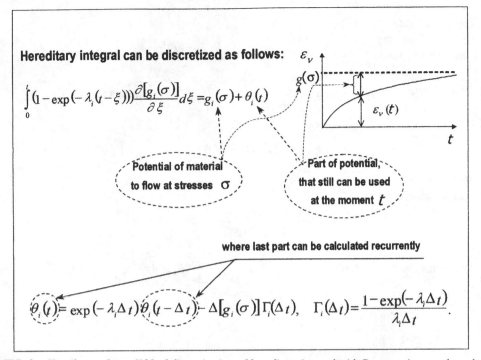

Hereditary integral can be discretized as follows:

$$\int_0^t (1-\exp(-\lambda_i(t-\xi)))\frac{\partial[g_i(\sigma)]}{\partial\xi}d\xi = g_i(\sigma) + \theta_i(t)$$

Potential of material to flow at stresses σ

Part of potential, that still can be used at the moment t

where last part can be calculated recurrently

$$\theta_i(t) = \exp(-\lambda_i\Delta t)\theta_i(t-\Delta t) - \Delta[g_i(\sigma)]\Gamma_i(\Delta t), \quad \Gamma_i(\Delta t) = \frac{1-\exp(-\lambda_i\Delta t)}{\lambda_i\Delta t}.$$

FIG. 3—*Henriksen scheme [13] of discretization of hereditary integral with Prony series type kernel.*

$$\Delta\bar{\varepsilon} = [(1 + v_0)\mathbf{M}_D + (1 - 2v_0)\mathbf{M}_H]\varphi'(\hat{\sigma})\Delta\bar{\sigma} + [(1 + v_1)\mathbf{M}_D + (1 - 2v_1)\mathbf{M}_H]\,\Delta\bar{\sigma}$$

$$\times \sum_{i=1}^n (\phi_i'(\hat{\sigma})g_i(\hat{\sigma}) + \phi_i(\hat{\sigma})g_i'(\hat{\sigma}))(1 - \Gamma_i(\Delta t)) + \phi_i'(\hat{\sigma})(g_i(\hat{\sigma}(t - \Delta t))\Gamma_i(\Delta t)$$

$$\tag{22}$$

$$+ \exp(-\lambda_i\Delta t)\theta_i(t - \Delta t)) + [(1 + v_1)\mathbf{M}_D + (1 - 2v_1)\mathbf{M}_H]\sum_{i=1}^n \phi_i(\hat{\sigma})$$

$$(\exp(-\lambda_i\Delta t) - 1)\bar{\theta}_i(t - \Delta t)$$

and inverted. Here again different variants are possible. The most obvious way is to factorize the stress increment $\Delta\bar{\sigma}$ and express it as a function of other parameters [9] (variant I in Fig. 4)

$$\Delta\bar{\sigma} = \{[(1 + v_0)\mathbf{M}_D + (1 - 2v_0)\mathbf{M}_H]\varphi'(\hat{\sigma}) + [(1 + v_1)\mathbf{M}_D$$

$$+ (1 - 2v_1)\mathbf{M}_H] \times \sum_{i=1}^n [(\phi_i'(\hat{\sigma})g_i(\hat{\sigma}) + \phi_i(\hat{\sigma})g_i'(\hat{\sigma}))$$

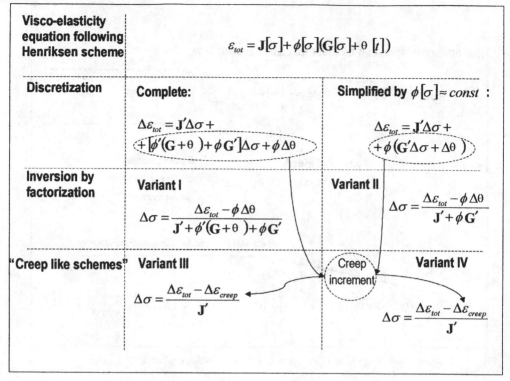

FIG. 4—*Different variants of numerical scheme for modeling of nonlinear viscoelasticity within the FEA package MARC.*

$$(1 - \Gamma_i(\Delta t)) + \phi_i'(\hat{\sigma})(g_i(\hat{\sigma}(t - \Delta t))\Gamma_i \Delta t) + \exp(-\lambda_i \Delta t)\theta_i(t - \Delta t))]\}^{-1} \quad (23)$$

$$\times \left(\Delta \bar{\varepsilon} - [(1 + v_1)\mathbf{M}_D + (1 - 2v_1)\mathbf{M}_H] \sum_{i=1}^{n} \phi_i(\hat{\sigma}) \right.$$

$$\left. (-\lambda_i \Delta t) - 1)\bar{\theta}_i(t - \Delta t) \right)$$

The vector θ_i can be calculated at the end of every time increment as follows

$$\theta_i(t) = \exp(-\lambda_i \Delta t)\theta_i(t - \Delta t) - \Delta[g_i(\hat{\sigma})]\Gamma_i(\Delta t) \quad (24)$$

while the value $\bar{\theta}_i$ will be estimated in analogy to the associated flow rule

$$\bar{\theta}_i(t) = \exp(-\lambda_i \Delta t)\bar{\theta}_i(t - \Delta t) - \left| \frac{\Delta[g_i(\hat{\sigma})]}{\Delta \hat{\sigma}} \right| \Gamma_i(\Delta t)\Delta \bar{\sigma} \quad (25)$$

The above scheme is recurrent. To estimate the viscous strains increment for the current time t, only the data for the stresses field σ and internal parameters θ_i and $\bar{\theta}_i$ from the previous step are needed. Although for the models with a large number of elements this numerical scheme can require large amounts of computer memory, the advantage in decreasing of computing time is, however, obvious.

The simplified variant of the above scheme, Eq 23, can be obtained assuming that the effective stresses are constant during the time increment [11] (variant II, Fig. 4)

$$
\Delta\bar{\sigma} = \{[(1 + v_0)\mathbf{M}_D + (1 - 2v_0)\mathbf{M}_H]\tilde{\varphi}(\hat{\sigma}) + [(1 + v_1)\mathbf{M}_D
$$

$$
+ (1 - 2v_1)\mathbf{M}_H] \sum_{i=1}^{n} \phi_i(\hat{\sigma})\tilde{g}_i(\hat{\sigma})(1 - \Gamma_i(\Delta t))\}^{-1}
$$

(26)

$$
\times \left(\Delta\bar{\varepsilon} - [(1 + v_1)\mathbf{M}_D + (1 - 2v_1)\mathbf{M}_H] \sum_{i=1}^{n} \phi_i(\hat{\sigma}) \right.
$$

$$
\left. (\exp(-\lambda_i\Delta t) - 1)\tilde{\theta}_t(t - \Delta t) \right)
$$

where

$$
\tilde{\theta}_i(t) = \exp(-\lambda_i\Delta t)\tilde{\theta}_i(t - \Delta t) - \Delta[\tilde{g}_i(\hat{\sigma})\overline{\sigma}]\Gamma_i(\Delta t)
$$

(27)

Another way to invert Eq 22 is to deduct first the viscous part from the total strain increment with the following inversion (variant III, Fig. 4)

$$
\Delta\bar{\sigma} = \{[(1 + v_0)\mathbf{M}_D + (1 - 2v_0)\mathbf{M}_H]\varphi'(\hat{\sigma})\}^{-1} \times \{\Delta\bar{\varepsilon} - [(1 + v_1)\mathbf{M}_D
$$

$$
+ (1 - 2v_1)\mathbf{M}_H] \sum_{i=1}^{n} [\Delta\bar{\sigma}(t - \Delta t)[(\phi_i'(\hat{\sigma})g_i(\hat{\sigma}) + \phi_i(\hat{\sigma})g_i'
$$

$$
\times (\hat{\sigma}))(1 - \Gamma_i(\Delta t)) + \phi_i'(\hat{\sigma})(g_i(\hat{\sigma}(t - \Delta t))\Gamma_i(\Delta t)
$$

(28)

$$
+ \exp(-\lambda_i\Delta t)\theta_i(t - \Delta t))] + \phi_i(\hat{\sigma})
$$

$$
(\exp(-\lambda_i\Delta t) - 1)\overline{\theta}_i(t - \Delta t)]\}
$$

Once more, Eq 28 can be simplified if one is to assume the effective stresses to be constant (variant IV, Fig. 4)

$$
\Delta\bar{\sigma} = \{[(1 + v_0)\mathbf{M}_D + (1 - 2v_0)\mathbf{M}_H]\tilde{\varphi}(\hat{\sigma})\}^{-1}
$$

$$
\times \{\Delta\bar{\varepsilon} - [(1 + v_1)\mathbf{M}_D + (1 - 2v_1)\mathbf{M}_H]
$$

(29)

$$
\times \sum_{i=1}^{n} [[\phi_i(\hat{\sigma})\tilde{g}_i(\hat{\sigma})(1 - \Gamma_i(\Delta t))]\Delta\bar{\sigma}(t - \Delta t) + \phi_i(\hat{\sigma})
$$

$$
\times (\exp(-\lambda_i\Delta t) - 1)\tilde{\theta}_i(t - \Delta t)]\}
$$

Verification of the Schemes Proposed

The above listed variants of the numerical scheme have been implemented in the FEA package MARC and tested. The tensile creep-recovery tests for several plastics have been simulated first. Due to strong nonlinearity of the material behavior, the numerical instabilities have been encountered while testing these numerical schemes. Several reasons for the instabilities have been distinguished:

1. When the loading history with *decreasing stresses* is simulated, it might appear that certain stress increments produce near-zero strain changes: $\Delta\varepsilon = G \cdot \Delta\sigma \approx 0$. Here G denotes the tangent compliance. In this case $G \approx 0$. Since the above stress-strain relation has to be inverted to the form of Eq 20, it comes into view that

$$\Delta\sigma = \frac{\Delta\varepsilon}{G} \approx \frac{0}{0} \qquad (34)$$

This type of instability occurs if the first variant of the numerical scheme, where the tangent compliance matrix depends on loading history, is used.

2. Another type of numerical instability that is also related to inversion of the tangent compliance matrix is caused by nonlinear elastic compliance J_0. It can lead to an ill-conditioned set of linear equations and, consequently, to loss of convergence. This problem is typical for all four schemes listed above, but might be avoided if one assumes $J_0 = $ const.

3. The instability can also emerge in any of the mentioned schemes when the material parameters $\alpha_{i,j}$ or $\gamma_{i,j}$ in Eqs 10 and 11 are found to be *negative*. Such parameters might give a good fit for experimental data, but will disturb the incremental numerical scheme. As mentioned above, a simpler form of Eqs 10 and 11 is preferable.

4. The last type of instability is encountered only for variants I or III. In other words, it can be avoided, if one is to assume that *the effective stresses are constant* during the time increment. The mechanism fo this type of instability is not clear yet; it might appear for one set of material parameters and will not emerge for other sets.

Thus, variants II and IV of the numerical schemes appear to be more stable. Usually they also show better convergence ability and, consequently, a higher computation rate than the schemes I or III. However, the assumption about *constant* (within the time increment) *effective stresses* $\hat{\sigma}$, which makes the schemes II and IV simpler and more stable, also has a disadvantage. It is obvious (Fig. 4) that it causes the material stiffness to be independent of loading history. In fact, relations II and IV correspond to the Leaderman equation [1] and, therefore, do not describe the recovery behavior of plastics properly. The latter can be clearly seen in Fig. 5 where the results of FE simulations of uniaxial creep and recovery of PMMA are compared with the corresponding experimental results. The calculations according to variants I and III give nearly the same results; thus they were not distinguished in Fig. 5. The same relates to schemes II and IV.

Unlike relations II and IV, schemes I and III give an adequate description for both creep and recovery behavior. Therefore, based on the above, it can be assumed that, if possible, *numerical scheme* III should be used. If the instability is emerging, one should *switch to scheme* IV.

Further, to compare two approaches, namely, the Schapery model, Eq 3 and the generalized representation, Eq 8, the behavior of simply supported thick HDPE plate ($60 \times 60 \times 3$ mm) under transversal loading has been simulated. Because of two planes of symmetry, only one-quarter of the plate has been modeled. The thickness of the plate has been divided into four tiers of elements. Each tier has been subdivided into a five-by-five element pattern, resulting in a mesh containing 180 nodes and 100 of 8-noded 3-D elements.

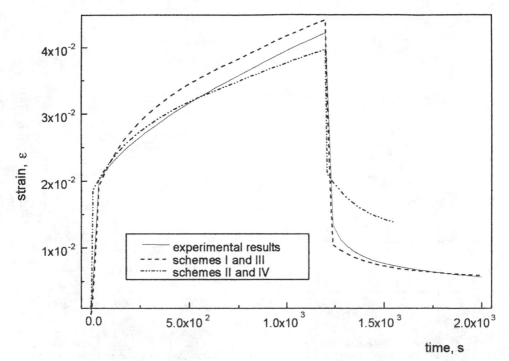

FIG. 5—*FE simulation of the creep and recovery behavior of PMMA (σ = 50 MPa) using different variant of numerical scheme.*

The upper surface of the plane plate has been loaded by uniform transversal pressure applied gradually (in 20 steps). The edges of the plate were supported against the downward displacement. The maximum level of loading reached was 0.2 MPa. After that, the loading was kept constant during time 1200 s.

The results obtained using these two models (Fig. 6) are nearly similar. The plate features simultaneously the relaxation of stresses and creep (increasing in time deflection of the plate). The redistribution of stress fields during the sustained loading is qualitatively the same for both models. The transversal displacement (deflection) at the middle of the plate differs for two models by 3%, while the effective von Mises stresses deviate by 5%. If one is to account for the accuracy of FE calculations (~5%) and deviation that is introduced by two different fittings to experimental data, this might be considered as a good agreement.

Conclusions

1. The nonlinear viscoelasticy model based on the Schapery approach is developed and implemented into the commercial FE package MARC as a user subroutine. This approach enables simulation of the creep and recovery behavior of engineering plastics with higher accuracy and easiness.

2. The newly proposed representation, Eq 8, produces similar results to the Schapery model, Eq 3, for the problems where both models are operational. On the other hand, Eq 8 is also able to describe the plastics behavior for the regions where experimental data are nonregular for some reason.

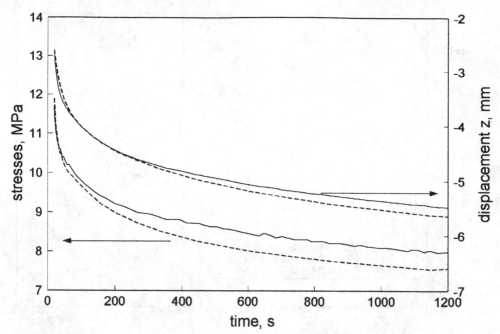

FIG. 6—*Evolution of the equivalent von Mises stresses and transversal displacement at the middle of HDPE plate, calculated using representation, Eq 8—(straight line) and Schapery model, Eq 3—(dashed line).*

3. Several numerical schemes for calculation of nonlinear viscoelastic behavior within an FE package were analyzed and compared from the point of view of their numerical stability and accuracy. It was shown that the variation of preintegral functions within the time increment is the main reason for numerical instabilities in these schemes. Restriction of preintegral functions within the time increment makes the routine stable, but decreases the accuracy of calculations for the stages of loading history where loading is decreasing.

References

[1] Leaderman, H., *Elastic and Creep Properties of Filamentous Materials and Other High Polymers,* Textile Foundation, Washington, DC, 1943.

[2] Tomlins, P. E., "Code of Practice for the Measurement and Analysis of Creep in Plastics," Technical Report NLP MMS 002: National Physical Laboratory, Teddington, U.K., Jan. 1996, 18 pp.

[3] Read, B. E. and Tomlins, P. E., "Time-Dependent Deformation of Polypropylene in Response to Different Stress Histories," Technical Report CMMT(A)16, NPL, March 1996, U.K., 21 pp.

[4] Schapery, R. A., "On the Characterization of Nonlinear Viscoelastic Materials," *Journal of Polymer Engineering Science,* Vol. 9, 1969, p. 295.

[5] Read, B. E. and Dean, G. D., *Polymer,* Vol. 25, 1984, p. 1679.

[6] Crissman, J. M. and McKenna, G. B. J., "Relating Creep and Creep Rupture in PMMA Using a Reduced Variable Approach," *Journal of Polymer Science, Part B—Polymers Physics,* Vol. 25, 1987, p. 1667.

[7] Crissman, J. M. and McKenna, G. B. J., "Physical and Chemical Aging in PMMA and Their Affect on Creep and Creep Rupture Behavior," *Journal of Polymer Science, Part B—Polymers Physics,* Vol. 28, 1990, p. 1463.

[8] Skrypnyk, I. D. and Zweers, E. W. G., "Non-linear Visco-elastic Models for Polymers Materials (Creep and Recovery Behavior)," Technical Report K345, Faculty of Industrial Design Engineering, Delft University of Technology, 1996, Delft, The Netherlands, 40 pp.

[9] Skrypnyk, I. D. and Spoormaker, J. L., "Modelling of Non-Linear Visco-elastic Behaviour of Plastic Materials," *Proceedings,* 5th European Conference on Advanced Materials/Euromat 97/, *Materials, Functionality and Design,* Vol. 4, pp. 4/491–4/495.

[10] Skrypnyk, I. D., Spoormaker, J. L., and Kandachar, P. V., "Modelling of the Long-Term Visco-elastic Behaviour of Polymeric Materials," Technical Report K369, Faculty of Industrial Design Engineering, Delft University of Technology, Delft, The Netherlands, 1997, 48 pp.

[11] Lai, J., "Non-linear Time-dependent Deformation Behavior of High Density Polyethylene," Ph.D. thesis, TU Delft, 1995, 157 pp.

[12] Zhang, L., "Time-dependent Behavior of Polymers and Unidirectional Polymeric Composites," Ph.D. thesis, TU Delft, The Netherlands, 1995, 172 pp.

[13] Henriksen, M., *Computers & Structures,* Vol. 18, 1984, p. 133.

[14] Press, W. H., et al. *Numerical Recipes (FORTRAN version),* Cambridge University Press, U.K., 1989.

[15] Beijer, J. G. J. and Spoormaker, J. L., "Viscoelastic Behaviour of HDPE Under Tensile Loading," *Proceedings,* 10th International Conference on Deformation, Yield and Fracture of Polymers, 1997, pp. 270–273.

[16] Findley, W. N., Lai, J. S., and Onaran, K., *Creep and Relaxation of Nonlinear Viscoelastic Materials,* Dover Publ. Inc., New York, 1989, 371 pp.

Gale A. Holmes,[1] Richard C. Peterson,[1] Donald L. Hunston,[1]
Walter G. McDonough,[1] and Carol L. Schutte[1]

The Effect of Nonlinear Viscoelasticity on Interfacial Shear Strength Measurements

REFERENCE: Holmes, G. A., Peterson, R. C., Hunston, D. L., McDonough, W. G., and Schutte, C. L., **"The Effect of Nonlinear Viscoelasticity on Interfacial Shear Strength Measurements,"** *Time Dependent and Nonlinear Effects in Polymers and Composites, ASTM STP 1357,* R. A. Schapery and C. T. Sun, Eds., American Society for Testing and Materials, West Conshohocken, PA, 2000, pp. 98–117.

ABSTRACT: Experimental evidence demonstrates that diglycidyl ether of bisphenol-A (DGEBA)/meta phenylenediamine (m-PDA) epoxy resin matrix used in the single fiber fragmentation tests exhibits nonlinear stress strain behavior in the region where E-glass fiber fracture occurs. In addition, strain hardening after the onset of yield is observed. Therefore, linear elastic shear-lag models and the Kelly-Tyson model are inappropriate for the determination of the interfacial shear strength for this epoxy resin system. Using a strain-dependent secant modulus in the Cox model, the calculated interfacial shear strength is shown to be relatively lower by at least 15% than the value determined using a linear elastic modulus. This decrease is consistent with numerical simulations which show the linear elastic approximation over predicts the number of fragments in the fragmentation test. In addition, the value obtained by the strain-dependent secant modulus is approximately 300% relatively higher than the value predicted by the Kelly-Tyson model.

KEYWORDS: single-fiber fragmentation test, viscoelasticity, nonlinear matrix behavior, strain-rate dependence, interfacial shear strength, epoxy resin, E-glass fiber, Cox Model, Kelly-Tyson model

In most composite interface research, the strength and durability of the fiber-matrix interface are estimated by interfacial shear strength measurements. To a large extent, estimates of the fiber-matrix interfacial shear strength are based on unit composite micromechanics models and experimental data from single fiber fragmentation (SFF) tests. In the single fiber fragmentation test, a dogbone is made with a resin having a high extension–to–failure and a single fiber embedded down the axis of the dogbone. The sample is pulled in tension and stress is transmitted into the fiber through the fiber-matrix interface. Since the fiber has a lower strain to failure than the resin, the fiber breaks as the strain is increased. This process continues until the remaining fiber fragments are all less than a critical transfer length, l_c. The critical transfer length is the length below which the fragments are too short for sufficient load to be transmitted into them to cause failure. This point is termed saturation. The fragment lengths at saturation are measured and a micromechanics model is used to convert the average fragment length into a measure of the interface strength or stress transfer efficiency.

The fiber-matrix interfacial shear strength is a critical parameter since it directly affects off-axis properties in unidirectional composites and the shear strength of laminates. One of

[1] National Institute of Standards and Technology, Polymers Division, Gaithersburg, MD 20899.

the central issues in predicting the long-term performance of a composite structure involves assessing the durability of the fiber-matrix interface. Although silane coupling agents are often added to the fiber to enhance the durability of this interface, there is no reliable method to measure the effectiveness of these treatments. The matrix properties play a strong role in determining the interfacial shear strength and the matrix properties usually change dramatically with environmental exposure, via plasticization. Therefore, the integrity of the fiber-matrix interface cannot be assessed without properly accounting for the changes in the matrix properties. In addition, delayed fracture events have been shown to occur during the testing procedure at times greater than 10 min after the implementation of a step-strain increment. This behavior is inconsistent with elastic and elastic-perfectly plastic based micromechanics models and suggests the presence of a time dependent failure process. The latter observation parallels research on full composites that relates time dependent fracture of composites to nonelastic properties of the matrix. These results and observations suggests that a critical look at the influence of matrix properties on the shear stress transfer process in micromechanics models is needed.

Background and Approach

Research by Galiotis et al. [1] on polydiacetylene fiber embedded in an epoxy matrix has demonstrated that the fiber stress distribution in a SFF test specimen can be approximated by the "classical" shear lag model derived by Cox [2]. Galiotis [3] also notes that the constant shear stress condition across the fiber matrix interface is seldom achieved in polymer matrix composites. Hence, the Kelly-Tyson model [4–6], based on elastic-plastic analysis, for determining interfacial shear strength is seldom applicable to polymer matrix composites. In addition, experimental data on the DGEBA/m-PDA epoxy resin will be presented which demonstrates the inappropriateness of the constant shear stress approximation for this DGEBA/m-PDA resin system. Therefore, in this paper, we will focus primarily on shear lag micromechanics models.

In developing the "classical" and "non-classical" shear lag models, researchers typically make simplifying assumptions about the shear-stress transfer process, the fiber-matrix interface, and the matrix material [7]. The common assumptions are as follows: (1) the matrix material is linear elastic, (2) a perfect bond exists between the fiber and the matrix, (3) the radius of the matrix, r_m, is unknown and typically assumed to be ½ the thickness of the test specimen, and (4) yielding of the matrix is not considered. In addition, as noted by Shioya et al. [8], these models make no assumption about the failure process occurring at the interface.

The "classical" shear lag models developed over the years [2,9–16] provide similar stress distribution profiles of the embedded fiber. However, these models violate the boundary conditions in the shear equation by predicting the maximum shear stress to be at the fiber ends. "Non-classical" shear lag models, developed recently by Whitney and Drzal [17], McCartney [18], and Nairn [19], overcome this violation. The shear stress profiles obtained from these "non-classical" models exhibit the appropriate zero shear stress at the fiber ends and fiber stress profiles similar to those predicted in the "classical" models. In addition, these models, along with a "classical" model developed by Amirbayat and Hearle [12–14], account for the radial pressure at the fiber-matrix interface that results from the difference in thermal expansion coefficients of the two materials. However, all of these models assume that the matrix material is linear elastic (the Kelly-Tyson model assumes elastic-perfectly plastic matrix behavior).

In applications that are strength and weight critical the matrix material, e.g., tetraglycidyl-4,4'-diaminodiphenylmethane (TGDDM)/4,4'-diaminodiphenylsulphone (DADPS) epoxy resin used in aerospace and military applications, is typically brittle and has a maximum elongation at room temperature of approximately 1.8% [20]. Hence, a linear elastic approximation may provide a reasonable estimate of the matrix material behavior in the composite, even though the initiation of failure in the composites may involve nonlinear matrix behavior on the microscopic scale. For commercial filament winding applications flexible epoxy resins are often used for impact resistance and greater elongation. These resins may have ultimate elongations of approximately 8% [21]. In the special case of interfacial shear strength measurements derived from the single fiber fragmentation (SFF) analysis, the epoxy matrix is also required to have a high extension–to–failure, greater than 7%. This is typically achieved by undercuring the resin [22] or using flexible hardeners to reduce the resin crosslink density [23]. Polymer materials exhibiting high extension–to–failure usually exhibit increased stress relaxation and nonlinear stress-strain behavior in the high strain region. In durability analyses, where the specimen has been exposed to moisture, the stress-strain behavior is often altered by moisture absorption, via plasticization. Hence, the linear elastic matrix assumption may be violated during the fragmentation process.

In this paper, the impact of matrix viscoelasticity on the SFF analysis procedure is investigated. The Cox model is utilized since it explicitly includes the effect of the matrix modulus in the equations for the fiber stress profile, shear stress profile at the fiber matrix interface, and the theoretical critical transfer length. As a first step toward understanding the effect of nonlinear viscoelasticity, we will extend the Cox model to accommodate linear viscoelasticity. Since research on nonlinear constitutive behavior in solids is complicated and still in its infancy, an engineering approximation of the nonlinear stress-strain behavior of the matrix will be presented. The impact of this engineering approximation on the analysis of data obtained from the single fiber fragmentation experiment will be considered.

Theory

The Cox model [2], developed in 1952, is a widely recognized and utilized linear elastic based unit composite model. This model affords the following equations for the stress profile of the embedded fiber, shear stress profile at the fiber-matrix interface, and maximum shear stress for a fiber at its critical length

$$\sigma_f\{z\} = (E_f - E_m)\varepsilon \left[1 - \frac{\cosh \beta \left(\frac{1}{2} - z\right)}{\cosh \beta \frac{1}{2}} \right] \tag{1}$$

$$\tau_{\text{interface}}\{z\} = \frac{d\beta\sigma_c}{4} \left(\frac{E_f - E_m}{E_m}\right) \frac{\sinh \beta \left(\frac{1}{2} - z\right)}{\cosh \beta \frac{1}{2}} \tag{2}$$

$$\tau_{\text{interface}_{\text{max}}}\{l_c\} = \frac{d\beta}{4} \frac{\sinh \beta \left(\frac{l_c}{2}\right)}{\cosh \beta \left(\frac{l_c}{2}\right) - 1} \sigma_f\{l_c\} \tag{3}$$

$$\beta = \frac{2}{d} \left[\frac{E_m}{(1 + v_m)(E_f - E_m) \ln \left(\frac{2r_m}{d} \right)} \right]^{1/2} \qquad (4)$$

where

$\sigma_f\{z\}$	is the tensile stress in the fiber
$\tau_{interface}\{z\}$	is the shear stress along the fiber-matrix interface
E_f	is the fiber elastic modulus
E_m	is the matrix elastic modulus
l	denotes the length of the fiber
z	is the distance along the fiber
ε	is the general or global strain
l_c	is the critical transfer length or ineffective length
σ_c	is the uniform stress applied to the composite
d	is the fiber diameter
r_m	is the radius of the matrix
v_m	is Poisson's ratio of the matrix

Of notable interest is the quantity β, which has been associated with the critical transfer length of the shear stress transfer process, $l_c/2 \cong 1/\beta$ [7,24]. With respect to the Cox model, β is used in determining the maximum shear stress in the interface (see Eq 3) at the critical transfer length. The determination of β has recently come under scrutiny because of its dependence on the nebulous radius of matrix, r_m, parameter (see Eq 4) [25–33]. Also of interest is the assumption by Cox, that the maximum stress transferred to the embedded fiber by shear is limited to the difference between the actual displacement at a point on the interface, at a distance "z" from the end of the fiber, and the displacement that would be observed if the fiber were absent. For an E-glass fiber, modulus 67.5 GPa, embedded in an elastic matrix, modulus 2.51 GPa, only 96% of the matrix shear stress is transferred to the fiber.

Linear Viscoelastic Composite Models

In 1968–1970, J. M. Lifshitz [29–33] published extensively concerning the effect of linear viscoelasticity on the longitudinal strength of unidirectional fibrous composites. His research focused primarily on investigating the impact of viscoelasticity on composite time dependent fracture or creep rupture. This research was driven by the experimental observation that the longitudinal strength of unidirectional fiber reinforced composites is time dependent. In addition, the elastic based cumulative weakening composite model of Rosen [10,11] assumed fracture in a single cross section of the composite, while experimental evidence showed a very complicated fracture surface. In his research, Lifshitz noted that in many cases of practical importance the matrix material is a polymer having time dependent properties that can be characterized by the laws of linear viscoelasticity. He hypothesized that the existence of breaks in the fibers results in local shear stresses in the matrix that can be expected to relax. Therefore, the length of the fiber fragment required to transmit the load from the matrix to the fiber (ineffective length) increases with time due to the relaxation of the matrix modulus. This sequence of events suggested to Lifshitz the likelihood of a time-dependent failure process for fibrous composites, even for unidirectional composites loaded in the fiber direction. In the development of his viscoelastic model, Lifshitz extended the cumulative weakening model of Rosen [10,11] to the linear viscoelastic regime by utilizing Schapery's Approximation (*Schapery's Correspondence Principle* [34,35]) of the *Elastic-Viscoelastic Correspondence Principle* [36].

$$\sigma_f\{z,t\} = \sigma_{f_o}[1 - \exp(\eta\{t\}z)] \tag{5}$$

$$\tau_f\{z,t\} = \left(\frac{G_m\{t\}}{2E_f}\frac{r_f}{r_m - r_f}\right)^{1/2} \exp(-\eta\{t\}z) \tag{6}$$

$$\eta\{t\} = \left(\frac{2G_m\{t\}}{E_f}\frac{1}{r_f(r_m - r_f)}\right)^{1/2} \tag{7}$$

$$\delta\{t\} = \left[J_m\{t\}\frac{E_f r_f(r_m - r_f)}{2}\right]^{1/2} \ln\left[\frac{1}{(1 - \Phi)}\right] \tag{8}$$

where

σ_{fo} is the tensile stress in the fiber at a large distance from the fiber end.

In this formulation, Lifshitz chose the ratio $(r_f/r_m)^2$ to be the same value as the volume fraction of the fibers within the composite. Using this approach, a composite with a fiber volume fraction of 40% and specific gravity of 1.40 has a r_f/r_m value of 0.4 [2]. Assuming the average fiber diameter is 15 μm, r_m has a value of 37.5 μm. In the single fiber fragmentation specimen, the fiber volume fraction is approximately 4×10^{-5}. Hence, r_m is 5.2 mm using this formula. Since the dogbone specimens are approximately 1 mm thick, this has led some to take r_m to be half the specimen thickness [7]. However, this would imply that the interfacial shear strength is dependent on specimen size! Therefore, researchers have focused on ways to accurately determining this parameter.

Equations 5 and 7 show that the stress profile in the fiber is time dependent, via the shear modulus of the matrix, $G_m\{t\}$. In addition, the ineffective length (denoted by $\delta\{t\}$ in this model) is defined as the distance required to transfer the stress from the matrix to the fiber is defined as the ineffective length. The ineffective length is time dependent through the matrix creep compliance, $J_m\{t\}$. From his viscoelastic model Lifshitz, noted that the stress at any point in this zone will relax with time. For a diglycidyl ether of bisphenol-A/bis(2,3-epoxycyclopentyl) ether/aromatic diamine blend system he calculated a relative relaxation of 13% of the initial stress at infinite time. The ratio of the stress at infinite time to the initial stress is given by Φ in Eq 8. To account accurately for the long-term time dependent strength of fibrous composites, Lifshitz suggested that the viscoelastic nature of the glass fibers be considered in addition to the matrix viscoelasticity. Based on his results, Lifshitz viewed the viscoelastic properties of the matrix as the main cause of time dependent strength of fibrous composites.

Making note of Lifshitz's research, Phoenix in 1988 [37] extended his "chain of bundles" model by including the linear viscoelastic effects of the matrix material. To incorporate viscoelastic effects, Phoenix made use of Hedgepeth's solution [38] from the linear elastic case. This solution showed that the geometric load transfer length (ineffective transfer length), δ, varied as $\sqrt{E_f/G_m}$ where E_f is the fiber tensile modulus and G_m is the matrix shear modulus. Harlow and Phoenix [39] showed that the geometric load transfer length, is related to the effective load transfer length, δ^*, and bounded by the following expression

$$1/(\zeta + 1) \le \delta^*/\delta \le 3/(\zeta + 1) \tag{9}$$

Hence, δ^*, with statistical effects included, is considerably shorter than δ and depends on the Weibull shape parameter, ζ, for fiber strength [40]. Phoenix et al. [37] quantified the time dependence of δ, resulting from the viscoelasticity of the matrix, by using a power-law creep function

$$J_m\{t\} = J_o[1 + (t/t_o)^\theta], \, t \ge 0 \tag{10}$$

where θ and t_o are the creep exponent and time constants, respectively, and $J_o = 1/G_m$.

Although the correlation with experimental data was not perfect, Phoenix's research indicated that the prediction capability of the "chain of bundles" model could be improved by careful characterization of fiber strength, matrix creep, and time-dependent debonding at the fiber-matrix interface. The research by Lifshitz and Phoenix indicates that matrix viscoelastic effects in the creep rupture of unidirectional fibrous composites are important. In addition to these researchers, Jansson and Sundstrom [41] noted that the viscoelastic properties of the matrix play an important role in the creep as well as in the creep rupture processes in composite materials. They indicate that composite matrix materials exhibit nonlinear anelastic behavior at low strains and that this behavior becomes pronounced above 1% strain.

Linear Viscoelastic Cox Model

Even though unit composite analyses, e.g., single fiber fragmentation (SFF) tests, have been used extensively to link microstructural interface research to composite performance, very little research has been done to quantify the impact of matrix viscoelasticity on unit composite analysis methods. One reason for this is that much of the early work assumed that the DGEBA/m-PDA epoxy resin system behaved in an elastic-perfectly plastic manner. Hence, interfacial shear strength values were obtained using the Kelly-Tyson model. This model contains no parameters related to matrix properties and so the inference was that matrix properties were not important.

It should be noted that the SFF test is performed in different ways. Some researchers strain the test specimen at a constant strain rate and continuously monitor and count break events by acoustic emissions (AE) and/or a video camera. The analysis proceeds until the fiber stops breaking with further extension. In principle, viscoelastic effects in this testing regime are exhibited when the strain rate of the test is changed. Therefore, results from tests performed at different strain rates may not be comparable unless the model accounts for changes in matrix properties due to viscoelasticity.

A second approach, utilized in this laboratory, is the manual application of sequential step strains (saw-toothed loading pattern) until saturation is reached. The step strains are made at constant time intervals, usually 10 min, and the number of breaks are counted after each step. The complete distribution of fragment lengths is obtained after saturation is reached. To perform a more detailed analysis of the fragmentation process, e.g., (1) obtain a map of the fragmentation process, (2) obtain fragment lengths at each strain increment, and (3) monitor the development of debond regions during the testing procedure, the time increment between successive strains must be allowed to increase to the time required to measure the fragment lengths after each strain increment. Since the matrix material is viscoelastic and the stress response of the matrix at a given time, t, depends on the previous stress history, the impact of matrix viscoelasticity on the fragmentation process and the ability to compare results obtained by different testing regimes can become a complex issue.

Because of the similarity between the linear elastic field equations (i.e., equilibrium equation, boundary conditions, strain equations, etc.) and the transform of the linear viscoelastic field equations, Laplace transformed viscoelastic solutions can be obtained from elastic solutions by replacement of the elastic moduli and elastic Poisson's ratio, v, by the Carson transform of the appropriate viscoelastic relaxation functions and viscoelastic Poisson's ratio (*Elastic-Viscoelastic Correspondence Principle*) [36]. This simple replacement holds if quasi-static and separation of variables, i.e., $\sigma_f\{z,t\} = \hat{\sigma}_f\{z\}g\{t\}$, conditions prevail [42]. The transformed viscoelastic Cox equations are readily written down from Eqs 1–4. The transformed expressions of Eqs 1 and 4 are given below

$$\overline{\sigma}_f\{z,\lambda\} = (E_f - \lambda\overline{E}_m\{\lambda\})\varepsilon \left[1 - \frac{\cosh \overline{\beta}\{\lambda\} \left(\frac{1}{2} - z\right)}{\cosh \overline{\beta}\{\lambda\} \frac{1}{2}} \right] \tag{11}$$

$$\overline{\beta}\{\lambda\} = \frac{2}{d} \left[\frac{\lambda\overline{E}_m\{\lambda\}}{(1 + \lambda\overline{v}_m\{\lambda\})(E_f - \lambda\overline{E}_m\{\lambda\}) \ln\left(\frac{2r_m}{d}\right)} \right]^{1/2} \tag{12}$$

To transform Eq 11 into the time domain $\overline{E}_m\{\lambda\}$ must be specified analytically. Unfortunately, the inversion of the resulting equation is not trivial. An approximate method of Laplace transform inversion has been given by Schapery, *Schapery's Correspondence Principle: Direct Method* [35]. Schapery notes that if a function $f\{t\}$ has a small curvature when plotted against $\log_{10}t$, the following condition applies

$$f\{t\} \cong [\lambda\overline{f}\{\lambda\}]_{\lambda=0.5t} \tag{13}$$

The "small curvature" restriction means that when $f\{t\}$ is plotted against $\log_{10}t$ and a tangent is drawn to $f\{t\}$ at any point, the net algebraic area, A_Δ, enclosed by the tangent line, the function $f\{t\}$, and about three-fourths to one decade on each side of the tangency point should be small relative to the area, A_T, under $f\{t\}$ in the same interval. Schapery indicates that if $f\{t\}$ has constant curvature over a 1.8 decade interval, one can show that the relative error in $f\{t\}$ is essentially equal to the area ratio, A_Δ/A_T, for this interval.

As a result of the above approximation, Schapery notes that with respect to moduli, a viscoelastic solution is obtained from an elastic solution by replacing all elastic constants with time-dependent relaxation moduli, *Schapery's Correspondence Principle: Quasi-Elastic Method* [35]. In a critique of Schapery's approximations, Christensen indicates that this method is applicable to quasi-static problems in viscoelasticity, for which the deformation history is rather smooth [42].

Krishnamachari [43] notes two addition engineering approximations that may be used to convert the Laplace transformed viscoelastic solutions to the time domain. The first assumes the bulk modulus in the *Elastic-Viscoelastic Correspondence Principle* is constant. This approach is based on the observation that of all the viscoelastic properties, the bulk modulus varies the least with time. Support for this assumption is found in the work of Tschoegl [44] where he notes that in many synthetic polymers the bulk relaxation modulus changes from the glassy to the equilibrium state by only a factor of about 2 to 3 while the shear modulus changes by 3 to 4 logarithmic decades. The second approach is closely related to Schapery's approximations and is called *pseudoelasticity* [43]. In *pseudoelasticity,* Poisson's ratio, v, is assumed to be a constant and the elastic moduli are replaced by their viscoelastic counterparts. This approach is based on the observation that v, although subject to variations of up to 35 to 40%, is still a weak variable in the expressions for stress and strain, and can be treated as constant in engineering calculations. Christensen [42] indicates that in most quasi-static cases where separation of variables conditions prevail v is indeed a constant.

Assuming the matrix material meets Schapery's small curvature approximation, the impact of embedding an elastic fiber into a linear viscoelastic matrix is readily seen by utilizing *Schapery's Correspondence Principle* on the Cox model. The linear viscoelastic version (quasi-elastic approximation) of this unit composite model is shown below

$$\sigma_f\{z,t\} = (E_f - E_m\{t\})\varepsilon \left[1 - \frac{\cosh \beta\{t\} \left(\frac{1}{2} - z\right)}{\cosh \beta\{t\} \frac{1}{2}} \right] \tag{14}$$

$$\tau_f\{z,t\} = \frac{d\beta\{t\}\sigma_c}{4} \left(\frac{E_f - E_m\{t\}}{E_m\{t\}}\right) \frac{\sinh \beta\{t\} \left(\frac{1}{2} - z\right)}{\cosh \beta\{t\} \frac{1}{2}} \tag{15}$$

$$\tau_{f\max}\{l_c\{t\},t\} = \frac{d\beta\{t\}}{4} \frac{\sinh \beta\{t\} \left(\frac{l_c\{t\}}{2}\right)}{\cosh \beta\{t\} \left(\frac{l_c\{t\}}{2}\right) - 1} \sigma_f\{l_c\{t\}\} \tag{16}$$

$$\beta\{t\} = \frac{2}{d} \left[\frac{E_m\{t\}}{(1 + v_m\{t\})(E_f - E_m\{t\}) \ln \left(\frac{2r_m}{d}\right)} \right] \tag{17}$$

Because of the time dependence of the viscoelastic matrix, we can see that the critical transfer length, the maximum interfacial shear stress, the shear stress profile of the matrix-fiber interface, and the stress profile in the fiber are time dependent. Equation 16 is of particular interest since its linear elastic counterpart, Eq 3, has been used to determine the interfacial shear strength at saturation. As a consequence of the time dependence of the matrix, the linear viscoelastic Cox model indicates that the interfacial shear strength at saturation will depend on what influence the matrix relaxation has on the critical transfer length.

Experimental

Fiber and Mold Preparation

To make single fiber fragmentation specimens, eight-cavity molds were made out of RTV-664 (General Electric[2]) following the procedure described by Drzal [45]. All molds were post cured at 150°C and rinsed with acetone prior to use. A 12″ long tow was cut from a spool of E-glass fibers (from Owens-Corning) previously shown to be bare with no processing aids. The tow was washed with spectrophotometric grade acetone, vacuum dried at 100°C overnight, and cooled prior to use. Single filaments of E-glass fiber were separated from the 12″ tow being careful to touch only the ends of the fiber. The fibers were aligned in the mold cavity via the sprue slots in the center of each cavity. The fibers were temporarily fixed in place by pressing them onto double-stick tape. Small strips of double-stick tape were placed over each fiber end to hold them in place until each fiber was permanently mounted with 5 min epoxy.

Embedding Procedure

100 grams of diglycidyl ether of bisphenol-A (DGEBA, Epon 828 from Shell Chemical Co.) and 14.5 g of meta-phenylene diamine (m-PDA, Fluka Chemical Company), were weighed out in separate beakers. To lower the viscosity of the resin and melt the m-PDA crystals, both beakers were placed in a vacuum oven (Fisher Scientific Isotemp Vacuum Oven, model 281 A) set at 75°C. After the m-PDA crystals were completely melted, the silicone molds containing the fibers were placed into another oven (Blue M Stabiltherm, model OV-560A-2) that is preheated to 100°C. With the preheated oven turned off, the silicone molds were placed in the oven for approximately 20 min. This last procedure dries the molds and minimizes the formation of air bubbles during the curing process.

At approximately 9 min before the preheated molds were removed from the oven, the m-PDA is poured into the DGEBA and mixed thoroughly. The mixture was placed into the Vacuum Oven and degassed for approximately 7 min. After 20 min, the preheated molds were removed from the oven and filled with the DGEBA/m-PDA resin mixture using 10 cc disposable syringes. The filled molds were then placed into a programmable oven (Blue M, General Signal, model MP-256-1, GOP). A cure cycle of 2 h at 75°C followed by 2 h at 125°C was used.

Fragmentation Test

The fiber fragmentation tests were carried out on a small hand operated loading frame similar to that described by Drzal [45] mounted on a Nikon Optiphot polarizing microscope. The image was viewed using a CCD camera (Optronics LX-450 RGB Remote-Head microscope camera) and monitor (Sony, PVM-1344Q). Before the test, the fiber diameter was measured with an optical micrometer (VIA-100 from Boeckeler) attached to the video system. The sample was scanned by translating the loading frame under the microscope with a micrometer. The position of the load frame is monitored by an LVDT (Trans-Tek, Inc. model

---

[2] Certain commercial materials and equipment are identified in this paper to specify adequately the experimental procedure. In no case does such identification imply recommendation or endorsement by the National Institute of Standards and Technology, nor does it imply necessarily that the product is the best available for the purpose.

1002-0012) connected to an A-to-D board (Strawberry Tree, Inc.) in a computer. To measure fragment lengths or other points of interest in the sample, the location was aligned with a cross hair in the microscope as seen on the video monitor, and the position of the LVDT was digitized into the computer. The standard instrument uncertainty in measuring a point is ± 0.3 µm. The standard uncertainty in relocating a point reproducibly is ± 1.1 µm. The load is also monitored during the experiment using a 2224 N (500 lb) load cell connected to a bridge (load cell and AED 9001A bridge, Cooper Instruments). The expected standard uncertainty of the load measurements is 3% of the load. The bridge is attached to the same computer via a serial connection. A custom program was developed to continuously record the load and any LVDT measurements that are made. The average application time of each strain step was (1.10 ± 0.17) s and the average deformation was (14.45 ± 3.11) µm. The strain was found to increase by 0.0034% for each 1 N change in load.

Results and Discussion

Viscoelastic Loading Profile

A typical load-time curve for a DGEBA/m-PDA epoxy resin SFF test specimen is shown in Fig. 1 (lower curve). Readily visible in this loading curve is the relaxation of the load after each strain increment. The relaxation of the stress with time is consistent with the constitutive law which governs the behavior of linear viscoelastic materials to step–strain responses, $\sigma\{t\}_{\text{output}} = E\{t\}\varepsilon_{\text{input}}$. The modulus, $E\{t\}$, is the relaxation modulus of the viscoelastic material and is the ratio of the time dependent stress to the initial applied strain. Initially this relaxation modulus has a maximum value, $E_{\text{unrelaxed}}$, which corresponds to the instantaneous elastic response. At a time $t \rightarrow \infty$ the modulus attains a minimum value E_{relaxed}. This contrasts with the stress response of a linear elastic material to a step strain input,

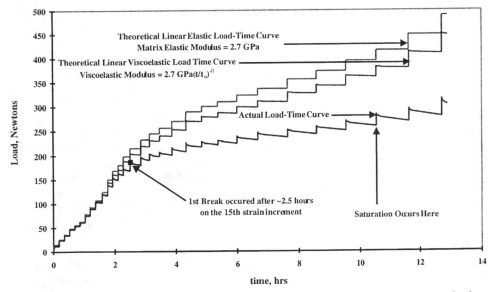

FIG. 1—*Actual load-time curve with linear elastic and linear viscoelastic approximations for bare E-glass fiber SFF test specimen.*

$\sigma_{\text{output}} = E_{\text{elastic}}\varepsilon_{\text{input}}$. In such a material the modulus, E_{elastic}, is constant with time and so the response to the applied strain is constant with time.

Using the small strain response, $\varepsilon < 0.1$, of the actual load-time curve, linear elastic and linear viscoelastic approximations of the load-time curve are also shown in Fig. 1. For the linear elastic approximation a modulus value of 2.7 GPa was used. For the linear viscoelastic approximation the time dependent modulus was approximated by a power law expression and Boltzmann's superposition principle [46]

$$E\{t\} = E_o \left(\frac{t}{t_o}\right)^{-\theta} \qquad (18)$$

From the step-strain response at small strains, the value of theta, θ, was found to be 0.016 when t_o was taken to be 100 s. Good agreement was obtained with the initial strain increments of the saw tooth loading pattern in Fig. 1 using both approximations. However, significant deviations arise before the first fiber break occurs, ~2.5 h. At saturation, ~3.6% strain, both approximations have deviated considerably from the actual load-time curve. This deviation is indicative of nonlinear stress-strain behavior. Thus fiber fracture is observed in the region where the matrix is exhibiting nonlinear viscoelastic behavior. In this test the time between strain increments is increased after the initial break and increases to the time required to measure the fiber fragments.

Viscoelastic Stress Strain Curves

"Pseudo-isochronal" stress versus strain plots from the data given in Fig. 1 are shown in Fig. 2. By this term we will mean the clock will be figuratively restarted after each loading step so we can compare loads after each loading step at the same time into that step. For example, the 10 min data will be the load recorded 10 min after each loading step was applied, i.e., the previous peak load [39,41]. Figure 2 shows data at 10 s (10 s stress) and 10 min (10 min stress) for the sample shown in Fig. 1. Since the time increment between step–strains varied, the stress immediately before the application of the next strain increment is also plotted. The measurement of fragment lengths began at 10 min after the step–strain was applied. Consequently, the separation between the 10 min stress plot and the stress before the next strain increment plot indicates the degree of stress relaxation and hence, matrix relaxation occurring during the measurement of the fragment lengths. In Fig. 2 the 10 s stress values from the theoretical linear viscoelastic fit in Fig. 1 are depicted by open circles. As expected the linear viscoelastic fit is intermediate between the linear elastic fit and the 10 s stress-strain data in the high strain region. From Fig. 2, it is clear that for the specific loading history in this experiment, the epoxy matrix exhibits nonlinear stress-strain behavior above 1% strain. Regression analyses of the data below 1% strain and above 1.8% strain shows that the tangent moduli for the 10 s stress data are (2.52 ± 0.01) GPa and (1.15 ± 0.01) GPa, respectively (see Fig. 2). Regression analyses of the 10 min stress-strain data resulted in tangent moduli values of (2.46 ± 0.01) GPa and (1.11 ± 0.01) GPa, respectively. However, the tangent moduli in the region above 1.8% strain are significantly less, ~55%, than the tangent moduli in the small strain region.

Based on these observations, a linear analysis of the interfacial shear stress via the Cox model, utilizing the small strain "linear elastic" modulus of the viscoelastic matrix, is not appropriate. From related research, numerical simulations of the fragmentation process using a linear elastic matrix results in an over prediction of the number of fragments that actually

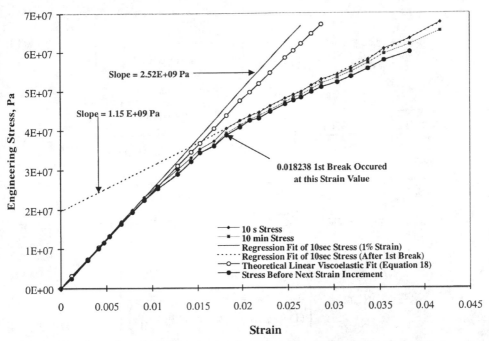

FIG. 2—*Pseudo-isochronal stress-strain plots for SFF test specimen containing a bare E-glass fiber (variable loading time increments).*

occur [47]. In addition, the onset of fragmentation in the unit composite specimens investigated by Feillard et al. [47] occurred in the nonlinear stress strain region, ~5% strain. Since the epoxy matrix exhibits increased load carrying capability after the onset of yield, the Kelly-Tyson model which assumes an elastic perfectly plastic matrix material is also a poor approximation.

Nonlinear Viscoelastic Model

Since the onset of fiber fracture occurs in the nonlinear stress-strain region, existing shear lag unit composite models must be extended to account for the nonlinear viscoelastic constitutive behavior of the matrix. Unfortunately, the development of nonlinear constitutive equations and models for solids is a complicated and active area of research. At present, there is no universal theory for developing nonlinear constitutive equations for melts or solids and, hence, all nonlinear models are somewhat empirical in nature. Motivated by the fact that matrix properties in the Cox and linear viscoelastic Cox model arise primarily through the matrix modulus and Poisson's ratio, it is plausible to assume that strain dependent non-linear effects involving the matrix are also manifested in these terms. Accepting this general format, one can make the engineering approximation that Poisson's ratio is a constant and replace the relaxation modulus, $E_m\{t\}$, in the linear viscoelastic Cox model with a modulus dependent on strain and time, $E_m\{\varepsilon,t\}$ Thus, the nonlinear viscoelastic Cox equations have the following functional forms and variable dependencies

$$\sigma_f\{z,\varepsilon,t\} = (E_f - E_m\{\varepsilon,t\})\varepsilon \left[1 - \frac{\cosh \beta\{\varepsilon,t\} \left(\frac{1}{2} - z\right)}{\cosh \beta\{\varepsilon,t\} \frac{1}{2}} \right] \qquad (19)$$

$$\tau_f\{z,\varepsilon,t\} = \frac{d\beta\{\varepsilon,t\}\sigma_c}{4} \left(\frac{E_f - E_m\{\varepsilon,t\}}{E_m\{\varepsilon,t\}}\right) \frac{\sinh \beta\{\varepsilon,t\} \left(\frac{1}{2} - z\right)}{\cosh \beta\{\varepsilon,t\} \frac{1}{2}} \qquad (20)$$

$$\tau_{f\max}\{1_c,\varepsilon,t\} = \frac{d\beta\{\varepsilon,t\}}{4} \frac{\sinh \beta\{\varepsilon,t\} \left(\frac{l_c}{2}\right)}{\cosh \beta\{\varepsilon,t\} \left(\frac{l_c}{2}\right) - 1} \sigma_f\{l_c\} \qquad (21)$$

$$\beta\{\varepsilon,t\} = \frac{2}{d} \left[\frac{E_m\{\varepsilon,t\}}{(1 + v_m)(E_f - E_m\{\varepsilon,t\}) \ln\left(\frac{2r_m}{d}\right)} \right] \qquad (22)$$

A plot of the relaxation modulus versus $\ln(t)$, Fig. 3, indicates that Schapery's small curvature

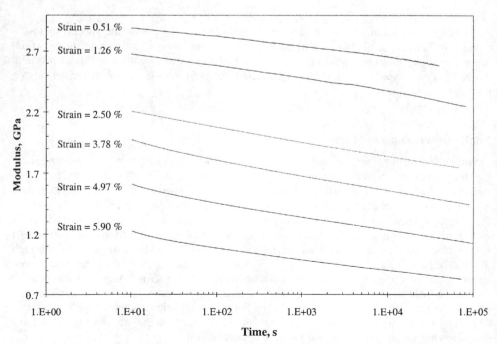

FIG. 3—*Single-step relaxation behavior of DGEBA/m-PDA dogbone specimens.*

approximation is met well into the nonlinear stress-strain region. The development of a nonlinear viscoelastic relaxation modulus will be the subject of future work. In the absence of an explicit expression for $E_m\{\varepsilon,t\}$, the matrix nonlinearity is further approximated by substituting into the nonlinear viscoelastic Cox model a "secant modulus," or average modulus, which is dependent on strain and time. In the nonlinear stress-strain region, this modulus is readily calculated from the experimental stress-strain data at each strain increment. This modulus should provide a conservative estimate of the impact of nonlinear viscoelasticity of the SFF test. Therefore, in the above equations

$$E_m\{\varepsilon,t\} = \langle E_m\{\varepsilon,t\}\rangle_{\text{secant}} \tag{23}$$

Additional motivation for the use of a strain dependent secant modulus is found in the work of Feillard et al. [47]. As noted previously, numerical simulations of the single fiber fragmentation process using a linear elastic matrix overpredicted the number of fragments that actually occur. These researchers found better agreement using a secant modulus.

Impact of Nonlinear Viscoelasticity on the Critical Transfer Length

From the research of Asloun et al. [7] the theoretical critical transfer length is given by the following expression

$$\frac{l_c}{2} \cong \frac{1}{\beta} = \frac{d}{2}\left[\frac{(1 + v_m)(E_f - E_m)}{E_m} \ln\left(\frac{2r_m}{d}\right)\right]^{1/2} \tag{24}$$

By replacing E_m in Eq 24 with Eq 23, the critical transfer length in a nonlinear viscoelastic material becomes time and strain dependent through the matrix modulus. The variation of the secant moduli, nonlinear and linear, with strain and the number of fiber breaks are shown in Fig. 4. At the final strain increment, the 10 s nonlinear secant modulus has decreased by 35% relative to the elastic modulus, whereas the linear viscoelastic secant modulus is predicted to decrease by only 6.5%. Assuming r_m to be ½ the thickness of the sample, the impact of these moduli variations on the critical transfer length are shown in Fig. 5. Except for the critical transfer length values determined by the theoretical equation of Asloun et al. (Eq 24), all transfer length values were determined graphically, at 96.5% of the maximum fiber stress, by substituting the appropriate viscoelastic secant moduli into Eqs 14 and 19.

Comparing the 10 s nonlinear secant modulus transfer length determined graphically, solid triangles, and transfer length determined theoretically from Asloun's equation extended to the nonlinear viscoelastic regime, solid circles, indicates that Asloun's equation captures the relative change in transfer length over the complete strain range, but is off by a constant factor of 1.684. Consistent with the moduli variations shown in Fig. 4, the transfer length for the 10 s nonlinear secant modulus at the last strain increment shows an increase of 25% relative to the elastic transfer length, $l_c/2 = 255$ μm. The transfer length of the linear viscoelastic secant modulus increases by only 3.6% over the same range. Analysis of the relaxation behavior within a strain step reveals that the relaxation of the secant modulus after 10 min into the strain step increases the transfer length by approximately 2%. Relaxation of the secant modulus up to 1 h increases the transfer length by less than 5% in each strain step. These results indicate that the variation in the transfer length from the experimental data is primarily due to the nonlinear behavior of the matrix material in the high strain region.

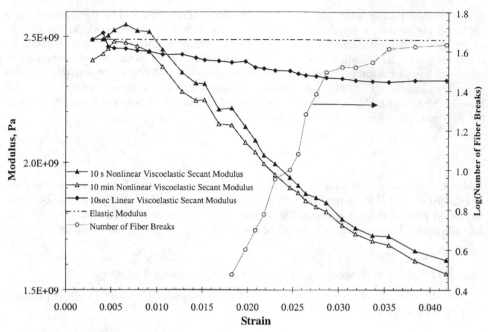

FIG. 4—*Variation of moduli with strain and number of fiber breaks.*

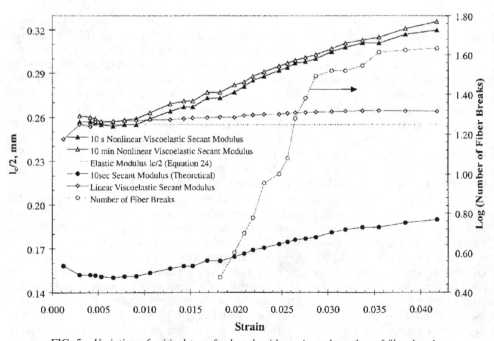

FIG. 5—*Variation of critical transfer length with strain and number of fiber breaks.*

Impact of Nonlinear Viscoelasticity on Interfacial Shear Strength Measurements

To determine the interfacial shear strength from the Cox model and its time and strain dependent variants (see Eqs 3, 16, and 21), the critical transfer length (l_c), the value of β, and the stress in the fiber at the critical transfer length ($\sigma_f\{l_c\}$) must be determined. Although the critical transfer length is obtained experimentally, the latter variables must be calculated. The stress at the critical transfer length is typically determined from Weibull parameters and is beyond the scope of this paper. However, Schultheisz et al. [48] have reported values for the stress in the fiber at the critical transfer length in the range of (2.5 to 2.8) GPa for fragments similar in size to those generated in this paper. Since this value is not essential for comparing the impact of nonlinear viscoelasticity on the analysis procedure, the value of 2.5 GPa will be used.

Although β is not constant for the viscoelastic models, the value of the matrix modulus at saturation is used to calculate β and determine the interfacial shear strength for all models. The use of this modulus value in the viscoelastic model is motivated by our desire to carry out durability analysis. Since the fiber strength has been shown to decrease with exposure to moisture, it is desirable to calculate the fiber strength value from the test data. Since the procedure devised by Schultheisz et al. [48] calculates the fiber strength at saturation, the value of the secant modulus at this value is appropriate. Recognizing that the secant modulus at saturation is dependent on the previous deformation history, the following guidelines were followed to minimize test variability in the measured secant modulus. (1) Multi-step loading was kept uniform, i.e., step-strains were nominally the same throughout the experiment with no premature unloads of the matrix. (2) The variability in intersample loading profiles was minimized. Intersample changes were limited to normal changes that are incurred from measuring or counting the fragments after each strain step. (3) The profile of the fragment distributions at the end of the test was checked in each sample and found to be nominally the same. (4) The incubation time, i.e., time before fragments are counted or measured, remained the same. For the DGEBA/m-PDA epoxy resin this time was 10 min.

In Table 1, the determined interfacial shear strength using the various moduli are shown. When the nonlinear secant modulus is used the determined interfacial shear strength is 20% lower than the value obtained when the elastic modulus value that is typically assumed for the DGEBA/m-PDA epoxy resin is used. The interfacial shear strength obtained by using the elastic modulus derived from the experimental data lowers the interfacial shear strength by approximately 8%. As expected from the data depicting the variation in moduli with strain (see Fig. 4), the interfacial shear strength obtained from the linear viscoelastic secant modulus is approximately the same as the value obtained from the experimental data elastic modulus. Utilizing the experimentally determined critical transfer length of 502 μm and the fiber diameter of 12.12 μm, the Kelly-Tyson model which has been widely used to analyze

TABLE 1—*Determined interfacial shear strength using various moduli.*

	Modulus, GPa	β, mm^{-1}	τ, MPa
Elastic Modulus, Assumed	3.06	14.48	116
Elastic Modulus, via Experimental Data	2.52	13.09	107
Linear Viscoelastic Secant Modulus	2.32	12.54	104
Nonlinear Secant Modulus	1.71	10.7	93
Kelly-Tyson Model	N/A	N/A	30

glass fibers embedded in this resin yields an interfacial shear strength value 3 to 4 times smaller than the elastic and secant moduli estimates.

Conclusions

Experimental data were presented for a single fiber fragmentation test specimen consisting of a bare E-glass fiber embedded in DGEBA/m-PDA epoxy resin. The data conclusively showed that fragmentation in the E-glass fiber occurred in the nonlinear viscoelastic region of the stress strain curve. In addition, the DGEBA/m-PDA epoxy resin was shown to exhibit increased load carrying capability after the onset of yield. Hence, the linear elastic Cox model and the Kelly-Tyson model were shown to be inappropriate for analyzing fragmentation data from this test specimen.

To accommodate the nonlinear viscoelastic behavior of the DGEBA/m-PDA resin, the Cox Equation was first extended to the linear viscoelastic regime by using *Schapery's Correspondence Principle*. Analysis of the DGEBA/m-PDA epoxy resin step-strain data revealed that the criteria for applying *Schapery's Correspondence Principle* was met even in the nonlinear viscoelastic region. Extending the linear elastic Cox model to the viscoelastic regime was shown to result in time dependent expressions for the fiber stress, shear stress, maximum fiber stress, and the parameter β, which has been associated with the critical transfer length. The nonlinear behavior of the matrix material had a major influence on the resulting transfer length.

The nonlinear viscoelastic Cox model was assumed to have a form analogous to the linear viscoelastic Cox model with the time dependent modulus and Poisson's ratio replaced by its time and strain dependent counterparts. In lieu of an explicit expression for the strain dependent modulus and motivated by a desire to do durability analyses, the secant modulus at saturation was chosen as an appropriate modulus to utilize in the nonlinear Cox model. In addition, v was assumed to be constant. With this approximation, the nonlinear behavior of the matrix was found to contribute significantly to the increase in the transfer length due to its dependence on the modulus. The modulus was found to decrease relative to the small strain modulus by approximately 55% during the experiment. This decrease in the modulus resulted in a 25% increase in the critical transfer length relative to the initial value during the experiment. As a result the interface strength calculated by the secant modulus at saturation was found to be 20% lower than the interface strength predicted by the elastic modulus. As noted previously, numerical simulations of the fragmentation process by Feillard et al. [47], assuming a linear elastic matrix was shown to over predict the degree of fragmentation in the specimen. These researchers found better agreement with the use of a "secant modulus." The Kelly-Tyson model which has been widely used to analyze glass fibers embedded in this resin yields an interfacial shear strength value 3 to 4 times smaller than the elastic and secant moduli estimates.

During the review process, it was observed that we have not addressed the impact of the change in fiber tensile strength as a function of fiber length on the determination of the interface strength. As pointed out by one reviewer, this effect will have a larger affect on the determined interfacial shear strength than changes in the matrix modulus and will make the values given in Table 1 much larger than the values that have been determined from measurements made on high volume fraction composites. The reviewer also noted that the number of fragments at saturation is experimentally determined and is relatively independent of the assumptions about matrix material behavior when slow straining rates are used. To the reviewer's first comment, we noted that no one has effectively dealt with the issue of change in fiber tensile strength as a function of fiber length and this issue remains an active research topic. Over the years, this issue has been discussed at length in our laboratory and because of its complexity, we decided that it should be dealt with as a separate topic. Hence, the values in Table 1 are not absolute values, but are given to illustrate only the effect of modulus changes on the derived value of the interfacial shear strength.

The driving force for the development of the nonlinear viscoelastic model has been the observation that the number of fragments at saturation from E-glass/DGEBA/m-PDA SFFT specimens depends on how the test is performed (test protocol). This observation is inconsistent with the assumption of elastic or elastic-perfectly plastic matrix behavior. Indeed, the reviewer's second observation alludes to this problem. The test protocol issue has also been of central concern in the development of a new round robin test procedure being administered by the Versailles Project on Advanced Materials and Standards (VAMAS). In the proposed VAMAS round robin test procedure, the DGEBA/m-PDA matrix is also used. Ongoing research in this laboratory has shown that the nonlinear viscoelastic model developed in this paper is useful in detecting differences in the fragmentation behavior of E-glass/DGEBA/m-PDA test specimens arising from changes in the test protocol. Since these observations require a detailed discussion of the fragmentation length distributions at saturation, future publications will cover the investigation of the effect of test protocol on the fragment length distribution at saturation and the application of the nonlinear viscoelastic model developed in this paper to this observed fragmentation behavior.

References

[1] Galiotis, C., Young, R. J., Yenug, P. H. J., and Batchelder, D. N., "The Study of Model Polydiacetylene/Epoxy Composites," Journal of Materials Science, Vol. 19, 1984, pp. 3640–3648.

[2] Cox, H. L., "The Elasticity and Strength of Paper and Other Fibrous Materials," British Journal of Applied Physics, Vol. 3, 1952, pp. 72–79.

[3] Melanitis, N., Galiotis, C., Tetlow, P. L., and Davies, C. K. L., "Interfacial Shear Stress Distribution in Model Composites. Part 2: Fragmentation Studies on Carbon Fibre-Epoxy Systems," Journal of Composite Materials, Vol. 26, No. 4, 1992, pp. 574–610.

[4] Kelly, A. and Tyson, W. R., "Tensile Properties of Fiber-Reinforced Metals: Copper/Tungsten and Copper/Molybdenum," Journal of Mechanics and Physics of Solids, Vol. 13, 1965, pp. 329–350.

[5] Kelly, A. and Tyson, W. R., "Fiber-Strengthened Materials," High Strength Materials: Proceedings of the 2nd Berkley International Materials Conference, V. F. Zackay, Ed., John Wiley & Sons, Inc., New York, Chapter 13, 1964.

[6] Kelly, A., "The Strengthening of Metals by Dispersed Particles," Proceedings of the Royal Society of London, A282, pp. 63–79.

[7] Asloun, El. M., Nardin, M., and Schultz, J., "Stress Transfer in Single-Fibre Composites: Effect of Adhesion, Elastic Modulus of Fibre and Matrix, and Polymer Chain Mobility," Journal of Materials Science, Vol. 24, 1989, pp. 1835–1844.

[8] Shioya, M., McDonough, W. G., Schutte, C. L., and Hunston, D. L., "Test Procedure for Durability Studies of the Fiber Matrix Interface," Proceedings of the Seventeenth Annual Meeting and the Symposium on Particle Adhesion, K. M. Liechti, Ed., The Adhesion Society: Orlando, Fl., 1994, pp. 248–251.

[9] Dow, N. F., "Study of Stresses Near a Discontinuity in a Filament-Reinforced Composite Metal," Report No. R63SD61, General Electric Space Sciences Laboratory, August 1963.

[10] Rosen, B. W., "Tensile Failure of Fibrous Composites," American Institute of Aeronautics and Astronautics Journal, Vol. 2, No. 11, 1964, pp. 1985–1991.

[11] Rosen, B. W., "Mechanics of Composite Strengthening," Fiber Composite Materials, S. H. Bush, Ed., American Society for Metals, Metals Park, Ohio, Chapter 3, 1965.

[12] Amirbayat, J. and Hearle, J. W. S., "Properties of Unit Composites as Determined by the Properties of the Interface. Part I: Mechanism of Matrix-Fiber Load Transfer," Fiber Science and Technology, Vol. 2, 1969, pp. 123–141.

[13] Amirbayat, J. and Hearle, J. W. S., "Properties of Unit Composites as Determined by the Properties of the Interface. Part II: Effect of Fiber Length and Slippage on the Modulus of Unit Composites," Fiber Science and Technology, Vol. 2, 1969, pp. 143–153.

[14] Amirbayat, J. and Hearle, J. W. S., "Properties of Unit Composites as Determined by the Properties of the Interface. Part III: Experimental Study of Unit Composites Without a Perfect Bond Between the Phases," *Fiber Science and Technology,* Vol. 2, 1969, pp. 223–239.

[15] Smith, G. E. and Spencer, A. J. M., "Interfacial Tractions in a Fiber-Reinforced Elastic Composite Material," *Journal of the Mechanics and Physics of Solids,* Vol. 18, 1970, pp. 81–100.

[16] Theocaris, P. S. and Papanicolaou, G. C., "The Effect of the Boundary Interphase on the Thermomechanical Behavior of Composites Reinforced with Short Fibers," *Fiber Science and Technology,* Vol. 12, No. 6, 1979, pp. 421–433.

[17] Whitney, J. M. and Drzal, L. T., "Axisymmetric Stress Distribution Around an Isolated Fiber Fragment," *Toughened Composites, ASTM STP 937,* N. J. Johnson, Ed., American Society for Testing and Materials, West Conshohocken, PA, 1987.

[18] McCartney, L. N., "New Theoretical Model of Stress Transfer Between Fiber and Matrix in a Uniaxially Fiber-Reinforced Composite," *Proceedings of the Royal Society of London,* A425, 1989, pp. 215–244.

[19] Nairn, J. A., "A Variational Mechanics Analysis of the Stresses Around Breaks in Embedded Fibers," *Mechanics of Materials,* Vol. 13, No. 2, 1992, pp. 131–154.

[20] Puglisi, J. S. and Chaudhari, M. A., "Epoxies(EP)," *Engineered Materials Handbook,* Metals Park, OH, Vol. 2, 1988, pp. 240–241.

[21] Shibley, A. M., "Filament Winding," *Handbook of Composites,* G. Lubin, Ed., Van Nostrand Reinhold Company, 1982, pp. 449–478.

[22] Drzal, L. T., Rich, M. J., and Lloyd, P. A., "Adhesion of Graphite Fibers of Epoxy Matrices: 1. The Role of Fiber Surface Treatment," *Journal of Adhesion,* Vol. 16, No. 1, 1982, pp. 1–30.

[23] Netravali, A. N., Henstenburg, R. B., Phoenix, S. L., and Schwartz, P., "Interfacial Shear Strength Studies Using the Single-Filament-Composite Test. 1: Experiments on Graphite Fibers in Epoxy," *Polymer Composites,* Vol. 10, No. 4, 1989, pp. 226–241.

[24] Galiotis, C., Young, R. J., Yeung, P. H. J., and Batchelder, D. N., "The Study of Model Polydiacetylene/Epoxy Composites. Part 1. The Axial Strain in the Fiber," *Journal of Materials Science,* Vol. 19, No. 11, 1984, pp. 3640–3648.

[25] Melanitis, N., Galiotis, C., Tetlow, P. L., and Davies, C. K. L., "Interfacial Shear Stress Distribution in Model Composites: The Effect of Fiber Modulus," *Composites,* Vol. 24, No. 6, 1993, pp. 459–466.

[26] Galiotis, C., "Interfacial Studies on Model Composites by Laser Raman Spectroscopy," *Composites Science and Technology,* Vol. 42, 1991, pp. 125–150.

[27] Melantis, N., Galiotis, C., Tetlow, P. L., and Davies, C. K. L., "Monitoring the Micromechanics of Reinforcement in Carbon Fibre/Epoxy Resin Systems," *Journal of Materials Science,* Vol. 28, No. 6, 1993, pp. 1648–1654.

[28] Li, Zong-Fu and Grubb, D. T., "Single Fiber Polymer Composites. Part 1. Interfacial Shear Strength and Stress Distribution in the Pull-Out Test," *Journal of Materials Science,* Vol. 29, No. 1, 1994, pp. 189–202.

[29] Lifshitz, J. M., "Time Dependent Fracture of Fibrous Composites," *Composite Materials. Vol. 5: Fracture and Fatigue,* L. J. Broutman, Ed., Academic Press: New York, Chapter 6, 1974.

[30] Lifshitz, J. M. and Rotem, A., "Time-Dependent Longitudinal Strength of Unidirectional Fibrous Composites," *Fiber Science and Technology,* Vol. 3, No. 1, 1970, pp. 1–20.

[31] Lifshitz, J. M. and Rotem, A., "An Observation on the Strength of Unidirectional Fibrous Composites," *Journal of Composite Materials,* Vol. 4, No. 1, 1970, pp. 133–134.

[32] Lifshitz, J. M., "Specimen Preparation and Preliminary Results in the Study of Mechanical Properties of Fiber Reinforced Material. Part I," AFML-TR-69-89, Air Force Materials Laboratory, Wright-Patterson Air Force Base, Ohio, July 1969.

[33] Lifshitz, J. M. and Rotem, A., "Longitudinal Strength of Unidirectional Fibrous Composites," AFML-TR-70-194, Air Force Materials Laboratory, Wright-Patterson Air Force Base, Ohio, September 1970.

[34] Schapery, R. A., "Approximate Methods of Transform Inversion for Viscoelastic Stress Analysis," *Proceedings of the 4th U.S. National Congress of Applied Mechanics,* ASME, 1962, p. 1075.

[35] Schapery, R. A., "Stress Analysis of Viscoelastic Composite Materials," *Journal of Composite Materials,* Vol. 1, No. 3, 1967, pp. 228–267.

[36] Lee, E. H., "Stress Analysis in Visco-Elastic Bodies," *Quarterly of Applied Mathematics,* Vol. 13, No. 2, 1955, pp. 183–190.

[37] Phoenix, S. L., Schwartz, P., and Robinson IV, H. H., "Statistics for the Strength and Lifetime in Creep-Rupture of Model Carbon/Epoxy Composites," *Composites Science and Technology,* Vol. 32, 1988, pp. 81–120.

[38] Hedgepeth, J. M., "Stress Concentrations in Filamentary Structures," NASA TN D-882, Langley Research Center, 1961.

[39] Harlow, D. G. and Phoenix, S. L., "Bounds on the Probability of Failure of Composite Materials," *International Journal of Fracture*, Vol. 5, No. 4, 1979, pp. 321–336.

[40] **Note:** The symbols have been changed in this article (see Ref *37*) to be consistent with previously referenced research results. Hence, δ denotes the geometric load (ineffective) transfer length and represents the distance along the broken fiber (counting both sides) where the stress is reduced below the maximum stress. δ^* denotes the "effective load transfer length" and represents the length of a fiber segment, i.e., fiber adjacent to the broken fiber, under a constant or uniform stress which would have the same probability of failure as a segment loaded under the assumed triangular overload load profile.

[41] Jansson, J.-F. and Sundstrom, H., "Creep and Fracture Initiation in Fiber Reinforced Plastics," *Failure of Plastics*, W. Brostow and R. D. Corneliussen, Eds., Hanser Publishers, Munich, Chapter 24, 1986.

[42] Christensen, R. M., "Theory of Viscoelasticity: An Introduction," Academic Press, New York, Chapter II, 1982.

[43] Krishnamachari, S. I., "Applied Stress Analysis of Plastics: A Mechanical Engineering Approach," Van Nostrand Reinhold, New York, Chapter 5, 1993.

[44] Tschoegl, N. W., "The Phenomenological Theory of Linear Viscoelastic Behavior," Springer-Verlag, Berlin, Chapter 11.3.5, 1989.

[45] Drzal, L. T. and Herrera-Franco, P. J., "Composite Fiber-Matrix Bond Tests," *Engineered Materials Handbook: Adhesive and Sealants, Vol. 3,* ASM International, Metals Park, Ohio, 1990, pp. 391–405.

[46] Dealy, J. M. and Wissbrun, K. F., *Melt Rheology and its Role in Plastics Processing: Theory and Applications,* Van Nostrand Reinhold, New York, 1990, pp. 128, 132, 146, and 188.

[47] Feillard, P., Desarmot, G., and Favre, J. P., "A Critical Assessment of the Fragmentation Test for Glass/Epoxy Systems," *Composites Science and Technology,* Vol. 49, 1993, pp. 109–119.

[48] Schulthcisz, C. R., McDonough, W. G., Kondagunta, S., Schutte, C. L., Macturk, K. S., McAuliffe, M., and Hunston, D. L., "Effect of Moisture on E-Glass/Epoxy Interfacial and Fiber Strengths," *Composite Materials: Testing and Design, ASTM STP 1242.*

Erhard Krempl[1] and Kwangsoo Ho[2]

An Overstress Model for Solid Polymer Deformation Behavior Applied to Nylon 66

REFERENCE: Krempl, E. and Ho, K., **"An Overstress Model for Solid Polymer Deformation Behavior Applied to Nylon 66,"** *Time Dependent and Nonlinear Effects in Polymers and Composites, ASTM STP 1357,* R. A. Schapery and C. T. Sun, Eds., American Society for Testing and Materials, West Conshohocken, PA, 2000, pp. 118–137.

ABSTRACT: Extensive experimental investigations at ambient temperature on commercial Nylon 66, PEI (Polyetherimide) and PEEK (poly(ether ether ketone)) have shown that the overstress model developed for viscoplasticity should be, in principle, capable of modeling for solid polymers the rate-dependent behavior, including creep, relaxation and cyclic motions. The viscoplasticity theory based on overstress was modified accordingly to allow for the modeling of typical solid polymer deformation behavior. Included are nonlinear rate sensitivity, curved unloading, significant strain recovery at zero stress and cyclic softening. The viscoplasticity theory based on overstress for polymers (VBOP) is introduced in uniaxial formulation. It is shown that VBOP can be thought of as a modified standard linear solid with overstress-dependent viscosity and nonlinear, hysteretic equilibrium stress evolution. VBOP consists of a flow law that is easily adopted to cases where the strain or the stress is the independent variable. The flow law depends on the overstress, the difference between the stress and the equilibrium stress with the latter being a state variable of VBOP. The growth law of the equilibrium stress in turn contains the kinematic stress and the isotropic or rate-independent stress, two additional state variables of VBOP. The material constants of VBOP are determined for Nylon 66 at room temperature and various tests are simulated by numerically integrating the set of nonlinear differential equations. The simulations include monotonic loading and unloading at various strain rates, repeated relaxation, recovery at zero stress that is dependent on prior strain rate, and cyclic strain-controlled loading. Finally, the stress-controlled loading and unloading are predicted with very good results. The simulations and predictions show that VBOP is competent at modeling the behavior of Nylon 66 and other solid polymers.

KEYWORDS: solid polymers, nylon, constitutive equations, viscoplasticity, viscoelasticity, numerical experiments, overstress model, rate dependence, creep, relaxation

Most of the modeling of solid polymer deformation behavior is via integral representations. An example is linear viscoelasticity where convolution integrals with strain or stress as the independent variable are employed. Nonlinear convolution integral formulations are used as well—for example, McKenna [1], Knauss and Emri [2], and Schapery [3], to name just a few investigators. Hasan and Boyce [4] model the viscoelastic viscoplastic behavior of glassy polymers using a mechanistically motivated differential formulation [4]. No attempt is made to summarize the extensive literature further. The purpose of this paper is to introduce the viscoplasticity theory based on overstress for polymers (VBOP) which was derived from a "unified" state variable theory for metallic materials.

[1] Professor of Engineering, Rensselaer Polytechnic Institute, Troy, NY 12180-3590.
[2] Now at Yeungnam University, Korea.

It is comparatively easy to invert a linear integral law with stress as the independent variable to an equivalent law with strain as the independent variable. However, it may be very difficult, if not impossible, to perform the same conversion for nonlinear behavior.

In experiments with servo-controlled systems a switch from strain control (strain is the independent variable) to stress control is easy and requires only the throw of a switch. Material response for stress and strain control is different. The experimental results have to be reproduced by material models and, therefore, easy inversion of the material model is very desirable.

Another form of a constitutive law is the differential representation. In the linear case the differential formulation is easily derived from the integral form. For nonlinear behavior the conversion from the integral to the differential form and vice versa can be very difficult.

Differential constitutive equations can be constructed so that inversion is simple. In addition, they need only the initial conditions to incorporate past history. For integral representations the entire previous history has to be "carried along" and deformation histories of the distant past contribute less and less as time progresses.

Strain and stress methods are used in stress analyses. Appropriately formulated differential constitutive laws can be immediately written for either the stress or the strain as the independent variable. No inversion is necessary. It is difficult, if not impossible, to invert nonlinear integral constitutive equations numerically or analytically.

The Overstress Model

The constitutive model to be introduced here is of a differential form and has its origin in the modeling of metal deformation behavior. With the experience in modeling the time-dependent inelastic behavior of metals and alloys, a room temperature test program was initiated for three solid polymers, Nylon 66, PEEK (poly(ether ether ketone)) and PEI (polyetherimide). Bordonaro and Krempl [5–8] report the results, without a model bias. Sure enough, differences between metal and solid polymer behavior were found. The recovery at zero stress is small for metals but large and dependent on prior loading rate for polymers. Also the unloading behavior is curved for polymers but nearly straight for metals. However, surprisingly many common features were found in the time-dependent behavior of solid polymers metals and alloys. Examples are the nonlinear rate sensitivity and the short-time creep and relaxation behavior when inelastic flow is fully established. Also transient experiments intended to confirm or falsify the presence of an equilibrium stress came out in favor of the existence of the equilibrium stress for metals and polymers [9].

Overall, these results suggested that it might be profitable to modify the viscoplasticity theory based on overstress (VBO), a "unified" state variable theory developed for metals and alloys [10], for application to polymers. The results of an initial modeling effort are presented here.

The modeling of solid polymers using the overstress concept was also encouraged by a series of publications by Kitagawa [11,12] and Ariyama [13,14] (only the first and last available papers are cited here for brevity). These authors report good modeling capabilities of an early overstress theory of strain rate sensitivity, creep and relaxation motions for monotonic but not for cyclic loading. It turns out that these authors use an early version of the overstress theory in which the equilibrium curve is only a function of strain. The equilibrium curve is therefore the same for loading and unloading and no hysteresis loop is formed. The model introduced below has overcome this limitation. An elastic/viscoplastic analysis that includes the overstress for polymeric composites was proposed by Gates and Sun [15]. The paper by Zhu and Sun [16] presented in this symposium also uses the overstress concept in identifying time-dependent behavior of polymeric composites.

The experimentally based method of modeling advocated by the senior author follows an article of faith of materials science: "The current response of a material depends on the present microstructure and *on the current input*" (text in Italics added). In this view the testing machine imposes a certain stress (strain) history, the input, and the material answers or responds with strain (stress) history, the output. It is evident that the response (answer/output) of a specimen includes the effects of the input and of the internal mechanisms. The specimen serves as a natural integrator of all active internal mechanisms. Therefore, the present, experimentally based, continuum approach includes the effects of the varying internal structure by letting the specimen do the integration and constructing the model from the comparison of input-output pairs.

The experiments include "standard" tests such as varying the strain rate from test to test that show nonlinear rate sensitivity. Other tests, such as short time creep during loading and unloading are specifically introduced to motivate and justify the use of the overstress. Details are given in Refs 5–9.

To show that the model to be introduced below has a familiar background, the standard linear solid (SLS) in the overstress formulation is discussed first. Included are the fast and the slow response of this model. The long-term behavior in a constant strain rate test is discussed and reveals linear rate dependence. As a linear viscoelastic solid it models full recovery of strain when unloaded to zero stress. The viscoplasticity theory based on overstress for solid polymers (VBOP) is, in broad terms, a modified SLS that reproduces, among other properties, nonlinear, loading rate sensitivity and partial recovery. This is accomplished by making the viscosity coefficient depend on the overstress and by making the equilibrium stress nonlinear with hysteresis so that different loading and unloading branches are obtained. After a general discussion of VBOP the material constants are determined for Nylon 66 at room temperature. Then uniaxial load and displacement controlled, transient tests are numerically simulated with good success. In essence the same tests are performed with the model and the real specimen. The model simulations involve numerical integration of the set of coupled, stiff, nonlinear differential equations using the LSODA program [17]. A three-dimensional theory that allows inelastic compressibility and finite deformation is presented [18] and will be submitted elsewhere.

The Standard Linear Solid

The simplest linear viscoelastic solid that shows creep, relaxation and loading rate sensitivity is the "standard linear solid," see Fig. 1. Accordingly, this model is used as a starting point by writing its constitutive equation in the "overstress" format

$$\dot{\varepsilon} = \frac{\dot{\sigma}}{E_1} + \frac{\sigma - aE_2\varepsilon}{a\eta} \tag{1}$$

A superposed dot denotes material time derivative. σ and ε are true (Cauchy) stress and true strain, respectively; E_1 and E_2 are the moduli of the spring in front and the spring in the Kelvin element, respectively. The quantity η is the viscosity coefficient with the dimension of stress $\times$ time; and $a = E_1/(E_1 + E_2)$. The effective modulus of the two springs in series is equal to aE_2.

The stress response for an infinitely slow loading is given by the two springs in series since the dashpot does not support any stress under these conditions. It is seen from Eq 1

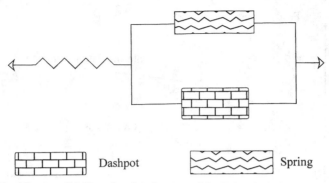

Dashpot   Spring

FIG. 1—*Standard linear solid. Note that the elements of the Kelvin body have special fills to indicate that these elements have different properties for the SLS and VBOP.*

that $\sigma = aE_2\varepsilon$ is the response in the limit as the strain rate approaches zero. Therefore the quantity $aE_2\varepsilon$ is called the equilibrium stress. The difference $\sigma - aE_2\varepsilon$ is the overstress.

The response of the SLS for infinitely fast strain rate is given by the response of the spring in front since the dashpot is rigid for infinitely fast loading. This can be demonstrated by writing

$$1 - \frac{d\sigma/d\varepsilon}{E_1} = \frac{\sigma - aE_2\varepsilon}{a\eta\dot{\varepsilon}} \tag{2}$$

and letting the strain rate go to infinity. For any other strain rate the response lies between the infinitely fast and infinitely slow response as shown in Fig. 2.

For loading with a constant strain rate the model predicts that the slope becomes equal to the slope of the equilibrium stress-strain curve for large time. To see this, the term $aE_2\dot{\varepsilon}/E_1$ is subtracted from each side of Eq 1 so that

$$\dot{\varepsilon}\left(1 - \frac{aE_2}{E_1}\right) = \frac{d}{dt}\left(\frac{\sigma - aE_2\varepsilon}{E_1}\right) + \left(\frac{\sigma - aE_2\varepsilon}{a\eta}\right) \tag{3}$$

This equation permits a solution for constant overstress. Then the first term on the right-hand side of Eq 3 vanishes. In this case the overstress is constant and linearly related to the strain rate.

When the expression of constant overstress is substituted into Eq 2 the final slope for constant strain rate is

$$\left\{\frac{d\sigma}{d\varepsilon}\right\} = aE_2 \tag{4}$$

where { } indicates the long-time solution. For the SLS the slopes of all stress-strain diagrams with constant strain rate are ultimately equal to that of the equilibrium curve.

For creep the stress rate is zero and Eq 1 yields

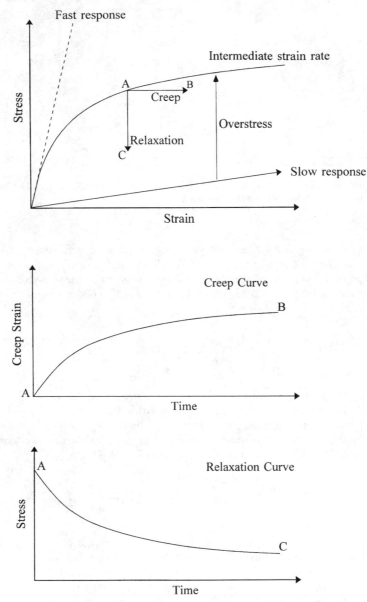

FIG. 2—*Infinitely fast, infinitely slow, and response for a constant intermediate constant strain rate test of the standard linear solid. Creep and relaxation paths are also sketched.*

$$\dot{\varepsilon} = \frac{\sigma_0 - aE_2\varepsilon}{a\eta} \tag{5}$$

For relaxation the strain rate is zero and we have

$$\dot{\sigma} = -E_1 \frac{\sigma - aE_2\varepsilon_0}{a\eta} \tag{6}$$

In the above equations the subscript $_0$ indicates that this quantity is constant. Examination of Eqs 5 and 6 shows that the creep and relaxation rates depend on the difference between the stress and the equilibrium stress, the so-called overstress. It is positive in tensile loading and is indicated in Fig. 2. It follows that for prior tensile loading the creep rate is positive and the relaxation rate is negative. Creep and relaxation motions, see the paths indicated in Fig. 2, end when the stress reaches the equilibrium stress. It can be demonstrated that the equilibrium stress is reached in infinite time.

Suppose, a numerical creep (relaxation) test is started when the long time solution holds and the overstress is constant. Although stress and strain are still increasing, the creep (relaxation) behavior is independent of the stress (strain) at which the test starts. In the regions where the long-time asymptotic solution holds, the curves of creep strain versus time and curves of relaxed stress (the stress measured from the starting point) versus time are congruent, respectively.

Viscoplasticity Theory Based on Overstress for Polymeric Solids (VBOP)

General

VBOP is in essence a modified SLS where the viscosity coefficient representing the dashpot is nonlinearly dependent on overstress and where the spring of the Kelvin model is changed to a nonlinear spring-like element with hysteresis. Many properties of the SLS are preserved. One of them is the existence of long-term solutions for constant strain rate loading.

The Equations

The VBOP is a state variable theory that consists of a flow law, three state variables with the dimension of stress and their growth laws. The flow law is given by

$$\dot{\varepsilon} = \dot{\varepsilon}^{el} + \dot{\varepsilon}^{in} = \frac{\sigma}{E} + \frac{(\sigma - g)}{Ek[\Gamma, A]} \tag{7}$$

and is similar to Eq 1. The first term on the right-hand side is the same in both cases and represents the instantaneous response of the spring in series in Fig. 1. The second term contains the equilibrium stress g and the nonlinear viscosity function k with dimension of time. It is positive, decreasing and $k[0] \neq 0$. Square brackets denote "function of."

The viscosity function normally depends only on $\Gamma = |(\sigma - g)|$, the overstress invariant which corresponds to the absolute value of the overstress in uniaxial loading. In this case there is a pure overstress dependence of the inelastic strain rate. The reasons for introducing the additional dependence on the isotropic stress A is discussed below.

The response for infinitely slow loading is $\sigma = g$. It is also evident that the strain rate depends nonlinearly on the overstress. VBOP represents nonlinear rate sensitivity.

The flow law, Eq 7, and the SLS share many properties. One phenomenon is the existence of long-term solutions for constant strain rate. All stress-strain curves can become ultimately equidistant and the separation depends nonlinearly on strain rate in VBO [10], and VBOP [18]. The dependence of the overstress is linear for SLS, see Eq 3. In the region of fully established inelastic flow, VBO and VBOP can exhibit identical relaxation properties [10,19].

Experimental evidence of the special relaxation behavior in the region of fully established inelastic flow has been reported in Figs. 4 and 5 of [5].

In the present case the viscosity function is also made to depend on the isotropic or rate-independent stress A, another state variable of the VBOP. This dependence on A is needed to model the merging of the stress-strain curves of Nylon 66 obtained at different strain rates for large strains, see Fig. 1 of Ref 5. The change of the isotropic stress A according to the differential equation given below is also a repository for modeling cyclic softening.

In the case of the SLS the growth of the equilibrium stress is linear, reversible and rate-independent. In VBOP the growth for the equilibrium stress is made almost rate-independent, nonlinear and hysteretic, it traces a different curve on loading and unloading. (In the early version of VBO the equilibrium stress depended on strain only and the equilibrium curve was the same for loading and unloading. Note that strain is not present in VBO and VBOP.) The growth of the equilibrium stress of VBOP is given by

$$\dot{g} = \psi \frac{\dot{\sigma}}{E} + \psi \left[\frac{(\sigma - g)}{Ek} - \frac{(g - f)}{A} \frac{\Gamma}{Ek} + \frac{(\dot{\sigma} - \dot{g})}{E} \right] + \left[1 - \frac{\psi}{E} \right] \dot{f} \tag{8}$$

It includes the positive shape function ψ, where $E_t < \psi < E$ that models the transition from quasi linear elastic behavior to inelastic flow. The quantity E_t is the tangent modulus based on inelastic strain at the maximum strain of interest. The first two terms on the right-hand side of Eq 8 are the elastic and inelastic hardening terms followed by the dynamic recovery term. The term $\dot{\sigma} - \dot{g}$ is introduced to emphasize transients for solid polymers; it is zero when the long-term solution is reached [18]. The last term involves the kinematic stress and its evolution equation, Eq 10, sets the slope of the stress-strain curve at large times.

The state variable A, with the dimension of stress, grows according to

$$\dot{A} = A_c[A_f - A] \frac{\Gamma}{Ek} \tag{9}$$

with A_f the final value of A, and A_c, is a dimensionless quantity that controls the speed with which the final value of A is reached.

The kinematic stress f has the growth law

$$\dot{f} = \left[\frac{|\sigma|}{\Gamma + |g|} \right] E_t \frac{(\sigma - g)}{Ek} \tag{10}$$

and sets the tangent modulus at the maximum strain of interest.

Similar to SLS, the four nonlinear differential equations of VBOP admit a long-term solution for a constant strain rate test for which $\{\sigma - g\} = constant$ where $\{ \}$ indicate the asymptotic long-term solution. Here the additional conditions $\{g - f\} = A_f$, $\{A\} = A_f$ and $\{\dot{\sigma}\} = \{\dot{g}\} = \{\dot{f}\}$ must be satisfied. This result indicates that in the region where the long-term solutions hold the stress grows equidistant to f, which in turn grows according to Eq 10. Note that the final tangent modulus E_t can be chosen positive, zero or negative to ultimately model hardening, perfectly plastic or softening behavior. For details, see Ref 10.

A comparison of the equations of the SLS and of VBOP with Fig. 1 shows that the linear spring in series is present in both models. However, the linear spring of the Kelvin model of the SLS becomes a nonlinear, hysteretic element in VBO and VBOP. It can be shown that the slope of the equilibrium curve before and after a sudden change in strain rate is different and this behavior gives rise to a loading and unloading branch of the equilibrium stress-strain curve. For VBO and VBOP, a damper depending nonlinearly on the overstress replaces the linear dashpot of SLS.

After determination of the material constants, the above differential relations can be exercised like a servo-controlled testing machine. A test condition needs to be specified, say, a constant total strain rate up to a certain strain and then unloading with a given strain rate until zero stress is reached. When this condition is put into the constitutive equations, they can be solved for the stress by numerical integration. Cross-plotting of the stress and the strain gives the loading/unloading stress-strain diagram.

Sample Response

A loading/unloading test at two strain rate magnitudes is depicted in Fig. 3. The thick continuous and dashed lines represent the evolution of the stress. The corresponding evolution of the equilibrium stress is in thin lines and dashes. For a given strain rate there is a stress evolution and an associated equilibrium stress evolution. It is noted that two equilibrium stress evolutions exist up to a strain of 6%. Then the two curves merge. The theory postulates that the long-term solution for the equilibrium stress is rate independent and so the two equilibrium stress-strain curves merge. This is a characteristic of the long-term

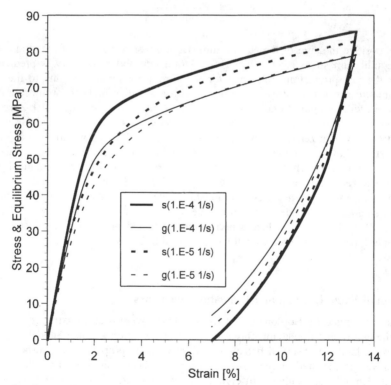

FIG. 3—*Simulation of stress-strain behavior during loading and unloading at two strain rates. In addition to the stress, the equilibrium stress is also plotted. Note the nonzero equilibrium stress when the stress reaches the strain axis. The equilibrium stress is different for the two tests and is largest for the high unloading strain rate. The overstress for the recovery test, Eq 11, is different for the two tests and this difference is the basis for modeling the influence of prior strain rate on recovery, see Fig. 5.*

solutions, see Ref *10*. Figures 2 and 3 show the unique and linear equilibrium stress-strain diagram of SLS and the nonlinear and the hysteretic and finally unique equilibrium stress-strain curve of VBOP, respectively. On unloading the linear equilibrium stress-strain diagram traces back on itself for the SLS whereas the equilibrium stress of VBO initiates an unloading branch due to the hysteretic evolution law. The stress follows this pattern.

The nonlinear rate sensitivity is evident; the strain rates differ by one order of magnitude, but the stresses differ only by approximately 5% at large strain. Initially the equilibrium stress curves are nearly indistinguishable from the stress curves but separate from them as strain increases. They show rate dependence, which ceases after 6% strain when the asymptotic condition is approached and the equilibrium stress-strain curves merge. The equilibrium stress-strain curves become rate independent at large strains as shown in Fig. 3.

At about 13% the simulation introduces unloading at the same strain rate magnitude until zero stress is reached. The two unloading stress-strain curves are almost identical and are curved. The hysteretic behavior of the equilibrium stress is evident as is the rate-dependent evolution upon unloading. The equilibrium stress associated with the high strain rate is always larger than the one associated with the low strain rate.

The computations leading to Fig. 3 were stopped at zero stress to show the behavior in a recovery test in which the stress is zero. Equation 7 is specialized for this condition to yield

$$\dot{\varepsilon} = \dot{\varepsilon}^{in} = \frac{(-g)}{Ek[\Gamma_{\sigma=0}, A]} \tag{11}$$

Since k is positive the sign of the equilibrium stress governs the sign of the strain rate. The positive equilibrium stress renders a negative strain rate and a recovery is predicted. It is larger for the fast prior strain rate than for the slow prior strain rate due to the different equilibrium stresses at zero stress, see Fig. 3. With a negative inelastic strain rate the equilibrium stress will continue to decrease as can be ascertained from Eqs 8–10. Zero equilibrium stress is reached for $\sigma = f = 0$ and $A = constant$ at some strain between zero strain and the strain at which zero stress was reached upon unloading, approximately at 7% in this case. The strain that is computed by integrating Eq 11 is the recovery strain, which must be added to the strain at which zero stress is reached during unloading. Mathematically it takes infinite time to reach this history-dependent equilibrium position. However, the approach to the equilibrium position is at an ever-diminishing rate so that it is possible to estimate the permanent strain. This will be demonstrated below.

The VBOP model can reproduce a permanent set that depends on prior history. For the SLS complete recovery is predicted for every equilibrium strain reached on unloading to zero stress.

Simulation of Some Loading and Unloading Behaviors

The data generated by Bordonaro and Krempl [5–7] were used to extract qualitative features of the deformation behavior. Based on previous experience a new model was invented that can reproduce the observed behavior qualitatively. Its properties had to be ascertained by qualitative analyses and by numerical experiments. They consist of numerically integrating the set of nonlinear, coupled, differential equations using the code LSODA. In essence, the model was "operated like a servo-hydraulic testing machine" using the same command functions.

Finally the present model was arrived at and consists of the previously postulated flow law, the growth laws for the state variables and the functions and material constants given in Table 1. Using these data and the LSODA numerical integrator the following test data were simulated.

The VBOP simulations are all performed using the true stress and strain quantities. The experiment uses displacement (engineering strain or strain rate) control and reports engineering stresses. To have a fair comparison the numerical results are converted to engineering quantities using the Poisson's ratio given in Table 1. It is important to know that there are, except for Fig. 9a, no predictions but only simulations in this paper. Other very reasonable predictions can be found in [18] for uniaxial and biaxial situations and these results will be published in the future.

Loading and Unloading at Different Constant Strain Rate Magnitudes

Figure 4 shows the numerical simulations of four different, constant, engineering strain rate tests with unloading at the same strain rate magnitudes. The strain rates differ by a factor of ten and start at the slow rate of 10^{-6} 1/s. The engineering stress-strain curves shown in Fig. 4 are simulations of Fig. 1 of Ref 5. The nonlinear rate sensitivity is apparent and is simulated very well, as is the merging of the stress-strain curves which is more pronounced in the experiments than in the simulation. The "overshoot," sometimes referred to as yielding, is reproduced at the highest strain rate. The overshoot for stress-strain curve at the strain rate of 10^{-4} 1/s is small compared with experimental results. Both the experiment and the simulations do not show an overshoot at the other two strain rates.

The curved unloading is a property of the experiments, see Fig. 1 of Ref 5. The simulations are fully capable of reproducing this behavior. After unloading the slope continuously decreases for each curve. All the simulated curves reach zero stress at roughly the same strain at about 7%, which is somewhat smaller than the strains found in the experiments. The experimental curve with the fastest strain rate reaches zero stress at a smaller strain (at about 7.5%) than the other stress-strain curves (at 7.8%). Overall the curved unloading behavior is captured rather well.

The loading and unloading simulations were repeated at 3% and 30% maximum strain. The same qualitative behavior was found.

Next the recovery process was simulated. The experimental curves are given in Fig. 2 of Ref 6. The relevant curves in this figure are those of specimens 21 through 24.

The simulations for these tests are given in Fig. 5. The absolute values of the recovery strains plotted on the ordinate start at the strain at which zero stress is reached during

TABLE 1—*Material constants for Nylon 66 at room temperature.*

Moduli	$E = 3700$ MPa; $E_t = 305$ MPa
Total Poisson's ratio	$\eta = 0.4$
Isotropic stress	
$\dot{A} = A_c(A_f - A)\dot{\phi}$	$A_0 = 60$ MPa; $A_f = 53$ MPa; $A_c = 3$
Viscosity function	
$k = k_1 \left[1 + \left(1 + \dfrac{A_0 - A}{A_0 - A_f} \right) \dfrac{\Gamma}{k_2} \right]^{-k_3}$	$k_1 = 1500$ s; $k_2 = 30$ MPa; $k_3 = 15$
Shape function	
$\psi = C_1^* + \dfrac{C_2 - C_1^*}{\exp(C_3 \varepsilon_{eff}^{in})}$	
$C_1^* = C_1 \left[1 + C_4 \left(\dfrac{g_{eff}}{A + f_{eff} + \xi \Gamma^\xi} \right) \right]$	$C_1 = 740$ MPa; $C_2 = 2670$ MPa; $C_3 = 100$ $C_4 = 2; \xi = 1$ (MPa)$^{-1}$; $\zeta = 2$

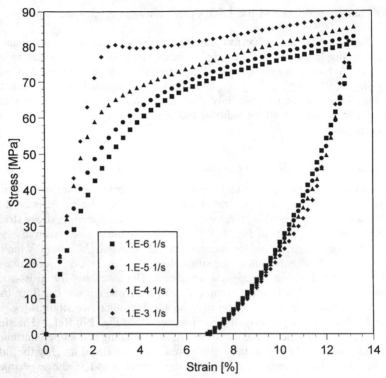

FIG. 4—*Simulation of loading and unloading of Nylon 66 at various strain rates. The nonlinear rate sensitivity, the merging of the stress-strain curves and the curved unloading behavior are simulated very well. This diagram is simulating Fig. 1 of Ref 5.*

unloading which is about 7% and nearly independent of the unloading strain rate, see Fig. 4. The magnitude of the recovered strain shown in Fig. 5 has to be subtracted from the 7% value to get the nearly permanent set. Figure 5 shows an initial fast recovery that increases with an increase of the prior unloading strain rate. The maximum strain recovery is for the test with the unloading strain rate of 10^{-3} 1/s. Although the recovered strains in the simulations are smaller than the experimental ones [6], the influence of prior strain rate and the ever-decreasing rate of recovery are excellently modeled. The methods by which VBO models recovery has been explained before, see Eq 11.

In both the experiments and the simulations the rate of recovery decreases continuously. Although the experiments lasted 12 h, it is reasonable to say that a full recovery is not expected. In light of the fact that the rate of recovery is continuously decreasing and that less than 3% recovery occurred in 12 h, it is unlikely that zero strains will be reached. Even if a full recovery were to take place for the test with the maximum prior strain rate, it would not occur at the smaller strain rates. The experiments and the simulations predict that the permanent set gets larger for a decrease in prior strain rate.

Repeated Relaxation

Figure 6 shows the simulation of repeated relaxation tests in the region of fully established inelastic flow. The relaxation periods last 1024 s and after the relaxation period is over,

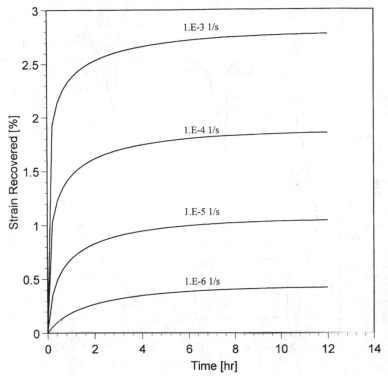

FIG. 5—*Simulation of the recovery behavior of Nylon 66 shown in Fig. 2 of Ref 5. The recovered strain increases nonlinearly with prior strain rate.*

loading resumes at the same strain rate. When the maximum strain is reached unloading takes place at the same strain rate magnitude. The experimental data are given in Fig. 4 of Ref *5* where only the curves for the fastest and the slowest strain rate are plotted.

It is seen that in both the experiment (see also Fig. 6 of Ref *5* where the forgetting of the prior history is clearly demonstrated) and in the simulation, the stress-strain curves without the interruption are reached quickly after loading commences. In the experiment a more pronounced overshoot is observed than in the simulation. The strains reached at zero stress after unloading are very close to those found in experiments.

It is interesting to observe that the relaxation drop in the constant relaxation time increases with an increase in prior strain rate. The terminal point of the relaxation test is lowest for the test with the fastest prior strain rate and is highest for the slowest strain rate. The model reproduces this unexpected behavior very well. For a detailed discussion of this behavior, see Ref *19*.

Figure 7 compares the relaxation curves at 4.9 and 9.5% and it is seen that the relaxation curves are nearly identical at the two slow strain rates but that there is a small deviation at the highest strain rate. This is a surprising result with which the experimental results fully agree, see Fig. 5 of Ref *5*. The overstress dependence together with the property of the asymptotic solution is the reason for the modeling capabilities of VBO and VBOP, see Ref *10* for an explanation.

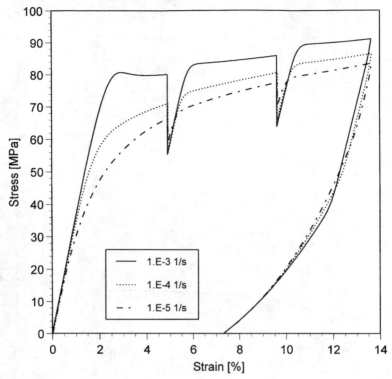

FIG. 6—*Simulation of relaxation behavior in 1024 s at two different strains and three different strain rates. Compare with Fig. 4 of Ref 5. Note that the relaxation drop increases nonlinearly with prior strain rate. The stress at the end of relaxation period is lowest for the fastest prior strain rate and is highest for the smallest prior strain rate.*

Completely Reversed Cyclic Loading

The next numerical experiment is a strain controlled, completely reversed test at a strain rate magnitude of 10^{-3} 1/s. The results are given in Fig. 8a. It is seen that the first half-cycle is distinct from the other ones and that cyclic softening takes place. The hysteresis loop has a propeller like shape with the middle section being somewhat concave.

The hysteresis loops found in the experiment with a tubular Nylon 66 specimen do not have a propeller-like shape but exhibit a separate first loading and strain softening, see Fig. 4.27 of Ref 20 reproduced here as Fig. 8b.

Propeller-shaped hysteresis loops, however, have been reported in Fig. 6 of Ref 21 and Fig. 6 of Ref 22 for polypropylene. After 50 cycles the shape of the hysteresis loop has returned to a normal shape. In the present simulation the propeller-like shape persists. Aside from the difference in material, Nylon 66 versus polypropylene, the strain rate used by Ariyama [21] is ten times the one used for testing Nylon 66.

Stress-Controlled Loading and Unloading

As a final check on the adequacy of the model, stress-controlled loading and unloading experiments were performed. Figure 9a shows the predicted results. In contrast to the pre-

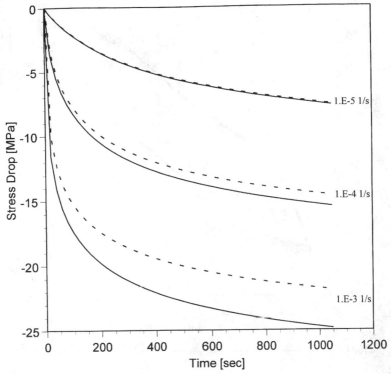

FIG. 7—*Relaxation curves for Nylon 66 at two different strains. Compare with Fig. 5 of Ref 5.*

viously reported results of numerical experiments, data from this stress-controlled experiment had not been used in determining the material constants. The experimental results are shown in Fig. 9*b*.

The prediction generally displays the same shape of the loading and unloading curves as the experiments. In both cases forward inelastic flow continues after reversal of the stress rate. After accumulation of additional strain, the tangent modulus increases, and reaches infinity before strain reduction starts. The behavior in stress control is completely different from the unloading characteristic in strain control. In both cases, the unloading is curved. The numerical and real experiments reach the zero stress axis at about the same strain. The difference is less than 0.5% strain.

Discussion

General

The constitutive equations of VBO and VBOP can easily be adapted for stress and strain control. In the form given they are written for stress control, and the strain rate has to be found. If, on the other hand, the strain rate is prescribed, Eq 7 is easily solved for computing the stress for a given strain input. The growth laws for the state variables contain only the stress rate, the stress and the state variables and their first time derivatives and need not be inverted. Because of these features VBOP can easily be adopted for strain or stress as the

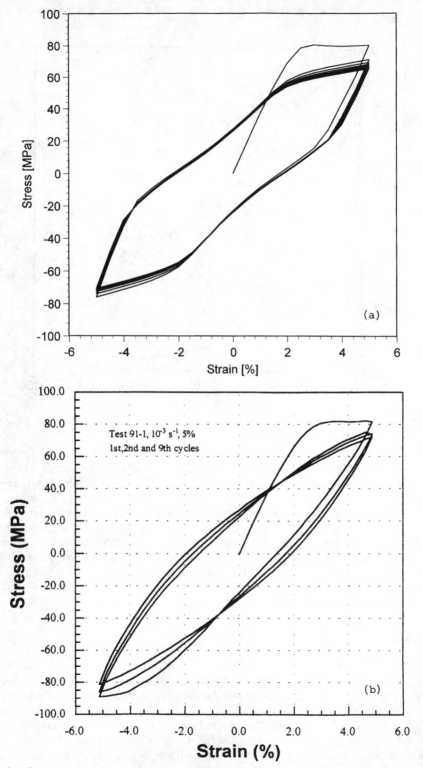

FIG. 8—*Completely reversed, strain-controlled loading of Nylon 66:* (a) *shows the simulation, and*
(b) *the experimental results from Ref 20. Note the propeller shape of the simulated hysteresis loops.*

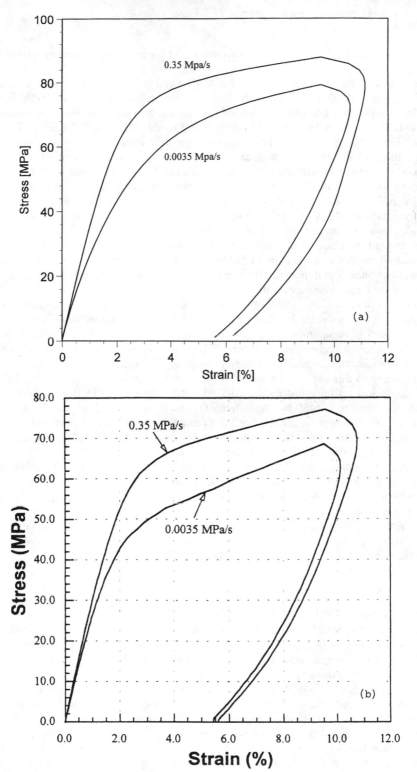

FIG. 9—*Stress-controlled loading and unloading of Nylon 66:* (a) *shows the predictions of the experimental results depicted in Fig. 9* (b) *and taken from Ref 20. Note that the strain still increases after the stress rate has changed sign.*

independent variable. This property is also displayed for the three-dimensional, finite strain model [18].

The results of the simulations show that VBOP can reproduce the nonlinear strain rate sensitivity, the unloading and recovery behavior, the relaxation and the cyclic softening behavior of Nylon 66. The stress-controlled loading and unloading behavior was successfully predicted. The experiments reported in Refs 5 and 6 and others on PEEK and PEI showed very similar behavior and there is no reason why their data could not be modeled. It is realized that PEI is amorphous whereas the other two polymers are semi-crystalline. Surprisingly, the three polymers do not show significant differences in the tests conducted so far. True enough the ductility of PEEK and PEI is much less than that of Nylon 66, but the behavior prior to fracture is not qualitatively different except that there was no extended neck formation in PEEK and PEI. However, the necking process was not modeled at all since all results pertain to homogeneous deformation behavior.

The simulations are all in the regions where inelastic flow is fully established. It remains to be seen how the modeling of VBOP would be at the transition from quasi-linear to nonlinear behavior. VBOP may not be able to reproduce the wide range of relaxation spectra that actual data can exhibit given its close relation to the SLS.

The model proposed here is that of a solid. At rest the model can accommodate a nonzero rest stress. The ever-decreasing inelastic strain rate in a relaxation test, see Fig. 7, will ultimately lead to equilibrium which can be at a nonzero stress. There may be situations where the relaxation proceeds to zero stress but a nonzero stress at rest is expected in general. For a discussion of the relaxation behavior found in experiments and in VBO refer to Ref 19. According to this paper the position of the equilibrium stress at the end of relaxation test is dependent on prior history. For the relaxation simulations in Figs. 6 and 7 history dependence manifests itself by the different stresses at the end of the 1024 s relaxation period. The stresses for the test with the prior fastest strain rate is smallest, see Figs. 6 and 7. In view of the similar relaxation curves indicative of nearly equal relaxation rates at 1024 s, it is likely that this difference prevails when equilibrium is reached.

The present formulation is for temperatures below the glass transition temperature. The behavior at temperatures approaching the glass transition temperature will introduce softening which can be modeled by introducing static recovery terms. In this case the stress at rest is the zero stress.

Due to the nature of the differential equation and in accordance with the properties of the SLS, it will take infinite time to reach equilibrium. For practical purposes equilibrium can be found at finite time with sufficient accuracy.

Differences between VBO and VBOP

The properties of VBO are explained by Krempl [10]. The system of nonlinear differential equations can be viewed as an interpolation between initial elastic behavior and fully established inelastic flow at large strains where an asymptotic, long-term solution may exist. The VBOP equations maintain these properties although changes have been made to every equation except for the growth law for the isotropic stress.

The viscosity function $k[\Gamma,A]$ in the flow law, Eq 7, depends on the overstress invariant Γ and on the isotropic stress A. The isotropic stress is responsible for modeling the merging of the stress-strain curves shown in Fig. 4. If there is no dependence of the viscosity function on the isotropic stress A the stress-strain curves are equidistant at long times.

The growth law for the equilibrium stress contains on the right side the time derivative of the overstress as a positive or hardening term, see Eq 8. There is no influence of this term on the long-term asymptotic solution since $\{\sigma - g\}$ = constant. It enhances the transients when there is a rapid change in the overstress.

The pre-factor on the right-hand side of the growth law for the kinematic stress, see Eq 10, is equal to one when the asymptotic solution holds. It was introduced to increase the recovery strain. At zero stress the kinematic stress rate is zero and has no influence on the recovery. When this pre-factor is absent in Eq 10 the kinematic stress rate would be negative and would enhance the decrease in the equilibrium stress during recovery thereby decreasing the recovered strain.

The previous two changes had an effect only on the transient region. Another contribution to the transient, curved unloading behavior is from the shape function ψ, which is different from that used for metals, see Table 1. This change does not affect the long-term asymptotic behavior.

It is seen that the general framework of VBO developed for metallic materials permits the specialization for polymeric solids. These changes are capable of reproducing special properties of polymeric materials.

Specific Properties

The overstress dependence ensures the existence of the long-term asymptotic solutions, the modeling of the unusual relaxation and the recovery behaviors. The relaxation rate increases with prior strain rate and the stress at the end of the relaxation tests is lowest for the curve pertaining to the fastest prior strain rate. These facts can be easily modeled, see Fig. 6. The dependence of the flow law on the overstress and the presence of the stress rate term in the growth law for the equilibrium stress are major contributors to these modeling capabilities. The overstress dependence and the modification of the growth law for the kinematic stress are responsible for the correct trend of the recovery strain evolutions.

The model and the simulations have shown that the overstress theory is capable of modeling the behavior of Nylon 66. The simulations did not always produce a match with the experimental data, but their trend is definitely captured. Also, the prediction of the stress-controlled test is satisfactory.

A three-dimensional theory has been developed [18] and has been applied to the corner path biaxial tests [8]. Again, the qualitative agreement and the trend of the simulations are excellent. The quantitative agreement could be improved.

The data reported [6,7] on PEI and PEEK were not qualitatively different from those of Nylon 66. There is ample reason to believe that the modeling of these two polymers can be easily accomplished using VBOP. It will be attempted.

Use of Overstress Models by Others

The behavior of polypropylene in torsion using thin-walled tubes during strain rate changes and relaxation periods has been investigated [11]. In their Fig. 6 the relaxation drop is shown to depend on prior strain rate and becomes nearly independent of strain at strains beyond 0.1. The relaxation drop is smallest for the fastest prior strain rate, a result that is opposite to the observations in this study. The text explains that the stress drop is plotted but the ordinate of Fig. 6 is labeled shear stress. If this stress is the residual stress after the 2000 s relaxation tests then this result is in full agreement with the present experimental results. This interpretation is most likely correct since the theoretical results plotted in Fig. 6 stem from an early overstress theory of the second author and his students. In this early version of VBO the equilibrium stress is nonlinear but not hysteretic. It has the same properties under monotonic loading as the present theory but is not a good model for cyclic loading. This version is called the deformation theory of VBO.

Kitagawa and Zhou [12] investigate amorphous and semi-crystalline solid polymers. Uniaxial and biaxial experiments were performed including unloading and creep experiments. Their results show no significant difference between the deformation behavior of amorphous and semi-crystalline, solid polymers. These results are completely in agreement with the experiments reported by Bordonaro [20] on three different solid polymers. That the unloading behavior also depends on the strain rate has been reported by Kitagawa et al. [11].

Polypropylene in monotonic and cyclic loading has been investigated where an influence of prior strain rate on relaxation behavior has been found (Fig. 7 of Ref 12). The results on polypropylene [14] regarding relaxation behavior (see Figs. 5–8 of Ref 14) are in agreement with Bordonaro's result on Nylon 66 [5]. In both cases the relaxation drop in a fixed period of time increases nonlinearly with prior strain rate.

These papers confirm the majority of the results reported by Bordonaro [20]. The good modeling capabilities of the old deformation theory of VBO reported by Kitagawa and Ariyama for monotonic loading can now be extended to cyclic loading since a hysteretic equilibrium stress has been proposed here. It has been shown that unloading and cyclic behaviors of solid polymers can be modeled with VBOP.

Acknowledgment

The support of the Department of Energy, Grant DE-FG02-96 ER 14603, is gratefully acknowledged.

References

[1] McKenna, G. B. and Zappas, L. J. "Nonlinear Viscoelastic Behavior of Poly(methyl Methacrylate) in Torsion," *Journal of Rheology*, Vol. 23, 1979, pp. 151–166.

[2] Knauss, W. G. and Emri, I. "Volume Change and the Nonlinearly Thermo-Viscoelastic Constitution of Polymers," *Polymer Engineering and Science*, Vol. 27, 1987, pp. 86–100.

[3] Schapery, R. A., "Nonlinear Viscoelastic and Viscoplastic Constitutive Equations Based on Thermodynamics," *Mechanics of Time-Dependent Materials*, Vol. 1, 1998, pp. 209–240.

[4] Hasan, O. A. and Boyce, M. C., "A Constitutive Model for the Nonlinear Viscoelastic Viscoplastic Behavior of Glassy Polymers," *Polymer Engineering and Science*, Vol. 35, 1995, pp. 331–344.

[5] Bordonaro, C. M. and Krempl, E., "The Effect of Strain Rate on the Deformation and Relaxation Behavior of 6/6 Nylon at Room Temperature," *Polymer Engineering and Science*, Vol. 32, 1992, pp. 1066–1072.

[6] Bordonaro, C. M. and Krempl, E., "The Rate-Dependent Behavior of Plastics: A Comparison Between 6/6 Nylon, Polyetherimide and Polyetheretherketone," *Use of Plastics and Plastic Composites: Materials and Mechanics Issues*, MD-Vol. 49, American Society of Mechanical Engineers, New York, NY, 1993, pp. 43–56.

[7] Krempl, E. and Bordonaro, C. M., "A State Variable Model for High Strength Polymers," *Polymer Engineering and Science*, Vol. 35, 1995, pp. 310–316.

[8] Krempl, E. and Bordonaro, C. M., "Non-Proportional Loading of Nylon 66 at Room Temperature," *International Journal of Plasticity*, Vol. 14, 1999, pp. 245–258.

[9] Krempl, E., "The Overstress Dependence of Inelastic Rate of Deformation Inferred from Transient Tests," *Materials Science Research International*, Vol. 1, 1995, pp. 3–10.

[10] Krempl, E., "A Small Strain Viscoplasticity Theory Based on Overstress," *Unified Constitutive Laws of Plastic Deformation*, A. S. Krausz and K. Krausz, Eds., Academic Press, 1996, pp. 281–318.

[11] Kitagawa, M., Mori, T. and Matsutani, T., "Rate-dependent Nonlinear Constitutive Equation of Polypropylene," *Journal of Polymer Science: Part B: Polymer Physics*, Vol. 27, 1989, pp. 85–95.

[12] Kitagawa, M. and Zhou, D., "Stress-Strain Curves for Solid Polymers," *Polymer Engineering and Science,* Vol. 35, 1995, pp. 1725–1732.

[13] Ariyama, T., "Viscoelastic-Plastic Deformation Behavior of Polypropylene after Cyclic Preloadings," *Polymer Engineering and Science,* Vol. 34, 1994, pp. 1319–1326.

[14] Ariyama, T., Mori, Y., and Kaneko, K., "Tensile Properties and Stress Relaxation of Polypropylene at Elevated Temperature," *Polymer Engineering and Science,* Vol. 37, 1997, pp. 81–90.

[15] Gates, T. S. and Sun, C. T., "An Elastic/Viscoplastic Constitutive Model for Fiber-Reinforced Thermoplastic Composites," *AIAA Journal,* Vol. 29, 1991, pp. 453–468.

[16] Zhu, C. M. and Sun, C. T., "A Viscoplasticity Model for Characterizing Loading and Unloading Behavior of Polymeric Composites," *Time Dependent and Nonlinear Effects in Polymers and Composites, ASTM STP 1357,* American Society for Testing and Materials, West Conshohocken, PA, 2000, pp. 266–284.

[17] Hindmarsh, A. C., *ODEpack a Systematized Collection of ODE Solvers,* Scientific Computing, R. S. Stepleman et al, Eds., North Holland, Amsterdam, 1983, pp. 55–64.

[18] Ho, K., "Application of the Viscoplasticity Theory Based on Overstress to the Modeling of Dynamic Strain Aging of Metals and to the Modeling of the Solid Polymers Specifically to Nylon 66," Ph.D. thesis, Rensselaer Polytechnic Institute, Troy, NY, Aug. 1998.

[19] Krempl, E. and Nakamura, T., "The Influence of the Equilibrium Stress Growth Law Formulation on the Modeling of Recently Observed Relaxation Behaviors," *JSME International Journal, Series A,* Vol. 41, 1999, pp. 103–111.

[20] Bordonaro, C. M., "Rate-Dependent Mechanical Behavior of High Strength Plastics: Experiment and Modeling," Ph.D. dissertation, Rensselaer Polytechnic Institute, Troy, NY, 1995.

[21] Ariyama, T., "Stress Relaxation Behavior after Cyclic Preloading in Polypropylene," *Polymer Engineering and Science,* Vol. 33, 1993, pp. 1494–1501.

[22] Ariyama, T. and Kaneko, K., "A Constitutive Theory for Polypropylene in Cyclic Deformation," *Polymer Engineering and Science,* Vol. 35, 1995, pp. 1461–1467.

Composites

*Thomas S. Gates,[1] L. Catherine Brinson,[2] Karen S. Whitley,[1]
and Tao Bai[3]*

Aging During Elevated Temperature Stress Relaxation of IM7/K3B Composite

REFERENCE: Gates, T. S., Brinson, L. C., Whitley, K. S., and Bai T., "**Aging During Elevated Temperature Stress Relaxation of IM7/K3B Composite,**" *Time Dependent and Nonlinear Effects in Polymers and Composites, ASTM STP 1357,* R. A. Schapery and C. T. Sun, Eds., American Society for Testing and Materials, West Conshohocken, PA, 2000, pp. 141–159.

ABSTRACT: An experimental and analytical study was performed on the use of tension stress relaxation to characterize the effects of elevated temperature and physical aging on the linear viscoelastic behavior of IM7/K3B. Isothermal stress relaxation tests on a $[\pm 45]_{2s}$ laminate were run over a range of sub-glass transition (T_g) temperatures. The sequenced test method most commonly employed for creep was successfully adapted to the stress relaxation test and from those sequenced tests, material parameters such as aging shift rates and momentary master curve coefficients were developed for use in the analytical model.

The analytical viscoelastic model was based on classical lamination theory, the hereditary integral formulation type constitutive law, and effective time theory. Time-aging time superposition, effective time theory, and viscoelasticity were used to determine the physical aging related material parameters from the relaxation tests. Results were compared to previously measured isothermal creep compliance results via known relationships for the convolution of compliance to modulus. Time-temperature superposition was also used to evaluate master curves and related shift factors. All of the results illustrated that the relative influence of temperature and aging must be considered when assessing long-term performance and that the loading mode may have to be considered when accurate predictions of viscoelastic behavior are required.

KEYWORDS: polymeric composites, stress relaxation, physical aging, creep

Prediction of the long-term mechanical behavior of advanced polymer matrix composites (PMC's) in structural applications such as those found in the biomedical, civil infrastructure, and aerospace disciplines requires development of accurate experimental characterization and advanced analytical tools. Comparing analysis results to test data for a variety of test environments provides verification of the methods. The intent of this research was to present the experimental and analytical results for the elevated temperature, constant strain (relaxation) response of an advanced PMC. The effects of physical aging were modeled and the constant strain, time-dependent behavior was compared to previous data on the elevated temperature, constant load (creep) behavior of the same material.

[1] Research Scientist, Mechanics of Materials Branch, NASA Langley Research Center, MS 188E, Hampton, VA 23681.

[2] Associate Professor, Department of Mechanical Engineering, Northwestern University, 2145 Sheridan Road, Evanston, IL 60208.

[3] Graduate Student, Department of Mechanical Engineering, Northwestern University, 2145 Sheridan Road, Evanston, IL 60208.

For graphite/thermoplastic composites, the long-term durability of the material can be affected by aging of the polymer matrix. In general, aging in polymers can be classified as either chemical or physical. The relative contribution of these aging behaviors to the material response is a function of mechanical and environmental exposure. Chemical aging is an irreversible process associated with such conditions as continued curing and oxidative stability and results in changes in the polymer chemistry. This work has assumed that the thermoplastic based composite does not exhibit chemical aging during the tests described here and that the total aging behavior is predominately of the type called physical aging. Physical aging (Hutchinson [1] and Struik [2]) starts after quench from above the glass transition temperature and will increase the relaxation times and decrease ductility in the polymer which in turn can alter fundamental engineering properties such as stiffness, strength, and toughness. Consequently, as the aging progresses in a PMC, the susceptibility of the composite to damage mechanisms such as matrix cracking or delamination will become more prevalent.

One method for establishing the rate and severity of aging in the polymer matrix is to use time-dependent loading and assess the relative changes in measured stiffness or compliance over a suitable aging time. To shorten the testing process, temperature can be used as an accelerant. These ideas have been explored extensively in the literature by Struik [2], McKenna [3] for neat resins over a range of test conditions. Results from those studies have led to developments in the concepts of effective time theory and relationships between aging and mechanical property evolution. These concepts, combined with the established methods for viscoelastic analysis of polymers (Ferry [4], Findley, et al. [5]), have provided the motivation for similar studies in PMC's.

Much of the work in physical aging of PMC's has focused on combining tension creep tests below the glass transition (T_g) temperature, linear viscoelasticity, and mechanics of composite materials. Examples are found in Gates and Feldman [6], Wang, et al. [7], Hastie and Morris [8], and Sullivan [9]. More recently, Veazie and Gates [10] extended these methods to characterize the aging performance of a PMC under isothermal, compressive creep loads.

Although the concepts of constant-strain (stress relaxation) testing has been well established, little data has been found in the literature on the use of a constant-strain test for measuring the effects of physical aging in a PMC. It is recognized that the stress relaxation test of a matrix dominated PMC is in itself a difficult task. As shown in Gates [11], for elevated temperatures, a constitutive model for stress relaxation at moderate strain levels may require an elastic/viscoplastic formulation to account for nonlinear behavior. However, Gates [12] found that the elastic/viscoplastic model by itself was not able to predict the effects of aging during extended holds at temperatures below the glass transition.

The current research outlines a viscoelastic model and presents the experimental method developed to measure the effects of aging on tension stress relaxation in the linear range. Results from these tests are compared to previously measured behavior of the same material under tension and compression creep. For the purposes of this study, experimental characterization of the in-plane shear behavior of the composite will suffice to illustrate the similarities and differences of creep and relaxation. The viscoelastic model was used to analyze the relaxation and creep data and to predict the long-term time-dependent behavior.

Viscoelasticity and Aging

An analytical model as provided in Brinson and Gates [13], Bradshaw and Brinson [14], [15] was used to invert creep compliance to relaxation modulus as well as predict the long-term tension relaxation modulus using as input the material properties developed from short term tests. A "long-term" test was defined as a test time at least 10 times greater than the time of the "short-term" material property tests.

Linear Viscoelasticity

Integral of differential equations give the relationships between material functions in linear viscoelasticity. For the case at hand, compliance is related to modulus by

$$\int_0^t S(t - \xi)Q(\xi)d\xi = t \tag{1}$$

where $S(t)$ and $Q(t)$ are the compliance and modulus respectively. In general, the constitutive relationship can be written as a hereditary integral

$$\sigma(t) = \int_0^t Q(t - \xi)\frac{d\varepsilon}{d\xi} d\xi \tag{2}$$

Where σ is the stress, ε is the strain, Q is the modulus, and ξ is a dummy variable.

Considering an orthotropic lamina for a fiber reinforced, laminated composite, the modulus Q and compliance S can be written in matrix form (Jones [15]). For viscoelastic behavior, the time-dependent form of these matrices are given as

$$S_{ij} = \begin{bmatrix} S_{11} & S_{12} & 0 \\ S_{12} & S_{22}(t) & 0 \\ 0 & 0 & S_{66}(t) \end{bmatrix} \tag{3}$$

$$Q_{ij} = \begin{bmatrix} Q_{11} & Q_{12}(t) & 0 \\ Q_{12}(t) & Q_{22}(t) & 0 \\ 0 & 0 & Q_{66}(t) \end{bmatrix} \tag{4}$$

where the subscripts 1, 2, and 6 refer to the material coordinate system aligned with the fiber, transverse to the fiber, and in-plane shear respectively. It was assumed that only the two matrix dependent compliance terms (S_{22} and S_{66}) are time dependent. Due to the form of these matrices, the relationship between shear modulus and shear compliance is uncoupled from that of the other terms and the Eq 1 holds between $S_{66}(t)$ and $Q_{66}(t)$.

To model the experimental data, the short-term modulus was fit with the three parameter Kohlrausch-Williams-Watts (KWW) (Shiro [16]) expression. The three parameters (Q_o, τ, β) are described as the initial modulus, relaxation time, and shape parameter respectively and describe the time-dependent modulus using

$$Q(t) = Q_o e^{(-t/\tau)^\beta} \tag{5}$$

Although the KWW function works quite well to fit experimental data, the double exponential form makes it difficult to work with analytically. Consequently, for the inversions shown in this paper, the KWW functions originally used to reduce the relaxation and creep data were converted to Prony Series expansions for both modulus and compliance, and written in the form

$$S(t) = S_0 + \sum_{i=1}^{N} S_i(1 - e^{-t/\tau_i}) \qquad (6)$$

$$Q(t) = Q_\infty + \sum_{i=1}^{N} Q_i e^{-t/\tau_i} \qquad (7)$$

such that S_0, Q_∞ are the initial compliance and rubbery modulus respectively.

The procedure for fitting viscoelastic functions with Prony Series and the subsequent interconversion is described in detail in Bradshaw and Brinson [14]. In Eqs 6 and 7, the relaxation/retardation times, τ_i, are chosen over a broad time spectrum and the magnitudes of S_i and Q_i are then determined as that which best fits the data.

If the modulus and compliance are both expressed in Prony Series forms, substitution of Eqs 6 and 7 into Eq 1 results in an expression that can be integrated and simplified readily, since it involves only the sums of simple exponentials. For the case where S_i are known and Q_i are to be found (inverting from known compliance to corresponding modulus), the expression results in

$$\sum_{i=1}^{N} Q_i \psi_i(t) = t \qquad (8)$$

where $\psi_i(t) = \psi_i(S_i, \tau_i, t)$ is a completely known function.

Because Eq 8 is linear in terms of the unknowns Q_i, it can then be solved by a general least squares fitting algorithm. The disadvantage of a simple least squares algorithm is that it results in some values of Q_i being negative, which are physically unrealistic. Thus, the algorithm used to solve Eq 8 was an iterative approach in which an additional constraint such that all $Q_i > 0$ was added (Bradshaw and Brinson [14]).

Time/Aging Time Superposition

The material parameters required by the model were found from sequenced, short-term relaxation and recovery tests. This sequenced test procedure is depicted in Fig. 1. Briefly, Fig. 1 illustrates how a series of short, sequenced relaxation tests were performed on the material while it is held at a constant temperature. The short relaxation segments during the sequenced test provide the data for calculating modulus curves as a function of time. Due to aging, these modulus curves were not coincident for the different times at temperature. These differences are shown in Fig. 2 for a sequenced test at 225°C. As demonstrated by Struik [2] and illustrated in the log-log plot of Fig. 2, horizontal separation of the sequenced relaxation modulus was characterized by the aging shift factor (a). This factor was simply defined as the factor required to shift a modulus curve horizontally to achieve the best match with a reference modulus curve.

The use of time/aging-time superposition on the sequenced modulus test data provided the means for the short-term data to be collapsed into a single momentary master curve (MMC) at each test temperature. Figure 3 shows a single MMC formed by collapse of the data in Fig. 2.

Plotting the shift factors versus aging time on a log-log plot (Fig. 4) implied a linear relation existed between (log a) and the logarithmic aging time (log t_e) where t_e is the aging time. A linear fit gave the aging shift rate (μ).

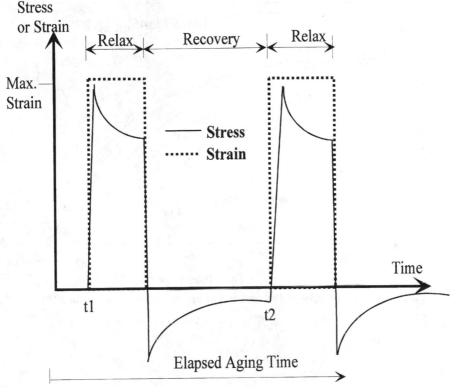

FIG. 1—*Schematic illustrating the sequenced relaxation test method.*

$$\mu = \frac{-d \log a}{d \log t_e} \tag{9}$$

The aging time reference modulus curve could be any of the sequenced curves and as shown in Bradshaw and Brinson [*17*], some small differences in aging shift rates may occur due to selection of a reference curve. For the relaxation tests, the momentary (short-term) sequenced relaxation/aging curves were collapsed into MMC's through a combined horizontal (time) and vertical (modulus) shift using the shortest aging time as the reference curve. To facilitate data reporting, the set of shifted curves (as in Fig. 3) at a 2 h reference aging time were best fit with a single curve which was then used as the MMC at that temperature.

The vertical shifts employed were small in comparison to the magnitude of the horizontal shifts and did not form any consistent trend. Although the use of vertical shifts may be associated with stress dependent effects or environmental (e.g., humidity) effects, it was felt that for this study the vertical shifts obtained were strongly associated with the physical aging. This assumption was based on the low strains, previous linearity checks, and laboratory conditions during the tests. All of the vertical shifts were small relative to the horizontal shifts.

For a horizontal (time) translation of a modulus curve, only the relaxation time parameter (τ) needs to be recalculated. Given the aging shift rate and reference curve parameters, the translation from one aging time to another was accomplished through the use of

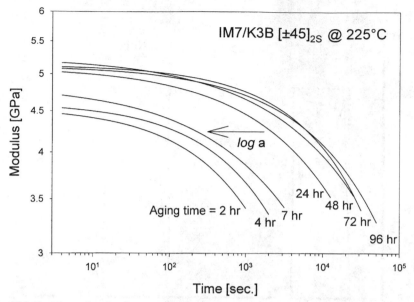

FIG. 2—*Example of sequenced relaxation modulus curves using test data at 225°C.*

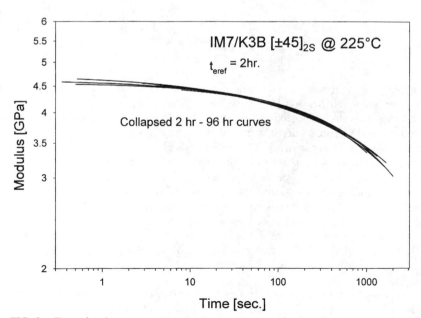

FIG. 3—*Example of momentary master curve using the sequenced test data at 225°C.*

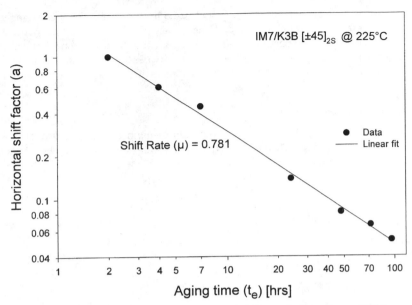

FIG. 4—*Example of aging shift factors and shift rate for test data at 225°C.*

$$\tau_e = \tau_{\text{ref}} \left(\frac{t_e}{t_{\text{eref}}} \right)^\mu \tag{10}$$

where t_{eref} is the reference aging time (Brinson and Gates [13]).

Long-term Modulus

For test periods that exceed the time required to collect the short term (momentary) data, the response can be expected to be influenced by the ongoing aging process. Struik [2] proposed an effective time that could be used to replace real time such that modulus can be written as

$$Q(t) = Q_o e^{(-\lambda/\tau(t_e^0))^\beta} \tag{11}$$

where λ is the effective time defined by

$$\lambda = \int_0^t a_{te}(\xi) d\xi = \int_0^t \left(\frac{t_e^0}{t_e^0 + \xi} \right)^\mu d\xi \tag{12}$$

which can be calculated according to

$$\lambda = t_e^0 \ln\left(\frac{t}{t_e^0} + 1 \right) \quad \text{for } \mu = 1$$

$$\lambda = \frac{t_e^0}{1 - \mu} \left[\left(1 + \frac{t}{t_e^0} \right)^{1-\mu} - 1 \right] \quad \text{for } \mu \neq 1 \tag{13}$$

where t_e^0 is the aging time prior to loading and at some time later the total aging time is $t + t_e^0$ where t is the relaxation test time. Brinson and Gates [13] demonstrated that the effective time theory is self-consistent and shift rates exceeding unity are both physically and mathematically permissible. The effective time expression was used with the MMC parameters, Q_o, τ, β, to calculate the long term response of the material.

Time/Temperature Superposition

The MMC's from the aging tests were a function of temperature. The use of time/temperature superposition (TTSP) (Ferry [4]) required that relaxation modulus be a function of temperature (T) and time (t) such that

$$Q = Q(T,t) \tag{14}$$

and that

$$Q(T,t) = Q(T_0, \zeta) \tag{15}$$

$$\zeta = t/a_T(T) \tag{16}$$

where ζ is the reduced time that is related to the real time t by the temperature shift factor $a_T(T)$ and T_o is the reference temperature. The use of TTSP implies that the effect of temperature on time-dependent behavior is equivalent to shrinking or stretching respectively, the real time for temperatures above or below the reference temperature. For thermorheologically simple materials, vertical shifts above the T_g are well-defined Ferry [4]; while vertical shifts are often needed to collapse TTSP data below T_g as well. This fact is especially true for PMCs, where the origin of these shifts is less well understood. However, as with the vertical shifts used for the time-aging time superposition, the vertical shifts obtained for TTSP in this study were also small relative to the horizontal shifts.

Test Material and Specimens

The composite material system chosen for this study was a continuous carbon fiber-reinforced thermoplastic polyimide fabricated by DuPont and designated IM7/K3B. The fiber, IM7, was an intermediate modulus carbon fiber manufactured by Hexcel. The unaged T_g in the composite as measured by TA Instruments Dynamic Mechanical Analyzer (DMA) G'' peak was 240°C.

The in-plane shear measurements were performed on rectangular test specimens similar to those described in ASTM Specification D3039-76. Each specimen measured 24.1 cm by 1.91 cm and were cut from laminated panels. All specimens consisted of 12 plies where each ply measured approximately 0.0135 cm thick. The in-plane shear (Q_{66}) relaxation modulus was measured using a $[\pm 45]_{3s}$ layup. To reduce experimental errors at least two replicates were tested at each test condition. Prior to testing, all specimens were dried for at least 24 h at 110°C in a convection oven.

Test Equipment

The sequenced test procedure required a return to zero strain condition to allow for recovery prior to the next loading (Fig. 1). Due to the slight compressive load imposed during the return to zero strain, the specimen had to be supported to prevent column buckling. In addition, provisions had to be made for heating the specimen as well as accommodating strain gages and axial extensometers. To meet all these requirements, a unique fixture, fabricated at NASA Langley, was used for all tests. This fixture, shown in Fig. 5, was made of aluminum and contained four resistance type heater rods that provided zone control of the specimen temperature. Thermocouple probes were used for monitoring and control functions while the heat was directly transferred to both sides of the test specimen through direct

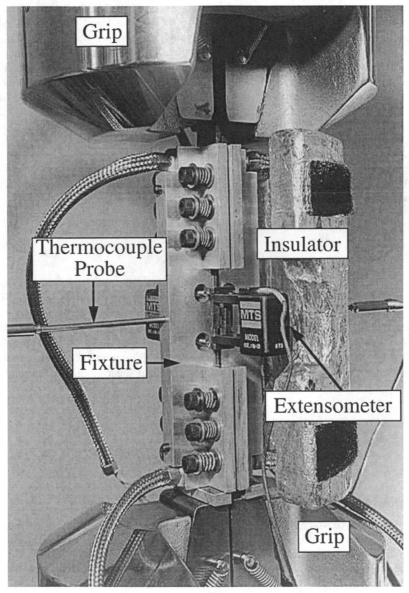

FIG. 5—*Stress relaxation test apparatus.*

contact of the fixture and specimen along the non-gripped section of the specimen. Both specimen and fixture were enclosed in a reflective, clamshell type insulator to control heat loss.

Buckling constraint was achieved by connecting the two fixture halves along their length with spring-loaded screws. Because of the continuous support along the length of the specimen, only a small amount of lateral force was needed to suppress buckling. The fixture was lined with Teflon™ tape to reduce friction.

Testing was performed with an electro-mechanical test machine capable of running predetermined strain history profiles. Axial strain was measured on the specimen by using extensometers mounted opposing each other along the specimen's edges. One of the extensometers was used to provide feedback control to the test machine. Transverse strain was measured by using back-to-back, center mounted, high temperature strain gages. The extensometers were connected to the test machine's internal signal conditioners while an external unit was used to condition the strain gage signals. Load, as measured by the load cell, was converted to nominal stress using the average cross-sectional area of the specimen measured prior to testing. If the applied stress is σ_x then the shear stress (σ_{12}) in a $[\pm 45]_{2s}$ laminate is

$$\sigma_{12} = \tfrac{1}{2}\sigma_{xx} \tag{17}$$

where the subscripts 1 and 2 refer to the directions aligned with and transverse to the fiber axis respectively. Therefore, the shear modulus for an elastic material is given by

$$Q_{12} = \frac{\sigma_{12}}{\varepsilon_{xx} - \varepsilon_{yy}} \tag{18}$$

where ε is the strain and the subscripts x and y refer to the directions aligned with and transverse to the loading axis respectively. For a viscoelastic material note that Eq 2 for this loading configuration is applied with $\varepsilon(t) = \varepsilon^{\text{eff}}(t) = 2[\varepsilon_{xx}(t) - \varepsilon_{yy}(t)]$ and $\sigma(t) = \sigma_{xx}(t)$; while the material is under constant axial (ε_{xx}) strain during the loading steps, ε_{yy} actually varies with time so that in order to obtain an expression similar to Eq 18, the approximation is made that $\varepsilon^{\text{eff}}(t) = \varepsilon^{\text{eff0}} H(t)$ (where $\varepsilon^{\text{eff0}}$ is a constant) to remove the convolution integral. With this "quasi-elastic" approximation the time-dependent shear modulus during relaxation is given as

$$Q(t) = \frac{\mu_{xx}(t)}{2(\varepsilon_{xx}(t) - \varepsilon_{yy}(t))} \tag{19}$$

Experimental Procedures and Data Reduction

As described above, the effects of physical aging may be measured by employing a well-documented technique (Struik [2]) that relies on creep compliance. This procedure was adapted for relaxation modulus and consisted of a sequence of relaxation and recovery tests while the specimen isothermally aged. The test temperatures selected for the study were 208, 215, 220, 225, and 230°C. These test temperatures were selected to ensure that measurable aging occurred within the test period.

Linearity

For all the sequenced tests, a single applied strain level of .0007 was chosen and subsequently used at all temperatures. Determination of the strain level necessary to stay within the linear viscoelastic range was made by ensuring the proportionality condition and Boltzman's superposition (Findley, et al. [5]) would be satisfied.

Single loading relaxation and recovery tests provided data for checking superposition. Proportionality checks were performed by plotting isothermal relaxation modulus versus test time for a specimen that was repeatedly rejuvenated (described below), quenched and loaded. Several strain levels were checked before arriving at the chosen level of .0007. The supposed transition from linear to nonlinear behavior would be evident by the vertical separation of the modulus curves with increasing strain. Because the highest test temperature is the worst case condition, linearity checks were made at the highest test temperatures thereby ensuring linearity of the total range of temperatures.

Short-term Tests

To provide for the required condition that all specimens start the test sequence in the same unaged condition, a means of rejuvenating the specimen was required. Rejuvenation was accomplished by a procedure based upon work by Struik [2] and others who showed that physical aging is thermoreversible and the excursion above T_g prior to quenching effectively rejuvenates the material. In the current tests, the gaged specimen was located in the fixture and heated in an oven to 250°C (10°C above T_g) for 30 min immediately before the start of any physical aging test sequence. Quenching from above T_g occurred when the specimen was removed from the oven for installation of the extensometers. Once the extensometers were mounted, the specimen was immediately heated to test temperature.

The duration of each relaxation segment was ¹/₁₀th the duration of the prior total aging time. The aging times (time after quench) selected for starting each relaxation segment were 2, 4, 7, 24, 48, 72, and 96 h. After each relaxation segment, the specimen was unloaded to zero strain and allowed to recover until the start of the next segment.

Momentary Master Curves

An individual MMC was found at each test temperature. These MMC's are shown in Fig. 6. All of these curves represent the best fit to collapsed data from all replicates used in the test program. The three-parameter KWW fit (Eq 5) was used to characterize each of the MMC's and these parameters are given in Table 1. It should be noted that the valid range of the KWW fit to the MMC data is governed by the extent of the test data. The length of the curves in Fig. 6 corresponds exactly to the range of experimental data.

Results and Discussion

Master curves, aging shift rates, and material parameters measured from tests for the in-plane, relaxation shear modulus of IM7/K3B are provided over the range of test temperatures. Where applicable, the creep compliance data of Gates and Feldman [6] and Veazie and Gates [10] was used to provide a comparison between creep and relaxation test results.

Momentary Master Curves

The momentary master curves (Fig. 6) that resulted from time/aging-time superposition demonstrated that in-plane shear relaxation of IM7/K3B was a strong function of test tem-

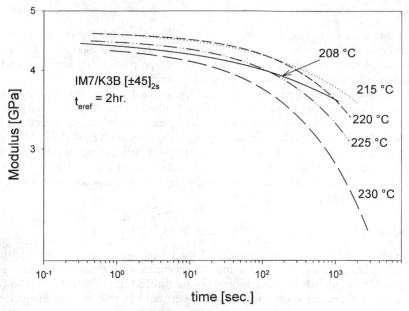

FIG. 6—*Stress relaxation momentary master curves (MMC's) for all test temperatures.*

perature with an increase in temperature resulting in a decrease in modulus and associated relaxation rate. Some crossing over of the MMC's was observed at the lower temperatures and the initial modulus appeared to be a weak function of temperature for all cases. It was expected that some small data scatter would be present due to testing procedures and constraints imposed on the specimen by the testing apparatus. However, in general, data from the replicate tests confirm that the test procedure was repeatable.

Aging Shift Rate

Based upon the known effects of physical aging in amorphous polymers, it was expected that aging shift rate would decrease as the temperature approached T_g. The measured rate from the relaxation (Fig. 7) reflects these expected trends. The sharp increase in shift rate at 208°C was unexpected and may be related to the same phenomena that caused the crossing over of the lower temperature MMC's.

A study by Nguyen and Ogale [18] showed that polyetheretherketone exhibited a shift rate that was a function of loading mode. A more recent study by Gates, et al. [19] showed

TABLE 1—*Momentary master curve parameters for relaxation modulus. (2 h ref)*

Modulus	T (°C)	Q_0 (GPa)	τ (sec.)	β	μ
Q_{66}	208	4.535	209 786.	0.274	0.891
Q_{66}	215	2.668	88 323.	0.344	0.726
Q_{66}	220	4.621	18 647.	0.472	0.793
Q_{66}	225	4.512	14 177.	0.438	0.806
Q_{66}	230	4.390	6 850.	0.433	0.652

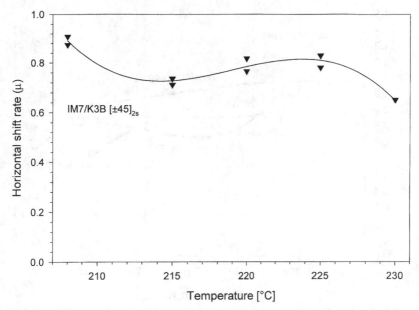

FIG. 7—*Shift rates for relaxation aging tests. Line represents best fit to trend of data.*

that shift rate in IM7/K3B depended on loading mode and loading direction. Comparing tension creep, compression creep, and tension relaxation (Fig. 8) it was apparent that the relaxation shift rates were similar to the compression shift rates.

Time-Temperature Superposition

The time-temperature superposition principle was used to collapse the sets of momentary master curves into single material master curves. These collapsed curves are shown in Figs. 9 and 10 for the relaxation and creep data respectively. The relaxation TTSP master curve (Fig. 9) as well as the creep TTSP master curves (Fig. 10) required small vertical shifts to achieve a satisfactory collapse. The master curve parameters for the curves in Figs. 9 and 10 are given in Table 2.

The TTSP shift factors used to collapse these curves are given in Table 3 and were plotted versus temperature in Fig. 11 for both creep and stress relaxation. A good linear fit to this data, over the entire range of values, was not possible. Although the two sets of creep data gave approximately the same shift factors, Fig. 11 indicates that the relaxation tests have a significantly different TTSP behavior than creep over the range of test temperatures. As noted by Schapery [20] these differences may be attributed to the temperature dependence of initial compliance or modulus.

Interconvertability of Compliance and Modulus

Figure 12 shows the modulus curve from the relaxation tests and the calculated modulus curves from tension creep tests and compression creep tests on the IM7/K3B material at test temperatures 208° and referenced to a 2 h aging time. The creep compliance were converted to the corresponding moduli using the procedure described at the beginning of

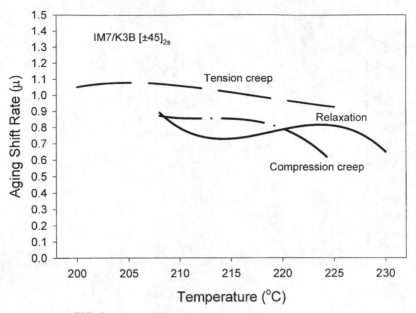

FIG. 8—*Aging shift rates for stress relaxation and creep.*

this paper, via inversion of Eq 1. This procedure has been well tested, and the results for the moduli as obtained via inversion of the creep data can be regarded as exact. For all temperatures, the three modulus curves show that the relaxation behavior is different from the creep behaviors but more closely resembles the tension creep than the compression creep.

It is clear that there are discrepancies between the viscoelastic material parameters determined from the three types of sequenced aging tests. The shift rates differ, as do the master curves. It is noted that although every attempt was made to make the experiments as equivalent as possible, variations exist between the tests and the specimens themselves and it is difficult to quantify how much of an impact such factors could have on the results. It is also possible that the material is nonlinear, therefore loading mode dependent. These discrepan-

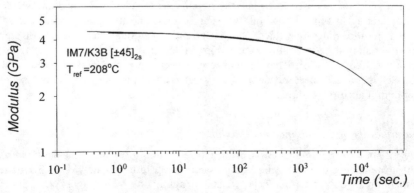

FIG. 9—*Time-temperature superposition master curve for relaxation formed using the MMC data.*

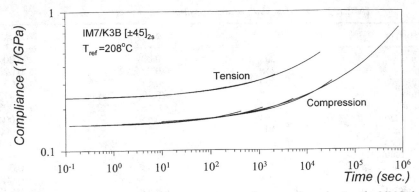

FIG. 10—*Time-temperature superposition master curves for creep formed using the MMC data.*

cies in the fundamental modulus of the material found here indicate that further work needs to be done to address this issue.

Long-term Predictions

Material constants from Table 2 were used along with the predictive model to perform parametric studies on the long-term relaxation behavior at elevated temperature. These predictions are shown in Fig. 13. Note that as is commonly done in long term predictions, only the horizontal shift factors are used in calculating the effective time (see Eqs 11–12); the small vertical shift factors used in data reduction are not included in the long term prediction. Although no data verifying extremely long-term behavior exists, the predictions were run out to approximately 3.2 years to provide parametric comparisons of long-term performance. These predictions illustrate that the relative effects of temperature are to decrease modulus as temperature increases. The relative effect of aging is to induce a gradual stiffening over time. This stiffening is less apparent at temperatures near T_g due to the increased ductility of the material at these temperatures and the decrease in the aging shift rate.

Concluding Remarks

It was established in previous investigations that the linear viscoelastic behavior of the advanced polyimide composite material IM7/K3B would be affected by physical aging below the glass transition temperature. These studies, which utilized tension and compression creep experiments to characterize the aging process, provided the short-term data necessary for predicting the long-term creep compliance.

The current study was concerned with the use of tension stress relaxation to characterize the effects of elevated temperature and physical aging on the linear viscoelastic behavior of

TABLE 2—*Time-temperature superposition master curve parameters. (208°C ref)*

Test Type	S_0 (1/GPa) or Q_0 (GPa)	τ (sec)	β
Tension Creep	0.234	47 693.	0.319
Compression Creep	0.147	127 568.	0.269
Relaxation	4.389	38 037.	0.422

TABLE 3—*Time-temperature superposition shift factors. (208°C ref)*

Test Type	T (°C)	Log(a_T)	Log(Vertical Shift)
Relaxation	208	0.00	0.000
Relaxation	215	0.07	0.020
Relaxation	220	−0.18	0.056
Relaxation	225	−0.39	0.013
Relaxation	230	−0.73	0.001
Tension Creep	200	0.45	−0.004
Tension Creep	208	0.00	0.000
Tension Creep	215	−0.50	−0.001
Tension Creep	225	−1.31	−0.079
Compression Creep	200	0.48	0.001
Compression Creep	208	0.00	0.000
Compression Creep	215	−0.61	−0.021
Compression Creep	220	−1.32	−0.045
Compression Creep	225	−2.36	−0.039

IM7/K3B. The test conditions covered a range of sub-T_g temperatures. An electro/mechanical test machine was used to run isothermal, strain-controlled tests on $[\pm 45]_{2s}$ laminated composite coupons. Tests on the $[\pm 45]_{2s}$ (in-plane shear) laminate illustrated the need for a supportive fixture to prevent buckling during the unloading segments of sequenced testing.

The sequenced test method most commonly employed for creep was successfully adapted to stress relaxation. A sequenced test procedure was then developed to provide the relaxation modulus curves as a function of aging time. From these sequenced tests, material parameters such as aging shift rates and momentary master curve coefficients were developed for use in the analysis model.

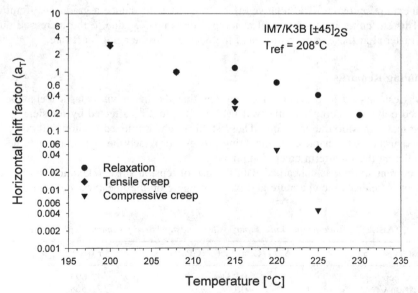

FIG. 11—*Time-temperature superposition shift factors for relaxation modulus and creep compliance data.*

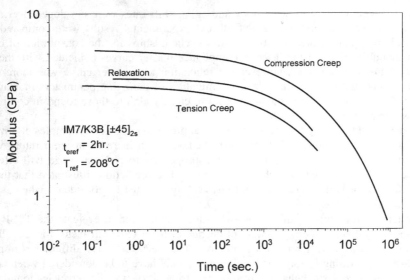

FIG. 12—*Relaxation MMC and calculated creep MMC's at 208°C.*

The analytical model used in the study provided a scheme for predicting relaxation modulus for a constant temperature condition. The model was based on classical lamination theory, the hereditary integral formulation type constitutive law, and effective time theory. Time-aging time superposition, effective time theory, and viscoelasticity were used effectively to determine the physical aging related material parameters from the relaxation tests.

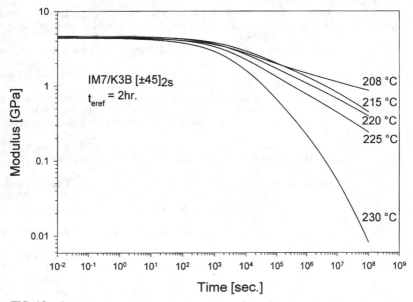

FIG. 13—*Long-term prediction of relaxation modulus based on short-term test data.*

Subsequently, momentary master curves and aging shift rates were developed over a range of sub-T_g temperatures and both analytical and experimental results were compared to the isothermal creep compliance results via known relationships for the convolution of compliance to modulus. None of the inverted compliance master curves coincided with the relaxation modulus master curves, however the relaxation test master curve was more closely akin to the inverted tension creep data. Examination of the aging parameters indicated that aging shift rates for stress relaxation were more closely akin to those found in compression creep as compared to tension creep.

Parametric studies using the measured material properties provided examples of how physical aging and temperature affect long-term behavior. An increase in temperature will increase the compliance and creep rate while a propensity to physically age will effectively decrease the rate of compliance change over time. The predictions illustrated that the combined influence of both temperature and physical aging must be considered when assessing long-term performance.

Time-temperature superposition was successfully employed to evaluate the TTSP master curve and related shift factors. The rate of change of the TTSP shift factor for the relaxation tests was significantly different than those from the creep tests. The differences imply that for an arbitrary loading history, the loading mode may have to be considered when accurate predictions of viscoelastic behavior are required. In addition, the discrepancies found in shift rate, shift factors and invertability indicate that further study of the impact of loading mode on PMC viscoelastic response should be investigated.

References

[1] Hutchinson, J. M., "Physical Aging of Polymers," *Progress in Polymer Science,* Vol. 20, 1995, pp. 703–760.

[2] Struik, L. C. E., *Physical Aging in Amorphous Polymers and Other Materials,* New York: Elsevier Scientific Publishing Company, 1978.

[3] McKenna, G. B., "On the Physics Required for the Prediction of Long Term Performance of Polymers and Their Composites," *Journal of Research of the National Institute of Standards and Technology,* Vol. 99, 1994, pp. 169–189.

[4] Ferry, J. D., *Viscoelastic Properties of Polymers,* 3rd ed., New York, John Wiley and Sons, Inc., 1980.

[5] Findley, W. N., Lai, J. S., and Onaran, K., *Creep and Relaxation of Nonlinear Viscoelastic Materials,* Toronto, North-Holland Publishing Company, 1976.

[6] Gates, T. S. and Feldman, M., "Effects of Physical Aging at Elevated Temperatures on the Viscoelastic Creep of IM7/K3B," *Composite Materials: Testing and Design (Twelfth Volume),* ASTM STP 1274, R. B. Deo and C. R. Saff, Eds., American Society for Testing and Materials, West Conshohocken, PA, 1996, pp. 7–36.

[7] Wang, J. Z., Parvatareddy, H., Chang, T., Iyengar, N., Dillard, D. A., and Reifsnider, K. L., "Physical Aging Behavior of High-Performance Composites," *Composites Science and Technology,* Vol. 54, 1995, pp. 405–415.

[8] Hastie, R. L. and Morris, D. H., "The Effect of Physical Aging on the Creep Response of a Thermoplastic Composite," *High Temperature and Environmental Effects in Polymer Matrix Composites,* ASTM STP 1174, C. Harris and T. Gates, Eds., American Society for Testing and Materials, West Conshohocken, PA, 1992, pp. 163–185.

[9] Sullivan, J. L., "Creep and Physical Aging of Composites," *Composites Science and Technology,* Vol. 39, 1990, pp. 207–232.

[10] Veazie, D. R. and Gates, T. S., "Compressive Creep of IM7/K3B Composites and the Effects of Physical Aging on Viscoelastic Behavior," *Experimental Mechanics,* Vol. 37, 1997, pp. 62–68.

[11] Gates, T. S., "Effects of Elevated Temperature on the Viscoplastic Modeling of Graphite/Polymeric Composites," *High Temperature and Environmental Effects on Polymeric Composites, ASTM STP*

1174, C. E. Harris and T. S. Gates, Eds., American Society for Testing and Materials, West Conshohocken, PA, 1993, pp. 201–221.

[12] Gates, T. S., "Matrix-Dominated Stress/Strain Behavior in Polymeric Composites: Effects of Hold Time, Nonlinearity, and Rate Dependency," *Composite Materials: Testing and Design* (*Eleventh Volume*), *ASTM STP 1206,* J. E. T. Camponeschi, Ed., American Society for Testing and Materials, West Conshohocken, PA, 1993, pp. 177–189.

[13] Brinson, L. C. and Gates, T. S., "Effects of Physical Aging on Long Term Creep of Polymers and Polymer Matrix Composites," *International Journal of Solids and Structures,* Vol. 32, 1995, pp. 827–846.

[14] Bradshaw, R. D. and Brinson, L. C., "A Sign Control Method for Fitting and Interconverting Material Functions for Linearly Viscoelastic Solids," *Mechanics of Time-Dependent Materials,* Vol. 1, 1997, pp. 85–108.

[15] Jones, R. M., *Mechanics of Composite Materials,* Washington, D.C., Scripta Book Company, 1975.

[16] Shiro, M., *Relaxation Phenomena in Polymers,* Munich, Hanser, 1992.

[17] Bradshaw, R. D. and Brinson, L. C., "Physical Aging in Polymers and Polymer Composites: An Analysis and Method for Time-Aging Time Superposition," *Polymer Engineering and Science,* Vol. 37, 1997, pp. 31–44.

[18] Nguyen, D. H. and Ogale, A. A., "Compressive and Flexural Creep Behavior of Carbon Fiber/PEEK Composites," *Journal of Thermoplastic Composite Materials,* Vol. 4, 1991, pp. 83–99.

[19] Gates, T. S., Veazie, D. R., and Brinson, L. C., "Creep and Physical Aging in a Polymeric Composite: Comparison of Tension and Compression," *Journal of Composite Materials,* Vol. 31, 1997, pp. 2478–2505.

[20] Schapery, R. A., "Inelastic Behavior of Composite Materials," *ASME Winter Annual Meeting, AMD-13,* C. T. Herakovich, Ed., Houston, ASME, 1975, pp. 122–150.

David R. Veazie[1] and Thomas S. Gates[2]

Tensile and Compressive Creep of a Thermoplastic Polymer and the Effects of Physical Aging on the Composite Time-Dependent Behavior

REFERENCE: Veazie, D. R. and Gates, T. S., **"Tensile and Compressive Creep of a Thermoplastic Polymer and the Effects of Physical Aging on the Composite Time-Dependent Behavior,"** *Time Dependent and Nonlinear Effects in Polymers and Composites, ASTM STP 1357,* R. A. Schapery and C. T. Sun, Eds., American Society for Testing and Materials, West Conshohocken, PA, 2000, pp. 160–175.

ABSTRACT: An experimental study was undertaken to compare the effects of physical aging on the viscoelastic behavior of IM7/K3B $[90]_{12}$ composite and the K3B neat resin loaded in tension and compression. The tests, run over a range of sub-glass transition temperatures, provided material constants, material master curves and aging related parameters. Comparing results from the short term K3B resin and the IM7/K3B $[90]_{12}$ composite behavior indicated that trends in the data with respect to aging time and aging temperature for tension loading are similar; however, trends in the IM7/K3B $[90]_{12}$ composite curves in compression appear more exponential than in the K3B resin curves, and show a more uniform creep rate as a function of temperature. Although most of the long term predictions made in this study physically aged with time as expected, the rate of aging in tension was more severely altered with increasing temperature than in compression.

KEYWORDS: fiber-reinforced composite laminates, viscoelasticity, physical aging, elevated temperature, compression, creep, polyimide matrix

Durability and performance are primary concerns when engineers employ advanced polymer matrix composites (PMCs) for use in modern applications such as those found in aerospace structures. Developing tools to predict long term performance and screen for final materials selection is the impetus for intensive studies at NASA and major industry based airframe developers. The intent of this research is to present the results of a recent experimental study on an advanced PMC and its matrix that applies to these concerns.

In polymer-based composites, Bank et al. [1] showed that the polymeric matrix is often the major constituent that experiences changes in physical and mechanical properties over time due to exposure to elevated temperatures. Their report also demonstrated that individual test methods combined in an integrated scheme will provide an accurate method for understanding the different contributions of various degradation mechanisms to durability. Physical aging, a thermoreversible process that occurs below the glass transition temperature (T_g), is

---

[1] Associate Professor, Department of Engineering, Clark Atlanta University, 223 James P. Brawley Drive, S. W., Atlanta, GA 30314.
[2] Senior Research Scientist, Mechanics of Materials Branch, NASA Langley Research Center, MS 188E, Hampton, VA 23681.

one of these mechanisms that can affect the properties of a polymer matrix. Physical aging emerges from the works of many investigators in the linear viscoelasticity regime and is relatively straight forward, indicating a structural dependence on the characteristic visco-elastic times [2]. Struik [3], Kovacs, Stratton and Ferry [4], among others, explain the structural changes in terms of free volume concepts; thus, we will not discuss them here. Changes in long term composite stiffness, strength and fatigue life can be a result of physical aging. Sullivan [5], as well as other investigators, showed that physical aging can affect the time dependent properties of composite, such as the creep behavior in unidirectional glass fiber reinforced thermosetting resin composites. Therefore, in order to develop time dependent phenomenological and/or micromechanical models to predict the long term properties of composites, the fundamental properties of the matrix material including the effects of the physical aging have to be studied.

The selection of creep testing to assess viscoelastic behavior in PMCs was based upon the previous work by many separate investigators. Scott et al. [6] found that temperature was a primary factor in creep behavior and that the use of time-temperature based super-position principles provided the type of parameters necessary for making accurate long-term predictions. Several investigators such as Struik [3] and Lee and McKenna [7] have used short term tensile creep tests of neat polymers to determine the effects of physical aging on the long term creep of these neat polymers or their composites. In these studies, investigations were made into the concepts of effective time theory and relationships between aging and free volume evolution. Brinson and Gates [8] determined that the physical aging shift rate was the most critical parameter in calculating the magnitude of aging's effect on long term performance. Studies by Gates and Feldman [9] and Veazie and Gates [10] showed that the sequence creep testing procedures for IM7/K3B laminates loaded in tension and compression produced repeatable test data. Time/temperature and time/aging-time superposition techniques provided the material properties required to make long-term predictions.

This paper extends these investigations by comparing the time-dependent effects of physical aging in tension versus compression of a composite material and its matrix constituent. To establish the background for the analytical methodology, a brief summary is given on linear viscoelasticity and physical aging. Neat K3B properties, along with the in-plane transverse (90°-direction) properties of the unidirectional IM7/K3B composite, are given and the experimental equipment and procedures are described. The methods of data reduction are explained including the use of superposition techniques. Results from the short and long term tests are discussed. Direct comparisons between tension and compression behavior of neat K3B and IM7/K3B composite are made by comparing aging shift rates, material constants, material master curves, long term data and predictions.

Physical Aging and Viscoelasticity

Brinson and Gates [8] provided the analytical model that was used in this study to predict the long term tension creep compliance using material properties developed from short term tests as input. A "long term" test was defined as a test time at least 10 times greater than the time of the "short term" material property tests. Using this definition, it was expected that the model would provide insights into the effects of physical aging on the long term viscoelastic behavior of advanced polymeric matrix materials and the polymeric composites that could be made from them. For the purposes of this discussion, the model is briefly recounted for tension loading only. Apart from the obvious sign differences between tension and compression loading, the tension based model was not modified for analysis of compression loading.

Creep Compliance and Time Based Superposition

The time dependent linear creep compliance was modeled with a three parameter expression given by

$$S(t) = S^0 e^{(t/\tau)^\beta} \tag{1}$$

where S^0, τ, and β are the initial compliance, retardation time and shape parameter respectively. It is recognized that some limitations may exist for the expression given in Eq 1 for extremely long times, however, the long term data collected in this study does not approach the limits of this function. Because long term predictions were not extrapolated beyond the available long term data, this function appeared to be usable over the time frames investigated in this work. The material parameters required by the model are found from sequenced short term creep and recovery tests, as depicted in Fig. 1. Time/aging-time superposition of the short term creep compliance test data provided the means for the sequenced, short term data to be collapsed into a single momentary master curve (MMC) at each test temperature. As demonstrated by Struik [3] and illustrated in the log-log plot of Fig. 2, horizontal separation of the sequenced creep compliance curves is due to aging and can be characterized by the aging shift factor ($-\log a$). This shift factor is simply defined as the horizontal distance required to shift a compliance curve to coincide with a reference compliance curve. A linear fit of all the shift factors versus the logarithmic aging time ($\log t_e$) for each MMC, Fig. 3 gave the aging shift rate

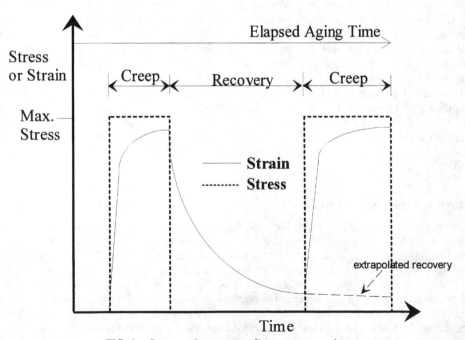

FIG. 1—*Sequenced creep compliance test procedures.*

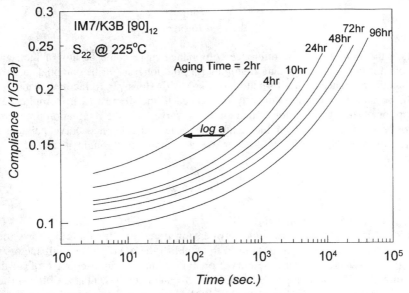

FIG. 2—*Typical transverse compliance momentary curves.*

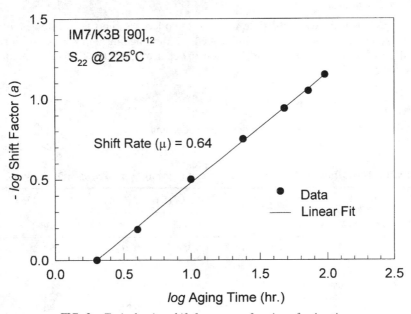

FIG. 3—*Typical aging shift factor as a function of aging time.*

$$\mu = \frac{-d \log a}{d \log t_e} \tag{2}$$

where t_e is the aging time. The reference compliance curve could be any of the sequenced curves, but for convenience of data manipulation, the longest (96 h) compliance curve was selected as the reference during formation of the MMC. However, to facilitate data reporting, all MMC parameters were subsequently referenced to the shortest (2 h) compliance curve. For a horizontal (time) translation of a compliance curve, only the retardation time parameter needs to be recalculated. Given the aging shift rate and reference curve parameters, the translation from one aging time to another was accomplished through the use of

$$\tau_e = \tau_{\text{ref}} \left(\frac{t_e}{t_{e_{\text{ref}}}} \right)^{\mu} \tag{3}$$

where $t_{e_{\text{ref}}}$ is the reference aging time [3].

Equation 2 implies a linear relation between log a and log t_e. Figure 3 shows this relationship for one of the replicate tests used to establish the MMCs. The shift factors for the data on Fig. 3 were found from the curves on Fig. 2 and are representative of all the neat K3B resin and IM7/K3B composite MMC data. To facilitate the collapse of the shifted data for a small number of the MMCs, vertical (compliance) shifts were also utilized. The vertical shifts for all data sets were small in comparison to the magnitude of the corresponding horizontal (time) shifts. A comparison of the typical horizontal versus vertical shift factors is shown in Table 1 for the neat K3B resin in tension. In this study, vertical shifts were only used to facilitate the collapse of the curves. Vertical shifting did not lend itself to the formation of vertical shift rates, nor was it used in the analysis method.

Long Term Compliance

For test periods that exceed the time required to collect the short term (momentary) data, the response can be expected to be influenced by the ongoing aging process. Struik [3] proposed an effective time that could be used to replace time such that the compliance in Eq 1 would be written as

$$S(t) = S^0 e^{(\lambda / \tau(t_e^0))\beta} \tag{4}$$

where λ is the effective time and is calculated according to

TABLE 1—*Horizontal and vertical shift factors (−log a) for K3B neat resin in tension.*

Aging Time (Hours)	200°C		215°C		225°C	
	Horizontal	Vertical	Horizontal	Vertical	Horizontal	Vertical
2	1.514	0.013	1.539	0.017	1.252	−0.018
4	1.277	0.015	1.281	0.018	1.016	−0.013
10	0.976	0.009	0.947	0.013	0.756	−0.017
24	0.640	0.005	0.584	0.011	0.520	−0.016
48	0.362	0.002	0.336	0.007	0.324	−0.010
72	0.159	0.001	0.118	0.010	0.190	0.000
96	0.000	0.000	0.000	0.000	0.000	0.000

$$\lambda = t_e^o \ln\left(\frac{t}{t_e^o} + 1\right) \qquad \text{for } \mu = 1$$

(5)

$$\lambda = \frac{t_e^o}{1 - \mu}\left[\left(1 + \frac{t}{t_e^o}\right)^{1-\mu} - 1\right] \text{ for } \mu \neq 1$$

where t_e^o is the aging time prior to loading and at some time later the total aging time is $t + t_e^o$ where t is the creep test time. Brinson and Gates [8] demonstrated that the effective time theory is self-consistent and shift rates exceeding unity are both physically and mathematically permissible. Use of the effective time expressions for the neat K3B resin, and in a laminated plate model given by Jones [11] for a single lamina under plane stress conditions for the IM7/K3B composite, allowed for the prediction of long term creep compliance. Input to the model were the material parameters measured from short term tests.

Time/Temperature Superposition

Because the MMCs from the aging tests were a function of temperature, the use of the time/temperature superposition principle (TTSP) (Findley, et al. [12]) required that creep compliance to be a function of temperature (T) and time (t) such that

$$S = S(T,t)$$

(6)

and that

$$S(T,t) = S(T_0,\zeta)$$

(7)

$$\zeta = \frac{t}{a_T(T)}$$

(8)

where ζ is the reduced time that is related to the real time t by the temperature shift factor $a_T(T)$ and T_0 is the reference temperature.

The collection of individual MMCs for tension and compression loading in both the neat K3B resin and the IM7/K3B composite can be collapsed into single material master curves using TTSP. The collapse is made using a single reference curve and horizontal (time) shifts only. Characterizing these master curves with an expression similar to Eq 1 along with the reference aging time and reference temperature allows the investigator to calculate the individual creep compliance curve for any test condition.

Test Materials and Equipment

The neat resin material chosen for this study was K3B, an amorphous thermoplastic polyimide fabricated by DuPont. The composite system consisted of the K3B matrix reinforced by unidirectional, intermediate modulus, carbon fibers manufactured by Hexcel and designated IM7. The unaged T_g as measured by Dynamic Mechanical Analyzer (DMA) G'' peak was 240°C. Change in the T_g from the unaged condition over extended aging times was measured by Boeing and found to remain within 3°C for 10 000 h. At Boeing, DMA testing was performed on individual samples that were removed at specific intervals from the isothermal aging condition (180°C). For this study, it was therefore assumed that chemical aging of the composite would not occur and the T_g would remain constant over the duration of the tests.

For in-plane transverse creep compliance of the composite, rectangular test specimens similar to those described in ASTM Standard Test Method for Tensile Properties of Fiber-Resin Composites (D 3039) measuring 24.1 cm by 2.54 cm for tension specimens, and 20.32 cm by 2.54 cm for compression specimens, were cut from laminated panels. For compression creep testing of the composite, specimens were shortened slightly to promote the use of the specially designed compression creep fixture described below. The composite specimens used in both tensile and compression tests were unidirectional 12-ply $[90]_{12}$, whereas each ply measured approximately 0.0135 cm thick. Rectangular test specimens measuring 15.11 cm by 1.78 cm by 0.635 cm thick were used for neat resin testing. To reduce experimental errors at least three replicates were tested at each test temperature. Although all the specimens came from the same material lot, many of the replicate specimens were cut from different panels. There was no evidence of differences in chemistry for the specimens tested in tension versus compression. Prior to testing, all specimens were dried for at least 24 h at 110°C in a convection oven.

Test Equipment

Testing was performed to understand the composite and matrix material behavior, develop material constants for the analytical model, and provide verification of the predictive model. This section will highlight some of the important test equipment and procedures. Specific procedures and techniques relating to the tensile and compressive testing may also be found in references [9] and [10].

The current study used the long, thin specimens described previously. A uniaxial constant load was applied through a dead-weight cantilever arm system. These tests were performed in convection ovens equipped with digital controllers. For the tensile tests, mechanical wedge grips held the specimen during the loaded or creep segments. High temperature tabs were attached to the specimen ends to prevent slipping. For compression testing, a unique apparatus was constructed to allow a tensile creep test frame to be used for application of a compressive load [10]. The compressive creep apparatus consisted of two rigid frames connected by steel rods running through linear bearings. To ensure stable compression, the specimen was supported from column buckling by lightweight knife edge guides. Column buckling was checked during loading by longitudinally aligned back-to-back strain gages, that would show a lack of parity in strain if bending occurred.

Strain in the specimen gage section was measured with high temperature foil strain gages applied in the center of the specimen, and thermal apparent strain was corrected. Laminate damage in the form of matrix cracks can alter the strain measurements, therefore, after each test sequence the specimens were inspected with an optical microscope for matrix cracks along their exposed edges. These inspections revealed no apparent damage after the sequenced tests.

Experimental Procedures and Data Reduction

To explore the effects of physical aging on the creep properties of the composite as well as its matrix constituent, a well-documented technique that measures the creep compliance as described in Struik [3] was used. This procedure consisted of a sequence of creep and recovery tests using a constant applied load while the specimen isothermally ages. The test temperatures selected for the study were 200, 215, and 225°C. These test temperatures were selected to ensure that measurable aging occurred within the test period.

Linearity

For all the creep tests, a single applied stress level (2.68 MPa) was chosen for the K3B neat resin and the IM7/K3B composite and subsequently used at all temperatures. To provide compatibility, the same stress level was used in both the tension and compression tests. Determination of the stress level necessary to stay within the linear viscoelastic range was made by checking that the proportionality condition and Boltzman's superposition would be satisfied [12]. Creep and creep/recovery tests provided data for checking superposition. Proportionality checks were performed by plotting isothermal, creep compliance versus test time for a specimen that was repeatedly rejuvenated (described below), quenched and loaded at various stress levels. The supposed transition from linear to nonlinear behavior would be evident by the vertical separation of the compliance curves with increasing stress. These linearity checks were made at the lowest and highest test temperatures thereby minimizing the effects of applied stress. This process also allowed a linear assumption to be used with assurance of reasonable accuracy.

Short Term Tests

To provide for the test condition that all specimens start the test sequence in the same unaged condition, a means of rejuvenating the specimen was required. Rejuvenation was accomplished by a procedure based upon work by Struik [3] and others who showed that physical aging is thermoreversible and the excursion above T_g prior to quenching effectively rejuvenates the material. In the current tests, the gaged specimens were heated to 250°C (10°C above T_g) for 30 min immediately before the start of any physical aging test sequence. High-pressure air was used to quench the specimen from above T_g to the aging temperature. Relaxation of thermal residual stresses during a tensile test after using this type of quenching procedure was investigated by Allen, et al. [13] using data from a resin similar to K3B. Although thermal residual stresses may play a role when trying to compare compressive and tensile static loading, Allen concluded that during creep the effect of apparent aging due to residual stress was much weaker than the effect of physical aging itself.

The duration of each creep segment was 1/10th the duration of the prior total aging time. The aging times (time after quench) selected for starting each creep segment were 2, 4, 10, 24, 48, 72, and 96 h. After each creep segment, the specimen was unloaded and allowed to recover until the start of the next creep test. To account for any remaining residual strain due to a lack of complete recovery, the strain measured in the creep segment was corrected by subtracting the extrapolated recovery strain from the prior creep curve as illustrated in Fig. 1.

The momentary (short term) sequenced creep/aging curves were collapsed into MMCs through a horizontal (time) shift using the longest aging time curve as the reference curve. In some cases, small (as compared to the horizontal shifts) vertical (compliance) shifts were also used in reduction of the IM7/K3B data.

An individual MMC was found for both transverse loaded IM7/K3B composite and the neat K3B resin at each individual test temperature. These MMCs are shown in Fig. 4 for tension loading and in Fig. 5 for compression loading. All of these curves represent the best fit to collapsed data from all replicates used in the test program. The three curve fit parameters used to characterize each tension and compression MMC are given in Tables 2 and

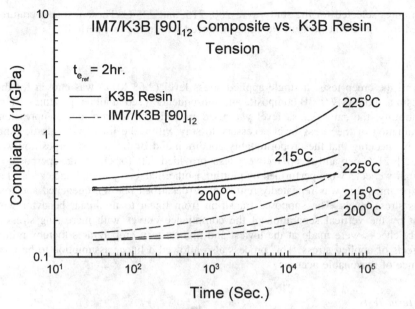

FIG. 4—*Isothermal momentary master curves for tension loading.*

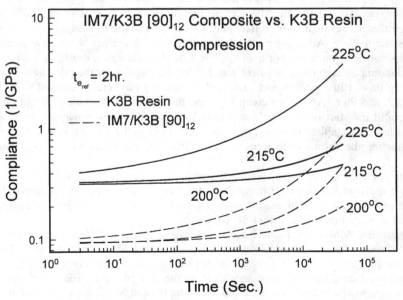

FIG. 5—*Isothermal momentary master curves for compression loading.*

TABLE 2—*Creep compliance momentary master curve parameters for both IM7/K3B composite and K3B neat resin in tension.*

Material	T (°C)	S^0 (GPa^{-1})	τ (sec.)	β	μ	$\Delta\mu$ (Std. Dev.)
	200	0.133	1.56E+5	0.423	0.864	0.139
IM7/K3B	215	0.127	7.69E+4	0.315	0.999	0.030
	225	0.134	1.76E+4	0.297	0.876	0.092
	200	0.344	9.21E+5	0.367	0.930	0.043
K3B Resin	215	0.348	1.28E+5	0.361	0.931	0.026
	225	0.355	9152.54	0.293	0.755	0.043

3, respectively. The parameters from this fit were termed the momentary master curve parameters for a given temperature.

Material Parameters

Aging shifts rates (μ) were calculated for the neat resin and the composite in tension and compression using the sets of master curves. These calculated values are given in Tables 2 and 3 and are plotted against test temperature for K3B resin and IM7/K3B composite in Figs. 6 and 7, respectively. It should be noted that, even though only one set of momentary master curve parameters is given in Tables 2 and 3, each of the replicate tests had their own shift rate (μ) found through the procedures described previously. Therefore, the shift rates given in Tables 2 and 3 are average values. The standard deviation of μ is also provided in Tables 2 and 3. The data points in Figs. 6 and 7 were connected with a third-order regression curve to illustrate the trends. Each set of master curves can be further reduced using TTSP by manually shifting to the reference temperature. Although TTSP was not used in the aging analysis, curves of creep compliance versus time resulting from the TTSP operation are given in Fig. 8. These curves show the contrast in the materials and the loading directions.

Results and Discussion

The momentary master curves that resulted from time/aging-time superposition demonstrated that both IM7/K3B composite and K3B neat resin varied with temperature as expected. Creep compliance in both the IM7/K3B composite and K3B neat resin was a strong function of test temperature with an increase in temperature resulting in an increase in

TABLE 3—*Creep compliance momentary master curve parameters for both IM7/K3B composite and K3B neat resin in compression.*

Material	T (°C)	S^0 (GPa^{-1})	τ (sec.)	β	μ	$\Delta\mu$ (Std. Dev.)
	200	0.094	7.76E+4	0.468	0.808	0.083
IM7/K3B	215	0.093	2.43E+4	0.559	0.742	0.033
	225	0.092	3378.87	0.316	0.644	0.058
	200	0.320	3.52E+4	0.448	0.832	0.002
K3B Resin	215	0.327	8.14E+4	0.386	0.819	0.050
	225	0.321	1078.98	0.244	0.797	0.028

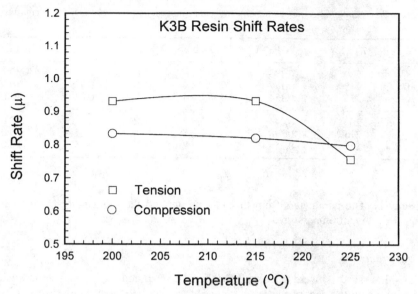

FIG. 6—*K3B neat resin shift rates as a function of temperature.*

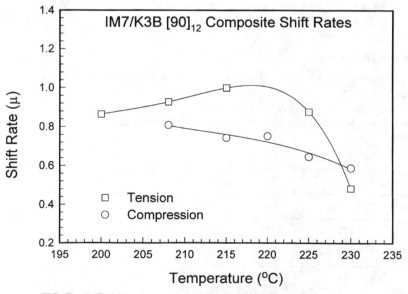

FIG. 7—*IM7/K3B composite shift rates as a function of temperature.*

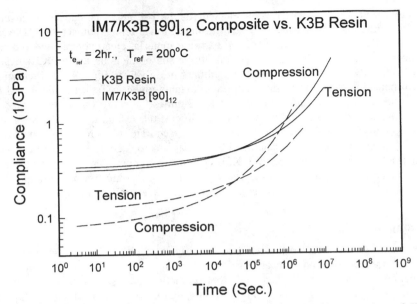

FIG. 8—*Time-temperature superposition master curves for the IM7/K3B composite and K3B neat resin.*

compliance and associated creep rate. The momentary master curves for the tension case showed almost identical trends for all temperatures, and a vertical (compliance) shift would allow the curves to virtually coincide. For the compression case however, trends in the IM7/K3B composite curves appear more exponential than in the K3B resin curves. Also, the IM7/K3B composite curves show a more uniform creep rate as a function of temperature. For both IM7/K3B composite and K3B neat resin, TTSP provided a means for collapsing the sets of momentary master curves into single material master curves. These curves, shown in Fig. 8, reveal that although similar trends exist in the tension and compression creep behavior, differences between the tension and compression loading direction in the IM7/K3B composite were more pronounced than in the K3B neat resin. The repeatability of the testing and the consistency of the data indicates that although theses differences may be slight, they do reflect real material behavior.

Aging Shift Rate

As expected, the aging shift rate was a function of test temperature, showing a decrease in shift rate as the temperature approached T_g (see Figs. 6 and 7). Tension shift rate trends behaved similarly for both the IM7/K3B composite and K3B neat resin. For compression loading, however, the shift rate trend for the K3B resin appeared somewhat level, whereas the shift rate for the IM7/K3B composite reduced continuously.

Long Term Test versus Prediction

Two sets of long term test temperatures, (200, 215°C) and (215, 225°C), were selected for tension and compression of K3B neat resin and IM7/K3B composite, respectively. The higher temperatures (215, 225°C) were selected for the IM7/K3B composite to ensure that measurable aging occurred during the tests. Similarly, lower temperatures (200, 215°C) were used for the K3B neat resin because of the extreme material creep compliance and rupture experienced at 225°C. The long term test results versus predictions for the K3B neat resin and the IM7/K3B composite at these temperatures are given in Figs. 9–12. Predictions of long term behavior were made using material constants given in Tables 2 and 3. All of the long term test times lasted approximately 11 to 12 times the short term tests.

Comparing the composite and its matrix constituent, the results indicated that prediction of creep compliance for the K3B neat resin is more accurate than predictions of IM7/K3B composite creep compliance. Expanding the comparison to include loading direction indicated that both K3B neat resin and IM7/K3B composite compression cases gave predictions that may diverge appreciably from the test data for long loading times, although the long term data for both tension and compression appear to have similar trends.

Concluding Remarks

Experiments were performed to determine the effects of physical aging on creep compliance of the IM7/K3B composite and K3B neat resin loaded in tension or compression. Experimental results and established analytical methods were used to investigate the similarities and differences between the composite and its matrix constituent.

The short term (96 h) tests, run over a range of sub-t_g temperatures provided material constants, material master curves and aging related parameters. The test data were consistent and repeatable over the entire range of test temperatures. Comparing results from the short term K3B resin and the IM7/K3B composite behavior indicated that trends in the data with respect to aging time and aging temperature for tension loading are similar; however, trends in the IM7/K3B composite curves in compression appear more exponential than in the K3B resin curves, and show a more uniform creep rate as a function of temperature.

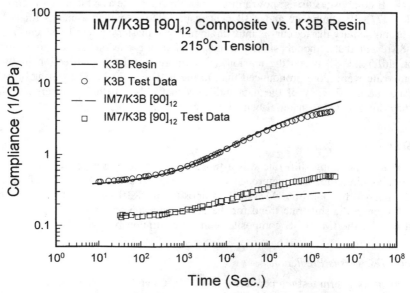

FIG. 9—*IM7/K3B composite and K3B neat resin tests versus prediction for long-term, tension creep tests at 215°C.*

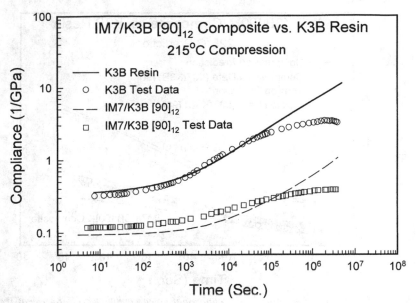

FIG. 10—*IM7/K3B composite and K3B neat resin tests versus prediction for long-term, compression creep tests at 215°C.*

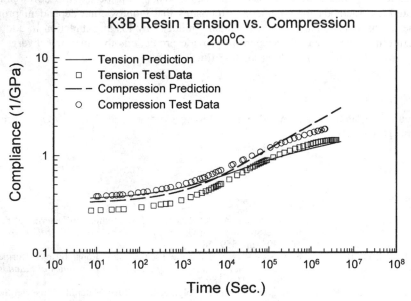

FIG. 11—*K3B neat resin tests versus prediction for long-term, tension and compression creep tests at 200°C.*

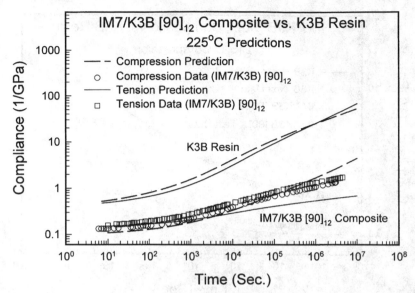

FIG. 12—*Predicted long-term tension and compression creep compliance for K3B neat resin and IM7/K3B composite at 225°C.*

The long term (1500+ hour) predictions compared favorably to the long term test data in tension with the model demonstrating more accuracy for the K3B resin as compared to the IM7/K3B composite. Although most of the long term predictions made in this study physically aged with time as expected, the rate of aging in tension was more severely altered with increasing temperature than in compression. When considering the development of accelerated test methods for polymeric composites that employ time dependent properties from the matrix constituent, aging mechanisms induced by loading direction in addition to temperature must be taken into account to provide predictions that may not diverge appreciably from the test data for long loading times.

Acknowledgment

The authors gratefully acknowledge The Boeing Company, St. Louis, MO for providing long term DMA data.

References

[1] Bank, L. C., Gentry, T. R., and Barkatt, A., "Accelerated Test Methods to Determine the Long-Term Behavior of FRP Composite Structures: Environmental Effects," *Journal of Reinforced Plastics and Composites,* Vol. 14, 1995, pp. 559–587.

[2] McKenna, G. B., "On the Physics Required for the Prediction of Long Term Performance of Polymers and Their Composites," *Journal of Research of the National Institutes of Standards and Technology,* Vol. 99, No. 2, 1994, pp. 169–189.

[3] Struik, L. C. E., "Physical Aging in Amorphous Polymers and Other Materials," Elsevier Scientific Publishing Company, New York, 1978.

[4] Kovacs, A. J., Stratton, R. A., and Ferry, J. D., "Dynamic Mechanical Properties of Polyvinyl Acetate in Shear in the Glass Transition Temperature Range," *Journal of Phys. Chem.,* Vol. 67, No. 1, 1963, pp. 152–161.

[5] Sullivan, J. L., "Creep and Physical Aging of Composites," *Composite Science and Technology,* Vol. 39, 1990, pp. 207–232.

[6] Scott, D. W., Lai, J. S., and Zureick, A.-H., "Creep Behavior of Fiber Reinforced Polymeric Composites: A Review of the Technical Literature," *Journal of Reinforced Plastics and Composites,* Vol. 14, 1995, pp. 588–617.

[7] Lee, A. and McKenna, G. B., "The Physical Aging Response of an Epoxy Glass Subjected to Large Stresses," *Polymer,* Vol. 31, 1990, pp. 423–430.

[8] Brinson, L. C. and Gates, T. S., "Effects of Physical Aging on Long Term Creep of Polymers and Polymer Matrix Composites," *International Journal of Solids and Structures,* Vol. 32, 1995, pp. 827–846.

[9] Gates, T. S. and Feldman, M., "Time Dependent Behavior of a Graphite/Thermoplastic Composite and the Effects of Stress and Physical Aging," *Journal of Composites Technology and Research, JCTRER,* Vol. 17, No. 1, 1995, pp. 33–42.

[10] Veazie, D. R. and Gates, T. S., "Compressive Creep of IM7/K3B Composites and the Effects of Physical Aging on Viscoelastic Behavior," *Experimental Mechanics,* Vol. 37, No. 1, 1997, pp. 62–68.

[11] Jones, R. M., Mechanics of Composite Materials, Scripta Book Company, Washington, D.C., 1975.

[12] Findley, W. N., Lai, J. S., and Onaran, K., "Creep and Relaxation of Nonlinear Viscoelastic Materials," North-Holland Publishing Company, Toronto, 1976.

[13] Allen, D. H., Zocher, M. A., and Groves, S. E., "Apparent Physical Aging of Polymeric Composites," *ASME Winter Annual Meeting,* November 1995.

S.-C. Hung[1] and K. M. Liechti[1]

Nonlinear Multiaxial Behavior and Failure of Fiber-Reinforced Composites

REFERENCE: Hung, S.-C. and Liechti, K. M., **"Nonlinear Multiaxial Behavior and Failure of Fiber-Reinforced Composites,"** *Time Dependent and Nonlinear Effects in Polymers and Composites, ASTM STP 1357*, R. A. Schapery and C. T. Sun, Eds., American Society for Testing and Materials, West Conshohocken, PA, 2000, pp. 176–222.

ABSTRACT: The objective of this study was to examine the nonlinear, three-dimensional behavior of a fiber-reinforced polymer. The material that was considered for this study was the carbon fiber composite, AS4/PEEK, graphite fibers in a semi-crystalline thermoplastic matrix. Arcan specimens were chosen for this study because the specimens are compact and biaxial stress states can be easily obtained. Strains were measured using geometric moiré, and specimens were subjected to shear and biaxial loading up to failure. The nonlinear behavior was examined in both the longitudinal and transverse planes of the composite. The effects of shear on the axial stress-strain behavior were obtained, as well as the effects of axial stress on the shear stress-strain behavior.

A one-parameter plasticity model was used to characterize the obtained nonlinear stress-strain curves. The original model was unable to characterize the nonlinear behavior. The effect of two-dimensional mean stress on the shear yield surface was then added to the model and a significant improvement was achieved. The work potential model with only one internal state variable was also used for characterizing the nonlinear behavior. It could not predict the observed difference of the nonlinear behavior of the composite under shear plus tension and shear plus compression.

KEYWORDS: fiber-reinforced polymer, AS4/PEEK, thermoplastic matrix, nonlinear behavior, shear plus compression

Nomenclature

E_{11}, E_{22}	Young's moduli
$G_{12}, G_{21}, G_{23}, G_{32}$	Shear moduli
$S_{11}, S_{22}, S_{12}, S_{33}, S_{44}, S_{66}$	Compliance
σ_{ij}	Stress
ε_{ij}	Strain
γ_{ij}	Shear strain
γ_n	Nominal shear strain
σ_e	Effective stress
ε_e^p	Effective plastic strain
θ	Loading angle
θ_p	Direction of principal tensile stress
ϕ	Angle between failure direction and y-axis of specimen
p	Hydrostatic pressure

[1] Research Center for the Mechanics of Solids, Structures and Materials, The University of Texas at Austin, Austin, TX 78712.

α, β, μ, κ	Constants
τ_{21}^y	Shear yield strength of 2-1 specimens
τ_{32}^y	Shear yield strength of 3-2 specimens
κ_{21}^c	Slope of curve—τ_{21}^y versus $-\sigma_{22}$
κ_{21}^t	Slope of curve—τ_{21}^y versus $+\sigma_{22}$
κ_{32}^c	Slope of curve—τ_{32}^y versus $-\sigma_{33}$
κ_{32}^t	Slope of curve—τ_{32}^y versus $+\sigma_{33}$
μ_{21}	Slope of curve—τ versus p for 2-1 specimens
μ_{32}	Slope of curve—τ versus p for 3-2 specimens
S	Internal state variable
$S_r = S^{\frac{1}{3}}$	Internal state variable
$\Gamma = a_{66}/a_{22}$	Yield function parameter
Γ_{21}	Γ for 2-1 specimen
Γ_{32}	Γ for 3-2 specimen

Many fiber-reinforced composites exhibit mechanically nonlinear behavior. The nonlinear behavior of fiber reinforced composites has been modeled using elasticity theory [1–3] or plasticity theory [4–10]. The plasticity models are mainly used for fiber reinforced metal matrix materials. Hahn and Tsai [1] employed a complementary strain energy density function to derive a stress-strain relation for a unidirectional composite lamina. By noting a much more severe nonlinearity in the longitudinal shear response than in the longitudinal and transverse tension or compression, an extra quadratic term with the longitudinal shear was added to the stress-strain relations. As a result, the stress-strain relation became nonlinear in shear but remained linear in uniaxial loading in both longitudinal and transverse directions. Hahn [2] extended the nonlinear elastic model of a unidirectional lamina to laminated composites. Sun et al. [3] derived a three-dimensional constitutive relationship for a fiber-reinforced composite with third-order strain terms in the stress-strain relations. The composite was modeled by a medium consisting of thin nonlinear matrix layers alternating with linearly elastic fibrous layers.

Dvorak et al. [9,10] used a micromechanical model consisting of elastic cylindrical fibers of vanishing diameters and an elastic-plastic matrix. The elastic-plastic behavior of the composite was described in terms of constituent properties, their volume fractions and their mutual constraints. However, difficulties were encountered [18] in attempting to apply the model to the composite being considered here due to its high fiber volume fraction.

Generalizing the original work by Hill [11] on anisotropic plasticity, Kenaga et al. [4] proposed an orthotropic elastic-plastic model to characterize the nonlinear behavior of a boron/aluminum composite. Off-axis tests were used to determine the three parameters contained in this orthotropic plasticity model. By taking advantage of the fact that most unidirectional fiber reinforced composites exhibit insignificant plasticity in the longitudinal direction, the orthotropic plasticity model [4] was further simplified by Sun and Chen [5]. They found that a plastic potential containing a single parameter was adequate for unidirectional fiber reinforced composites. Thus the number of parameters required in the model was reduced from three to a single one, i.e., a yield function that depends on stress through one constant. This parameter can be obtained from a simple uniaxial tension test of an off-axis specimen.

Very good agreement was reported between the experiments and the predictions by the one-parameter plasticity model for both boron/aluminum and graphite/epoxy composites [5]. Thus the one-parameter plasticity model was considered very suitable for describing the nonlinear behavior of composites exhibiting little plasticity in the fiber direction [7]. The application of this one-parameter plasticity model to characterize the effect of temperature variation on the elastic-plastic behavior of a unidirectional fiber reinforced composite was made in the shear-tension quadrant [6]. It was also used to characterize the elastic-viscoplastic properties of a unidirectional fiber reinforced composite [7]. Furthermore, it was found that the one-parameter plasticity model predicts the elastic-plastic behavior of a thermoplastic composite in the shear-compression quadrant at various temperatures very well [8].

Schapery [12,13] developed a model for nonlinear material characterization and analysis that uses the same mathematical formalism for nonlinear elastic and inelastic behavior and for growing damage and its effect. The model is based on thermodynamics with internal state variables (ISV), fracture mechanics and the experiment-based observation that the applied work is not sensitive to many details of the deformation history. This path insensitivity leads to the use of a work potential which is analogous to strain energy for a nonlinear elastic material. For example, for a two-dimensional unidirectional fiber reinforced composite, the model uses the linear elastic stress-strain constitutive equations with a generalization in which the material parameters (S_{11}, S_{22}, S_{12}, and S_{66}) vary with one or more internal state variables. These ISVs reflect changes in the microstructure, such as void growth, crystalline slip, shear banding and microcracking in the matrix and fiber/matrix debonding. As a result, they are called structural parameters. The dependence of material properties on the structural parameters is then determined experimentally. This work potential model has been shown theoretically and experimentally to be descriptive of both proportional and nonproportional loadings [12–16]. Existing applications to fiber reinforced plastics have used one ISV, denoted by S, to account for intrinsic nonlinearity; additional ISVs have been used to characterize effects of transverse cracking and delamination [15].

Experimental Procedures

Specimen Preparation

One of the principal advantages of the Arcan specimen is its relatively small size. This allows specimens to be cut in all directions so that any three-dimensional aspects of the material behavior can be determined. In this respect, the specimen width is the critical dimension, which in our case was 19.1 mm (0.75 in.). A unidirectional plate of AS4/PEEK with a 19.1 mm thickness was laminated and specimens were sliced from it in the various directions shown in Fig. 1. After considering the shear and normal stress distributions in Arcan specimens with various fiber orientations, 2-1 specimens (fibers running from notch to notch) and 3-2 specimens (fibers aligned with the normal to the specimen plane) were chosen [17,18] to investigate the nonlinear behavior in the longitudinal and transverse planes of the composite, respectively. Geometric moiré was used to measure the displacement fields in Arcan specimens. These were then differentiated to obtain strain distributions. In order to apply geometric moiré, a procedure of reproducing specimen gratings, attaching them to the Arcan specimen and applying the master grating to form the moiré fringes needed to be developed. Several preliminary tests were conducted in order to determine the range of displacements produced under various biaxial loading conditions. The criterion for the selection of the master grating was based on the fact that the grating should be sensitive enough to measure the smallest displacement, but not produce too many fringes to handle for the largest displacement induced in the biaxial loadings. As a result, a master grating with 120 lines/mm in the x-direction and 20 lines/mm in the y-direction was adequate. However, a crossed-line grating with above frequencies resulted in poor resolution in the x-direction due

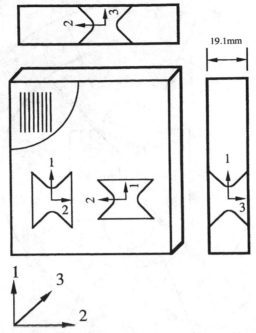

19.1mm

FIG. 1—*Schematic of a plate of unidirectional AS4/PEEK showing the orientations of specimens that were extracted from it.*

to the coarse pitch in the y-direction. Instead, two single-line gratings were chosen, which meant that two experiments had to be conducted in order to extract both displacement components.

The specimen gratings were reproduced via a master grating. Contact prints of the master grating were made to serve as specimen gratings in this work. Kodak Ektagraphic HC slide film (HCS 135) was chosen for making the contact prints. The procedure of reproducing specimen gratings by Kmiec [19] has been modified [18]. Before attaching the specimen grating onto the surface of specimens, a thin layer of white paint was sprayed over the specimen surface. This increased the visibility of the fringes. After the paint dried, the specimen grating was then glued onto the specimen with a liquid adhesive (M-bond 200), being careful not to glue the emulsion side of the specimen grating to the specimen. The bond was later found to be strong enough to endure shear strains up to 15%. At this stage, the specimens were ready for testing.

Experimental Setup

The specimen geometry and that of the connecting grips are shown in Fig. 2. Specimens were glued to the steel grips using epoxy adhesive (Hysol XEA9359.3). The specimens were also slightly thicker than the gap provided in the grips so that the specimens were also clamped. The grips were attached (Fig. 3) to the load cell and actuator of a servo hydraulic universal testing machine through a clevis and rod end arrangement that minimizes bending moments due to misalignments. Checks of bending and twisting effects on the specimens were made by mounting two strain gage rosettes (45°/0°/−45°) on both sides of the speci-

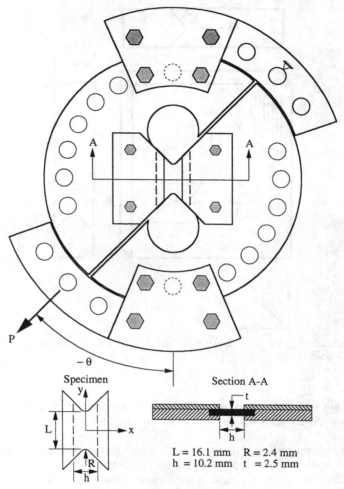

FIG. 2—*Arcan specimen and grips.*

mens. Both bending and twisting effects have been shown to be negligible [*18,20*] in Arcan specimens. For the biaxial loadings the external load was placed on the off-axis holes in the grips.

The geometric moiré fringes were formed by placing a glass reference grating in contact with the specimen grating. The best contrast was achieved by ensuring that the emulsion sides of the reference and specimen gratings were in contact. The reference grating was 9.9 mm wide and 25.4 mm long. The left edge of the reference grating was glued to the stationary grip which was connected to the load cell. Just before testing, a thin layer of mineral oil was applied between the reference and specimen gratings in order to make better contact, thereby increasing the contrast of the fringes. A pair of springs was used to ensure good contact between the reference and specimen gratings and to keep the reference grating in place after the specimen failed.

For each experiment, carrier fringes were introduced at zero load in order to reduce experimental error due to fewer fringes at lower loads. The introduction of carrier fringes was

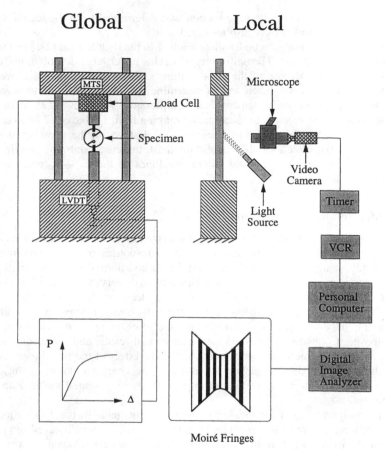

Moiré Fringes

FIG. 3—*Schematic of geometric moiré and the specimen mounted in a servohydraulic testing machine.*

made by aligning the reference grating at a small angle to the specimen axes. The specimen was illuminated by a fiber optic light source. The fringes were viewed by a video camera through a microscope and recorded on video tape (500-line resolution) for subsequent analysis (Fig. 3).

Strain Extraction

Since two single-line gratings were used, two specimens were used under the same conditions for one experiment. One test was used for obtaining u displacements while the other one was obtaining v displacements. The normal strain ε_{xx} was obtained by taking the derivative of the u displacements with respect to the x-axis from a single specimen. However, the shear strain γ_{xy} was determined by obtaining $\partial u/\partial y$ from the u displacement field and $\partial v/\partial x$ from the v displacement field.

The distribution of normal strain ε_{xx} along the specimen gage section was determined at various strain levels. The strain distributions were then used to determine correction factors for calculating the average strain between the notches. A similar procedure was applied to the shear strain. However in this case, $\partial u/\partial y$ along the gage section was determined from one specimen and $\partial v/\partial x$ along the gage section was determined from the second specimen. The shear strain distributions were also averaged.

The responses of the composite under shear loading in the longitudinal (2-1) and transverse (3-2) planes were studied first. The uniformity of strains under large deformations, nonlinear material behavior under shear and the shear failure stresses of the composite were determined. The next step that was taken was to determine the longitudinal and transverse shear response under the influence of tension and compression. A series of experiments with various biaxialities (normal stress to shear stress) ranging from 1.73 to -1.73 was conducted. From these experiments, the effects of transverse axial stress on the shear stress-strain response, as well as the effects of shear stress on the transverse axial stress-strain response were examined. The yield surfaces and failure envelopes on the σ_{21}—σ_{22} and the σ_{32}—σ_{33} planes were then obtained.

Experimental Results and Discussion

In this section, experimental results and characterization of the nonlinear behavior of the composite are presented. The nonlinear stress-strain responses of Arcan specimens under shear and biaxial loading were first determined. Shear and normal strain distributions along the gage section were determined at different strain (load) levels in order to obtain correction factors for strain measurements. The effects of shear stress on the normal stress-strain behavior and the effects of normal stress on the shear stress-strain behavior were also determined. The one-parameter plasticity model was used to characterize the nonlinear behavior of the composite. Comparisons between the experimental results and the predictions by the model are made. These motivated an examination of the effect of the two-dimensional mean stress on the nonlinear shear behavior. In addition, the characterization of the nonlinear behavior of the composite under biaxial loading by the work potential model with one ISV was also made.

Geometric moiré was used for measuring displacement fields in these experiments. The displacements were differentiated to obtain the strains. Ideally, it would have been preferable to obtain lengthwise averaged strains for all experiments. However, it was not always possible to obtain the strains from notch to notch at higher load levels due to diminishing fringe visibility. However, since fringes were always visible in the central region and it has been shown [17,18] that the shear strain and the normal strain (ε_{xx}) always had similar distributions irrespective of biaxiality, correction factors that were derived from 0° and $-45°$ experiments were used for all other orientations. As a justification for this step, it will be shown that the shear strain correction factors were insensitive to load level.

Longitudinal Plane (2-1 Specimens)

It was shown [17,18] that the 2-1 specimen (fibers were aligned with loading direction) is a better candidate than the 1-2 specimen (fibers were perpendicular to the loading direction) for determining shear properties in the longitudinal plane of the composite. It was assumed that this would extend to the nonlinear range. Therefore only 2-1 specimens were used to determine the nonlinear behavior in the longitudinal plane of the composite under shear and biaxial loading.

Strain Correction Factors—In Ref *17* and *18*, it was shown that the normalized shear stress distribution under biaxial loading with different biaxialities is all the same (even under pure shear loading). Therefore, shear strain correction factors obtained under shear loading at different strain levels will be applied to other experiments conducted under different biaxialities. In Fig. 4*a*, seven shear strain distributions along the gage section in the notch-

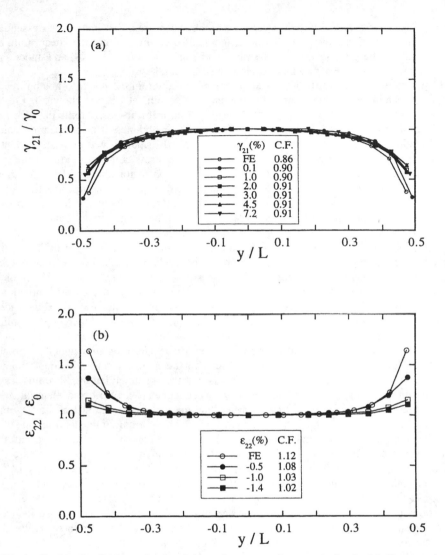

FIG. 4—*Strains distributions along a 2-1 specimen at various strain levels.* (a) *Shear strain,* (b) *normal strain.*

to-notch direction and their corresponding correction factors are shown. The average strain along the gage section can be obtained by multiplying the correction factor with the value of strain measured at the center point of the specimen. The first shear strain distribution in Fig. 4a was obtained [17] from a linear-elastic finite element analysis, while the second one was obtained [20] from moiré interferometry measurement at a 0.1% strain level. The correction factors for these two cases are 0.86 and 0.90, respectively. The difference between the measurement and analysis is about 4%. The shear strain distributions at strain levels 1, 2, 3, and 4.5% were obtained from a specimen under shear loading, while the strain distribution at strain level 7.2% was obtained from a specimen under biaxial loading in the −45° direction. The corresponding correction factors at these strain levels are all very close

(0.90~0.91). Therefore, it can be concluded that the shear strain distribution was essentially independent of the load level. Furthermore, it was also confirmed that the shear strain distributions along the gage section are the same under shear loading ($\theta = 0°$) and under shear plus compression ($\theta = -45°$) under large, as well as small strains.

The normal strain (ε_{22}) distributions along the gage section in the notch-to-notch direction at different strain levels and their corresponding correction factors are shown in Fig. 4b. Again, the small strain solution was obtained [17] from a linear-elastic finite element analysis. The distributions at 0.5, 1.0 and 1.4% strain were obtained from a specimen under biaxial loading in the $-45°$ direction. The normal strain distribution was more dependent on the load level than the shear strain distribution was. The stress concentration near the specimen notch roots reduced as load increased. The correction factor obtained from the linear-elastic finite element analysis was 1.12. From the moiré measurements, the correction factors at strain levels 0.5, 1.0 and 1.4% were 1.08, 1.03 and 1.02, respectively. Again, the correction factors obtained from this experiment were applied to other experiments conducted under different biaxialities.

Figures 5a and b depict the strain distributions along the x-axis ($y = 0$) at different strain levels. The shear strain distributions Fig. 5a at strain levels of 1, 2 and 3.7% were obtained from a specimen under shear loading, while the strain distributions at strain levels of 7.2 and 10.5% were obtained from a specimen under biaxial loading in the $-45°$ direction. It can be seen that, at lower strain levels, the shear strain was more uniformly distributed across the specimen. However, the shear strain distribution changed as the load increased. At higher load levels, the shear strain in the gage section became much larger than elsewhere in the specimen. This feature is not surprising since the Arcan specimen is a double V-notched specimen with the minimum cross-sectional area in the gage section. All double-notched specimens, including those whose notch radius is half the specimen height, have this feature of strain amplification *across* the center of the specimen. As a result, the material in the gage section reaches its nonlinear range earlier than the material in the rest of the specimen. The same feature was apparent (to a lesser extent) in the normal strain distributions along the x-axis (Fig. 5b). The normal strain (ε_{22}) distributions across the gage section at different strain levels were obtained from a specimen under biaxial loading in the $-45°$ direction. The lowest level strain distribution was obtained [17] from the finite-element analysis.

The advantage of the material first reaching its nonlinear range in the gage section is that any stress concentration will be reduced. Therefore, it can be expected that any stress concentrations in the gage section of an Arcan specimen will be reduced as the load increases. Since the shear strain distribution along the gage section of a 2-1 specimen exhibited no stress concentrations in the linear range, little change was noticed in the strain distributions (Fig. 4a) at higher strain levels. However, the normal stress concentrations near notch roots decreased significantly as the load increased which was already seen in Fig. 4b. This improves the stress uniformity in the gage section of Arcan specimens.

However, since the displacement is more localized in the gage section, the strain rate in the gage section cannot be expected to be constant, even in a constant displacement rate experiment. In Fig. 5c, the measured nominal shear strain (based on grip displacement) (γ_n) and lengthwise averaged shear strain (γ_{21}) were normalized with respect to their values at specimen failure, respectively. It was seen that the nominal shear strain rate was nearly constant. However, this was not the case for the lengthwise averaged shear strain rate which was initially small and increased continuously as the composite response became nonlinear. The rates ranged from 0.34 of the nominal strain rate initially to 1.9 close to failure. The question then arises as to how large an effect this has on the measured response, given that the AS4/PEEK may be time dependent.

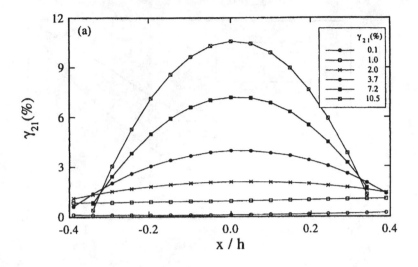

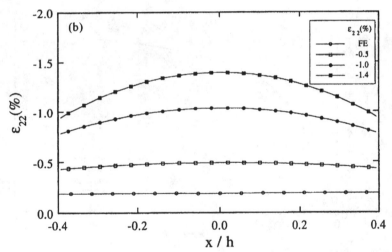

FIG. 5—(a) *Shear and* (b) *normal strain distributions across a 2-1 specimen at various strain levels.* (c) *The normalized strain rates.* (d) *The shear responses under constant displacement and strain rates.*

An experiment was therefore conducted where the central strain was measured via a strain gage and the resulting signal was used to control strain rate. The shear response (Fig. 5d) was the same for nominal strain rate and local strain rate control. Therefore, it can be concluded that, for the room temperature tests conducted in this study, controlling the nominal strain rate was acceptable.

Nonlinear Stress-Strain Curves—After correction factors were applied to the measured strains, average shear and normal stress-strain curves were obtained for all loading cases. Figure 6a shows the shear stress-strain curves, while the corresponding normal stress-strain

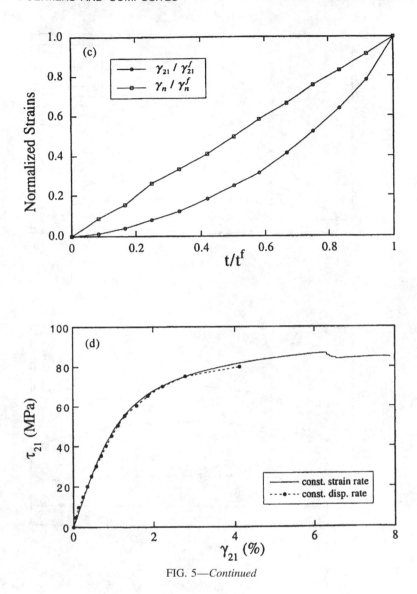

FIG. 5—*Continued*

curves are shown in Fig. 6*b*. The initial shear modulus obtained from each case was approximately the same. However, this was not the case for yield and failure strengths. It should be noted that the failure strength obtained in this work can only be associated with Arcan specimens. It is most likely not the real strength of the composite due to remaining normal stress concentrations in the notch region. For specimens under shear plus tension ($\theta > 0$), it was found that the larger the tensile stress, the lower the shear yield and failure strengths. Figure 6*b* shows the corresponding normal stress-strain curves. It was found that the specimen endures higher normal stresses and strains under shear plus compression than under shear plus tension. In both cases, higher biaxiality leads to higher normal strengths and failure strains.

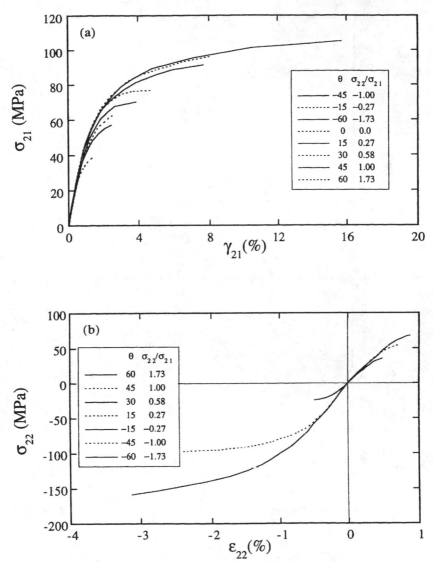

FIG. 6—*Averaged stress-strain curves of 2-1 specimens under biaxial loadings.* (a) *Shear stress-strain curves,* (b) *normal stress-strain curves.*

The shear yield strength of each specimen was determined by a 0.2% strain offset procedure. The shear yield and failure strengths of specimens at different normal stress levels are shown in Fig. 7a. Under shear plus tension, the measured shear yield and failure strengths were all lower than those obtained under simple shear loading. Compression had little effect on the shear yield strength (Fig. 7a). For specimens under shear plus compression ($\theta < 0$), it was found that the larger the compressive stress, the higher the shear failure strength except for the specimen loaded at $-60°$. Therefore, it can be concluded that the existence of normal stress has a significant effect on shear response in the longitudinal plane of unidirectional composites.

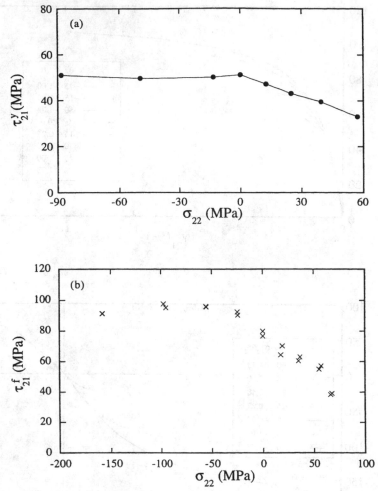

FIG. 7—*Shear strength versus normal stress for a 2-1 specimen.* (a) *Yield strength,* (b) *failure strength.*

Failure and Fracture Surfaces—Specimens 2-1 failed in a catastrophic manner. The specimen was suddenly torn apart along the fiber direction at the maximum load. Failed specimens for each loading direction are shown in Fig. 8. For all the cases studied, failure initiated at a point on the circular part of the notch. Failure was due to splitting along the fiber direction, except for the $-60°$ case, where the failure plane made a small angle with the fiber direction. Since the cracks initiate and grow in the y-direction, the normal stress σ_{xx} (σ_{22}) is important. Tensile stresses promoted the initiation of cracks, thereby reducing the strength while compressive stresses delayed the initiation of cracks, making the strength higher. Furthermore, as was already shown in Fig. 7, the existence of tensile stress not only affects shear failure strengths but also affects shear yielding strengths. Since tensile stresses promote the onset of shear yielding, crack initiation might be delayed due to plasticity effects whereas no mitigation can be expected on the compressive side because compressive stresses have little effect on the shear yield strength.

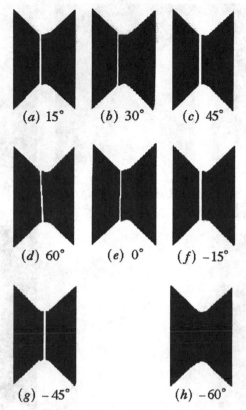

(a) 15°   (b) 30°   (c) 45°

(d) 60°   (e) 0°   (f) −15°

(g) −45°   (h) −60°

FIG. 8—*Failure of 2-1 specimens.*

A lot of work has been conducted on the interlaminar fracture of fiber reinforced composites and much of this has included fractographic features. The most predominant and most discussed of these features are "hackle" markings [21]. An examination of the fracture surfaces of all 2-1 specimens tested here showed that hackle markings existed on all specimens. The hackle markings are seen in the matrix (especially in the resin-rich areas) between fibers and appear as a more or less regular stacking of platelets. It has been suggested that the formation of hackles is due to matrix microcracks which are perpendicular to the direction of maximum principal tensile stress [22–24].

A scanning electron microscope (SEM) fractograph of the fracture surface of a 2-1 specimen under shear loading is shown in Fig. 9d. Note that the 10 μm scale marker in Fig. 9a applies to all the other fractographs in Fig. 9. Many hackle markings and fiber pull-outs can be seen. The tilt angle of the hackles is about 45° into the matrix which is normal to the direction of maximum principal tensile stress. This is in agreement with the suggestions [22–24] on the formation of hackles. The fracture surfaces of specimens under biaxial loading in the −60, −45 and −15° directions are also shown in Figs. 9a, b and c, respectively. It was found that, under shear plus compression, the sharp edges of hackles that were present at 0° are no longer seen. Instead, they appear to have been flattened, most likely due to the existence of compressive stresses. Apparently, the sharp edges of hackles seen in the shear loading case were erased due to the compression and friction between the two fracture surfaces. It was also noticed that higher compressive stresses reduced the number of hackles.

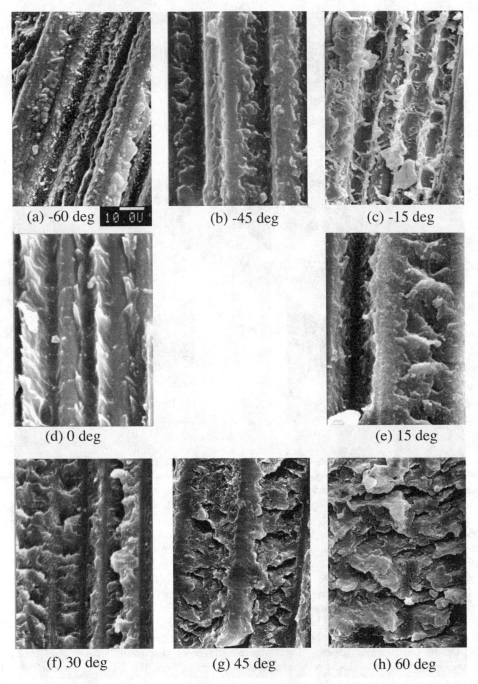

FIG. 9—*Fractographs of 2-1 specimens loaded in* (a) −60°, (b) −45°, (c) −15°, (d) 0°, (e) 15°, (f) 30°, (g) 45°, *and* (h) 60° *directions.*

Under shear plus tension, hackles were much more easily identified on the failure surfaces than was the case in shear plus compression. Figures 9e, f, g, and h are the failure surfaces of specimens loaded in the 15°, 30°, 45°, and 60° directions, respectively. Due to different biaxialities, the inclination of the maximum principal tensile stress varied with loading angle. As a result, the tilt angles of hackles also varied. It can be seen that the hackles were much more obvious and bigger than in the case of shear plus compression. Since the tensile stresses tend separate crack faces, it can be expected that the hackles are more likely to exist in more intact and larger form.

Transverse Plane (3-2 Specimens)

The 3-2 type of specimen was used to determine the nonlinear behavior in the transverse plane of the composite under biaxial loading. Again the biaxialities (σ_{33}/σ_{32}) ranged from -1.73 to 1.73.

Strain Correction Factors—Four shear strain distributions along the gage section were obtained at different strain levels and their corresponding correction factors are shown in Fig. 10a. The first shear strain distribution was again obtained [17] from a linear-elastic finite-element analysis which gives a correction factor of 1, while the second one was obtained [20] from moiré interferometry measurement at strain level 0.1% which gives a correction factor of 1.02. The difference between the measurement and analysis is only 2%. The shear strain distributions at strain levels 1% and 2% were obtained from a specimen under shear loading and the displacement measurements were made by geometric moiré. The corresponding correction factors for shear strain measurement at these two strain levels are 0.98 and 0.97, respectively. The biggest difference between the measurements and analysis is 3% for the cases considered here. The shear strain distribution along the gage section of 3-2 specimens again has virtually no dependence on the load level.

The normal strain (ε_{33}) distributions along the gage section at different strain levels and their corresponding correction factors are shown in Fig. 10b. Again, the small strain profile was obtained [17] from a linear-elastic finite-element analysis. The others were obtained from a specimen under biaxial loading in the $-45°$ direction. The stress concentration near the specimen notch reduced as the load increased in the same way that the 2-1 specimen did. However, the tendency is small. The correction factor obtained from the linear-elastic finite-element analysis was 1.10. From moiré measurements, the correction factors at strain levels 0.3%, 0.6%, 1.0%, and 1.5% were 1.07, 1.06, 1.06, and 1.06, respectively.

Nonlinear Stress-Strain Curves—After correction factors were applied to the measured strains, average shear and normal stress-strain curves were obtained for each loading case. Figure 11a shows the shear stress-strain curves of 3-2 specimens under various biaxial loadings, while the corresponding normal stress-strain curves are shown in Fig. 11b. The initial shear modulus obtained from each case was indistinguishable. For specimens under shear plus tension ($\theta > 0$), larger tensile stresses again led to lower shear yield and failure strengths. Here the shear yield strength of each specimen was determined from a 0.1% strain offset procedure instead of 0.2% due to the low failure strengths of 3-2 specimens. The shear yield and failure strengths were plotted against their corresponding normal stresses as shown in Figs. 12a and b. The yield and failure strengths under shear plus tension are all lower than those under shear loading. This is the same as was the case for 2-1 specimens. For shear plus compression ($\theta < 0$), it was found that the larger the compressive stress, the higher the failure strength. However, no noticeable differences in shear yield strength were observed. From Fig. 11a, it can be seen that the effect of compressive stress on shear yield strength is very small, which can also be verified from Fig. 12.

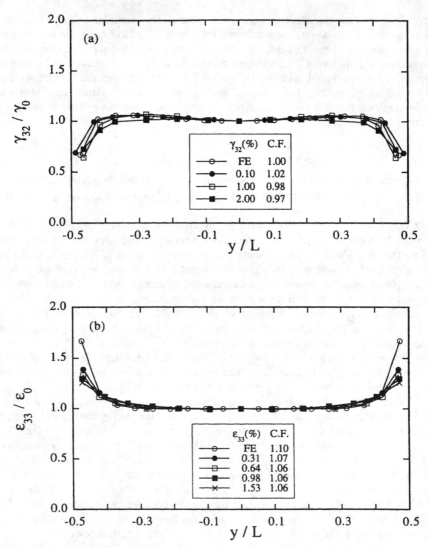

FIG. 10—*Strain distributions along a 3-2 specimen at various strain levels.* (a) *Shear strain,* (b) *normal strain.*

Figure 11*b* shows the corresponding normal stress-strain curves. It was seen that specimens endure higher stress and strain under shear plus compression than under shear plus tension. For both cases, the higher biaxiality, the higher the normal strength and failure strain, the same behavior as the 2-1 specimens. It again suggested that the tensile stress promotes the initiation of cracks while the compressive stress delays the initiation of cracks, leading to higher strength under shear plus compression.

Failure and Fracture Surfaces—The failure of specimens was also catastrophic for the 3-2 specimens. Each specimen broke when its maximum load was reached. However, the failed specimen was still held in the grips since it did not break into two pieces in the same way that the 2-1 specimen had. Figure 13 shows a failure specimen for each biaxial loading

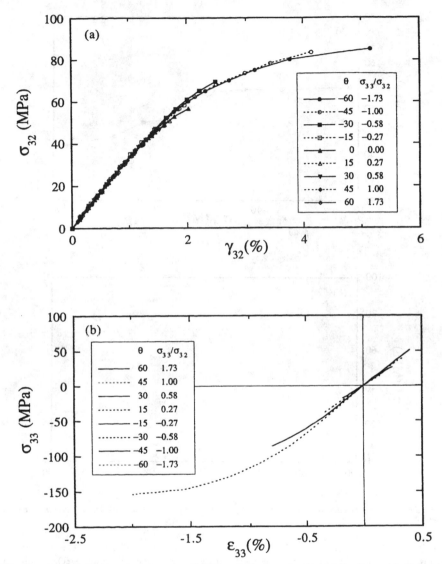

FIG. 11—*Averaged stress-strain curves of 3-2 specimens under biaxial loadings.* (a) *Shear stress-strain curves,* (b) *normal stress-strain curves.*

case. For each specimen, failure was due to cracks that initiated in the notch roots and propagated into the specimen with an angle ϕ (see Fig. 14) relative to the *y*-axis. The failure direction of each specimen is obviously different.

For specimens loaded in the −45° and −60° directions, the angle ϕ was found to be positive, while all the others were negative. A gradual change in angle ϕ of each specimen was also noticed. Therefore, an examination of maximum principal tensile stress and its direction along the notch flank for each biaxial loading case was conducted using the finite-element analysis procedures outlined in [17]. All the loading cases were examined. Three

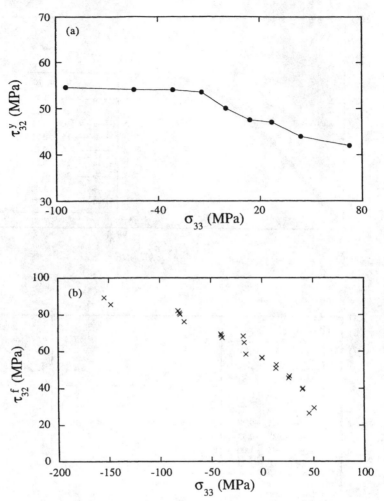

FIG. 12—*Shear strengths versus normal stress for a 3-2 specimen.* (a) *Yield strength,* (b) *failure strength.*

cases are now shown in Fig. 14 to represent the cases of shear plus compression, pure shear and shear plus tension. For each case, the angle ϕ was measured from the failure specimen, while the direction (θ_p) of the principal tensile stress at the location where crack initiated was determined from the analysis.

For the specimen loaded in the $-60°$ direction (Fig. 14a), the angles θ_p and ϕ were found to be 28° and 31°, respectively. Since they are so close, the cleavage plane is roughly normal to the direction of principal tensile stress. It suggests that the failure of 3-2 specimens (transverse plane) was due to tension. For the specimen under shear loading (Fig. 14b), two cracks were found on antisymmetric locations because the stresses at the top and bottom notch were antisymmetric. In this case, the angles θ_p and ϕ were $-41°$ and $-38°$. Again, the cleavage plane was found to be normal to the direction of maximum principal tensile stress. Therefore, it can be concluded that the 3-2 specimen failed in tension even under shear loading. Similar results were reported by Daniel et al. [25] when a silicon carbide

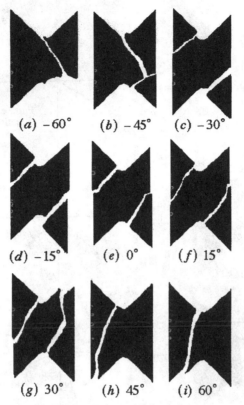

FIG. 13—*Failure of 3-2 specimens.*

particulate/aluminum composite was tested in the Iosipescu fixture. Figure 14c shows a case of shear plus tension ($\theta = 60°$), the angles θ_p and ϕ were $-24°$ and $-21°$, respectively, therefore, the cleavage plane was again found to be normal to the direction of maximum principal tensile stress. In fact, this statement holds for all loading cases considered, as can be seen in Fig. 15 where the corresponding angles θ_p and ϕ for each loading case are shown.

Fractographs of the failure surfaces are shown in Fig. 16. Note that the 10 μm scale marker in Fig. 16a applies to all the other fractographs in Fig. 16. They were made very close to the failure initiation sites. All the fracture surfaces were viewed in the x-direction which means (Fig. 13) that viewing was made at an angle to the fracture surface. The fiber direction is also the through-thickness (z) direction. In most cases, the fracture path did not exactly follow the fiber matrix interface. There was much evidence of matrix cracking and the surface were rough. However, no hackles were formed perpendicular to the fibers indicating an absence of σ_{32} as has been suggested in the previous paragraph.

Model Development for Characterizing the Nonlinear Behavior

In this section, a one-parameter plasticity model and work potential model with only one internal state variable were used to characterize the obtained nonlinear stress-strain curves. Modification of the one-parameter plasticity model is also made and discussed. The derivation of each model used is now given here.

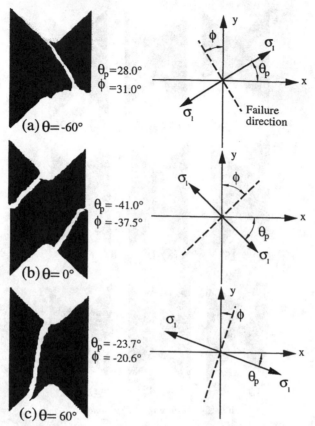

FIG. 14—*Direction of principal tensile stress and the orientation of cleavage in 3-2 specimens.*

One-Parameter Plasticity Model

The derivation of the one-parameter plasticity model was originally given in Ref 5. A detailed derivation of this model applied to Arcan specimens is given in Ref 18. A yield function that is quadratic in stresses is assumed for a general three-dimensional fiber reinforced composite. The function is a generalization of von Mises' yield criterion with nine coefficients (characteristic parameters) associated with the state of anisotropy. For unidirectional fiber reinforced composites, since the material behavior in the fiber directions is elastic up to failure, it is reasonable to assume that there is no plastic strain in the fiber direction, so that $d\varepsilon_{11}^p = 0$. In addition, the plane with the normal in the fiber direction is assumed to be isotropic. Therefore there exists a rotational symmetry about the fiber direction (1-axis). Furthermore, it is assumed that there is no plastic dilatation ($d\varepsilon_{ii}^p = 0$), which means that the material is incompressible under plastic deformation. Using the above three conditions, the original yield function (with nine coefficients) reduces to a form with only two coefficients, a_{22} and a_{66}. Without loss of generality, the coefficient a_{22} is set to 1. A parameter Γ is further introduced, which is defined as $\Gamma \equiv a_{66}/a_{22}$. The yield function can therefore be rewritten as

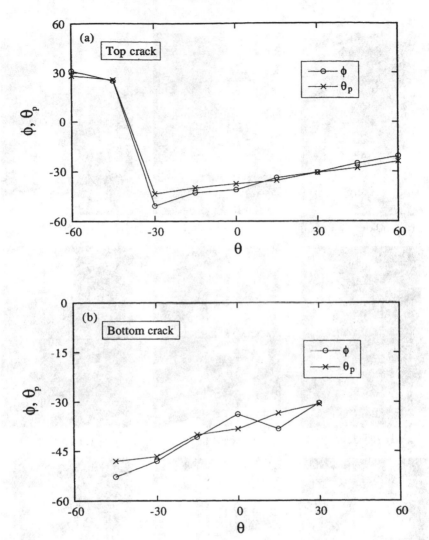

FIG. 15—*Direction (θ_p) of principal tensile stress and the direction (ϕ) of cleavage plane of 3-2 specimens.* (a) *Top crack, and* (b) *bottom crack.*

$$2f(\sigma) = (\sigma_{22} - \sigma_{33})^2 + 4\sigma_{23}{}^2 + 2\Gamma(\sigma_{12}^2 + \sigma_{13}^2) \tag{1}$$

The associated effective stress and strain become

$$\begin{cases} \sigma_e = \left\{ \dfrac{3}{2} \left[(\sigma_{22} - \sigma_{33})^2 + 4\sigma_{23}^2 + 2\Gamma(\sigma_{12}^2 + \sigma_{13}^2) \right] \right\}^{1/2} \\[2ex] d\varepsilon_e^p = \sqrt{\dfrac{2}{3}} \left\{ (d\varepsilon_{22}^p)^2 + \dfrac{1}{4} (d\gamma_{23}^p)^2 + \dfrac{1}{2\Gamma} \left[(d\gamma_{12}^p)^2 + (d\gamma_{13}^p)^2 \right] \right\}^{1/2} \end{cases} \tag{2}$$

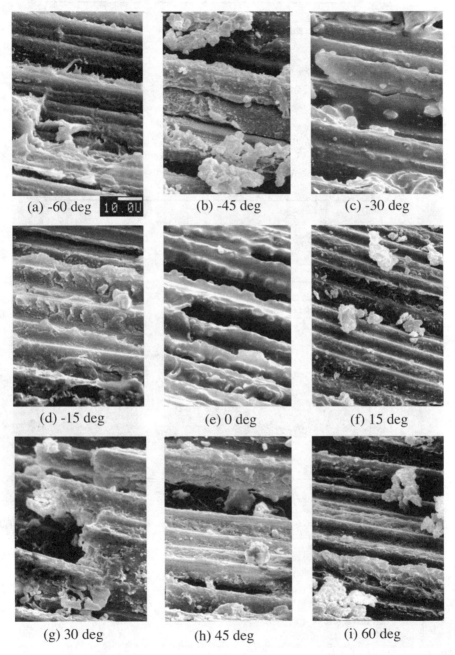

FIG. 16—*Fractographs of 3-2 specimens loaded in* (a) *−60°,* (b) *−45°,* (c) *−30°,* (d) *−15°,* (e) *0°,* (f) *15°,* (g) *30°,* (h) *45°, and* (i) *60° directions.*

Longitudinal Plane (2-1 Specimens)—When Arcan specimens with a 2-1 fiber orientation are used, a state of plane stress ($\sigma_{3i} = 0$) exists in the 2-1 plane. The state of plane stress was also confirmed by a three-dimensional finite-element analysis. For 2-1 specimens, the normal stress $\sigma_{yy} = v_{yx}\sigma_{xx}$ [17,18]. As a result, the yield function, effective stress and strain, respectively, reduce to

$$2f(\sigma) = \sigma_{22}^2 + 2\Gamma_{21}\sigma_{12}^2$$

$$\sigma_e = \left[\frac{3}{2} (\sigma_{22}^2 + 2\Gamma_{21}\sigma_{12}^2) \right]^{1/2} \tag{3}$$

$$d\varepsilon_e^p = \left\{ \frac{2}{3} \left[(d\varepsilon_{22}^p)^2 + \frac{1}{2\Gamma_{21}} (d\gamma_{12}^p)^2 \right] \right\}^{1/2}$$

The above equations are used to characterize the nonlinear behavior of unidirectional fiber reinforced composites (i.e., to determine a master curve). Once the master curve is determined, the process can be reversed and predictions can be based on it. The shear and normal responses of 2-1 specimens can be predicted by the following equations.

$$\begin{cases} \sigma_{21} = \dfrac{\sigma_e}{\sqrt{\dfrac{3}{2}} (\tan^2 \theta + 2\Gamma_{21})} \\ d\gamma_{21}^p = \dfrac{3\Gamma_{21}}{\sqrt{\dfrac{3}{2}} (\tan^2 \theta + 2\Gamma_{21})} d\varepsilon_e^p \end{cases} \quad \begin{cases} \sigma_{22} = \dfrac{\sigma_e \tan \theta}{\sqrt{\dfrac{3}{2}} (\tan^2 \theta + 2\Gamma_{21})} \\ d\varepsilon_{22}^p = \dfrac{3 \tan \theta}{2\sqrt{\dfrac{3}{2}} (\tan^2 \theta + 2\Gamma_{21})} d\varepsilon_e^p \end{cases} \tag{4}$$

Transverse Plane (3-2 Specimens)—When Arcan specimens with a 3-2 fiber orientation are used, a state of plane stress ($\sigma_{1i} = 0$) exists in the 3-2 plane. In the same way, the nonlinear behavior of 3-2 specimens can be characterized using

$$2f(\sigma) = (1 - \kappa)^2\sigma_{33}^2 + 4\sigma_{32}^2,$$

$$\sigma_e = \left\{ \frac{3}{2} [(1 - \kappa)^2 \sigma_{33}^2 + 4\sigma_{32}^2] \right\}^{1/2}, \tag{5}$$

$$d\varepsilon_e^p = \left\{ \frac{2}{3} \left[(d\varepsilon_{33}^p)^2 + \frac{1}{4} (d\gamma_{32}^p)^2 \right] \right\}^{1/2}$$

where κ is a constant in $\sigma_{yy} = \kappa\sigma_{xx}$ that was obtained from a linearly elastic finite-element analysis. The nonlinear stress-strain curves can be predicted using the following equations.

$$\begin{cases} \sigma_{32} = \dfrac{\sigma_e}{\sqrt{\dfrac{3}{2}} [(1 - \kappa)^2 \tan^2 \theta + 4]} \\[4mm] d\gamma^p_{32} = \dfrac{6}{\sqrt{\dfrac{3}{2}} [(1 - \kappa)^2 \tan^2 \theta + 4]} \, d\varepsilon^p_e \end{cases} \qquad \begin{cases} \sigma_{33} = \dfrac{\sigma_e \tan \theta}{\sqrt{\dfrac{3}{2}} [(1 - \kappa)^2 \tan^2 \theta + 4]} \\[4mm] d\varepsilon^p_{33} = \dfrac{\dfrac{3}{2} (1 - \kappa) \tan \theta}{\sqrt{\dfrac{3}{2}} [(1 - \kappa)^2 \tan^2 \theta + 4]} \, d\varepsilon^p_e \end{cases} \tag{6}$$

Modified One-Parameter Plasticity Model

For many polymers, the shear yield stress increases when they are subjected to a state of hydrostatic pressure [26,27]. For example, when polymers are under biaxial loading, normal stress components affect the shear yield strength. In such a case, the shear yield surface can be expressed as $\tau = \tau_0 + \mu p$ [26,27], where τ is the shear yield strength and τ_0 is obtained under shear loading. A modification of the yield function is needed to take this effect into account. One of the modifications suggested is that the value of k (in the yield function) increases linearly with the hydrostatic pressure of the stress system [30]. It takes the form, $k = k_0 + 6\alpha p$, where α and k_0 are constants and k reduces to k_0 under shear loading. The physical argument behind this assumption is that the hydrostatic pressure changes the state of the material and produces a structure that has a higher yield strength. Therefore, the yield function becomes $2f(\sigma) + 2\alpha I = k_0$. This form was originally suggested by Schofield and Wroth [31] for soil and later by Bowden and Jukes [30] for polymers. Since for transversely isotropic composites, the hydrostatic pressure affects the matrix (PEEK) of the composite and since, in the fiber direction, the material behavior is fiber dominant, it is reasonable to suggest that there is no hydrostatic pressure effect in the fiber direction, thus, the hydrostatic pressure effect reduces to two-dimensional mean stress effect. The yield function can therefore be written as

$$2f(\sigma) + 2\alpha(\sigma_{22} + \sigma_{33}) = k_0 \tag{7}$$

Longitudinal Plane (2-1 Specimens)—The modified yield function is now specialized for the stress state that exists in 2-1 specimens. The yield function, effective stress and strain were derived [18] and are listed as

$$2f(\sigma) = \sigma^2_{22} + 2\hat{\Gamma}_{21}\sigma^2_{12} + 2\hat{\alpha}_{21}\sigma_{22}$$

$$\sigma_e = \sqrt{\frac{3}{2}} \, [\sigma^2_{22} + 2\hat{\Gamma}_{21}\sigma^2_{12} + \hat{\alpha}_{21}\sigma_{22}]^{1/2} \tag{8}$$

$$d\varepsilon^p_e = \sqrt{\frac{2}{3}} \left\{ (d\varepsilon^p_{22})^2 + \frac{1}{2\hat{\Gamma}_{21}} (d\gamma^p_{21})^2 - \frac{\hat{\alpha}_{21}}{\sigma_{22} + \hat{\alpha}_{21}} (d\varepsilon^p_{22})^2 \right\}^{1/2}$$

where $\hat{\alpha}_{21} = 2/3 \, \mu \tau^y_{21} \hat{\Gamma}_{21}$. The τ^y_{21} is the shear yield strength under pure shear loading. The shear and normal responses can then be predicted by

$$
\begin{cases}
\sigma_{21} = \dfrac{-\dfrac{3}{2}\,\hat{\alpha}_{21}\tan\theta + \left[\left(\dfrac{3}{2}\,\hat{\alpha}_{21}\tan\theta\right)^2 + 2(6\hat{\Gamma}_{21} + 3\tan^2\theta)\sigma_e^2\right]^{1/2}}{6\hat{\Gamma}_{21} + 3\tan^2\theta} \\[4mm]
d\gamma_{12}^p = \left(3\hat{\Gamma}_{21}\,\dfrac{\sigma_{12}}{\sigma_e}\right)d\varepsilon_e^p
\end{cases}
$$

$$
\begin{cases}
\sigma_{22} = \sigma_{12}\tan\theta, \\[3mm]
d\varepsilon_{22}^p = \left(\dfrac{3}{2}\,\dfrac{\sigma_{22} + \hat{\alpha}_{21}}{\sigma_e}\right)d\varepsilon_e^p
\end{cases}
$$

$$(9)$$

Transverse Plane (3-2 Specimens)—In the same way, the nonlinear behavior of 3-2 specimens can be characterized using the following equations.

$$2f(\sigma) = (1 + 2\kappa - 2\kappa\tilde{\Gamma}_{32} + \kappa^2)\sigma_{33}^2 + 2\tilde{\Gamma}_{32}\sigma_{23}^2 + 2\hat{\alpha}_{32}(1 + \kappa)\sigma_{33} = k_0$$

$$\sigma_e = \left\{\frac{3}{2}\left[\left(1 + 2\kappa - 2\kappa\tilde{\Gamma}_{32} + \kappa^2\right)\sigma_{33}^2 + 2\tilde{\Gamma}_{32}\sigma_{23}^2 + \tilde{\alpha}_{32}(1 + \kappa)\sigma_{33}\right]\right\}^{1/2} \qquad (10)$$

$$d\varepsilon_e^p = \sqrt{\frac{2}{3}}\left\{[\beta\sigma_{33}^2 + \tilde{\alpha}_{32}(1 + \kappa)\sigma_{33}]\left(\frac{d\varepsilon_{33}^p}{[1 + \kappa(1 - \tilde{\Gamma}_{32})]\sigma_{33} + \tilde{\alpha}_{32}}\right)^2 + \frac{(d\gamma_{32}^p)^2}{2\tilde{\Gamma}_{32}}\right\}^{1/2}$$

where

$$\beta = 1 + 2\kappa - 2\kappa\tilde{\Gamma}_{32} + \kappa^2$$

The nonliner behavior of 3-2 specimens can be predicted based on

$$
\begin{cases}
\sigma_{32} = \dfrac{-\tilde{\alpha}_{32}(1 + \kappa)\tan\theta + \left\{[\tilde{\alpha}_{32}(1 + \kappa)\tan\theta]^2 + \dfrac{8}{3}(2\tilde{\Gamma}_{32} + \beta\tan^2\theta)\sigma_e^2\right\}^{1/2}}{2(2\tilde{\Gamma}_{32} + \beta\tan^2\theta)} \\[4mm]
d\gamma_{23}^p = 3\tilde{\Gamma}_{32}\,\dfrac{\sigma_{23}}{\sigma_e}\,d\varepsilon_e^p
\end{cases}
$$

$$
\begin{cases}
\sigma_{33} = \sigma_{32}\tan\theta \\[3mm]
d\varepsilon_{33}^p = \dfrac{3}{2}\,\dfrac{(1 + \kappa - \kappa\tilde{\Gamma}_{32})\sigma_{33} + \tilde{\alpha}_{32}}{\sigma_e}\,d\varepsilon_e^p
\end{cases}
$$

$$(11)$$

Work Potential Model

The nonlinear behavior of fiber reinforced polymers may arise from a variety of mechanisms, including void growth, crystalline slip, shear yielding, microcracking in the matrix and fiber/matrix debonding. Therefore a classical plasticity theory may not be fully suitable for describing the nonlinear behavior of fiber-reinforced composites. Schapery [12,13] developed model for material characterization and analysis that uses the same mathematical formalism for nonlinear elastic and inelastic behavior and for growing damage and its effect. The model is based on thermodynamics with internal state variables (ISV), fracture mechanics, and the experimental observation which finds that the stresses and mechanical work are practically independent of deformation history for suitably limited paths [32]. This limited path-independence leads to a mathematical description of material behavior in terms of a work potential which is analogous to that used for predicting stable crack growth and its effect on global structural response.

Longitudinal Plane (2-1 Specimens)—This model chooses the work of structural change (or damage work), which is the work per unit initial volume required to change the material's microstructure, as an internal state variable S. The engineering constants are generalized to be a function of this internal state variable S. Since the composite exhibits little nonlinearity in the fiber direction, the Young's modulus E_{11} and Poisson's ratio v_{12} are assumed to be independent of S (this is reasonable as long as the fibers do not break). Therefore, only the Young's modulus E_{22} and shear modulus G_{12} vary with the internal state variable S. The detailed derivation of this model applied to Arcan specimens is given in Ref *18*.

The characterization of material behavior is made by expressing material compliances in terms of damage work. The following equations are used to characterize the nonlinear behavior in the longitudinal plane of unidirectional fiber reinforced composites:

$$\begin{cases} S_{22} = \dfrac{\varepsilon_{11} - S_{12}\sigma_{11}}{\sigma_{22}} = \dfrac{\varepsilon_{22} + v_{12}^2\sigma_{22}/E_{11}}{\sigma_{22}} \\[2ex] S_{66} = \dfrac{\gamma_{12}}{\sigma_{12}} \\[2ex] S = \displaystyle\int_0^{\varepsilon_2} \sigma_{22}d\varepsilon_{22} + \int_0^{\gamma} \sigma_{21}d\gamma_{21} - \dfrac{1}{2}\left(\sigma_{22}\varepsilon_{11} + \sigma_{21}\gamma_{21}\right. \end{cases} \tag{12}$$

The nonlinear shear and normal responses can be made using

$$\frac{1}{2}\left[\sigma_{22}^2\frac{dS_{22}(S_r)}{dS_r} + \sigma_{12}^2\frac{dS_{66}(S_r)}{dS_r}\right] - 3S_r^2 = 0 \tag{13}$$

where $S_r = S^{1/3}$.

Transverse Plane (3-2 Specimens)—The nonlinear behavior of 3-2 specimens can be characterized using

$$\begin{cases} S_{33} = \dfrac{\varepsilon_{33}}{\sigma_{33} - v_{23}\sigma_{22}} = \dfrac{\varepsilon_{33}}{(1 - \kappa v_{23})\sigma_{33}} \\[2ex] S_{44} = \dfrac{\gamma_{23}}{\sigma_{23}} \\[2ex] S = \displaystyle\int_0^{\varepsilon_3} \sigma_{33}d\varepsilon_{33} + \int_0^{\gamma} \sigma_{32}d\gamma_{32} - \dfrac{1}{2}(\sigma_{33}\varepsilon_{33} + \sigma_{32}\gamma_{32}) \end{cases} \tag{14}$$

The nonlinear shear and normal responses can be made using

$$\frac{1}{2}\left[(1 + \kappa^2 - 2\kappa\upsilon_{23})\sigma_{33}^2 \frac{dS_{33}(S_r)}{dS_r} + \sigma_{23}^2 \frac{dS_{44}(S_r)}{dS_r}\right] - 3S_r^2 = 0 \qquad (15)$$

Results and Discussion

The results of applying the developed models to characterize the nonlinear behavior of the AS4/PEEK composite are given here. The comparisons of the predictions and the measurements are also shown and discussed.

One-Parameter Plasticity Model

The characterization of the nonlinear stress-strain curves of the AS4/PEEK composite in the longitudinal and transverse planes according to the one-parameter plasticity model is presented. The derivation of this model and its application to Arcan specimens with 2-1 and 3-2 fiber orientations and the procedures of making predictions of nonlinear behavior of the composite based on this model were presented in previous section. The procedures and results of applying this one-parameter plasticity model to characterize the nonlinear stress-strain curves obtained from the AS4/PEEK composite are now given here.

Longitudinal Plane (2-1 Specimens)—The effective stress σ_e and effective plastic strain ε_e^p were determined for each loading case from the measured average stress-strain curves using Eq 3. Since both the effective stress and effective plastic strain depend on the parameter Γ_{21}, the next step that was taken was to find a value of Γ_{21} that collapses all effective stress-strain curves obtained from experiments into one master curve. Once the master curve is determined, the process can be reversed and predictions of the nonlinear behavior of the composite can be made based on the master curve.

After several trails, it was concluded that a value of Γ_{21} which collapses all the experimental curves into one master curve could not be formed. For illustrative purposes (Fig. 17a, a value of 1.6 was assigned to Γ_{21}. This was the average of the values (1.5 and 1.7) that were used for tensile and compressive data obtained with AS4/PEEK by Sun et al. [7,8]. It is very clear that $\Gamma_{21} = 1.6$ did not collapse the data into one effective stress strain curve. When the value of Γ_{21} was increased, the stress-strain curves for shear plus compression became closer, however the curves for shear plus tension spread further apart. The opposite trend was encountered when the value of Γ_{21} was decreased.

Predictions of nonlinear stress-strain curves of 2-1 specimens were made (Eq 4) based on the master curve. The comparisons of the predictions and experimental results are shown in Figs. 17b and c. From the shear stress-strain curves in Fig. 17b, it was observed that this model performs well in the linear range but poorly in the nonlinear range. The composite under shear loading exhibited the stiffest shear response and became softer as the loading angle increased. This tendency is consistent with the experimental results only for the shear plus tension cases, but not for the shear plus compression cases. Furthermore, the predicted shear stress-strain curves are the same for loading in the directions of $\theta = -\theta$ ($\theta = 0°\sim 60°$). This feature is also inconsistent with the experimental measurements which show that the compressive stress increases shear strength and the tensile stress decreases shear strength. The predicted normal stress-strain curves are shown in the Fig. 17c. The predictions differ significantly from the measurements.

Transverse Plane (3-2 Specimens)—When the general form of the one-parameter plasticity model was reduced to the transverse (isotropic) plane, the dependence of the yield function on Γ_{32} was lost. Therefore, the yield function, effective stress and effective plastic strain in Eq 5 contain no parameter at all. The effective stress-strain curves obtained from experiments

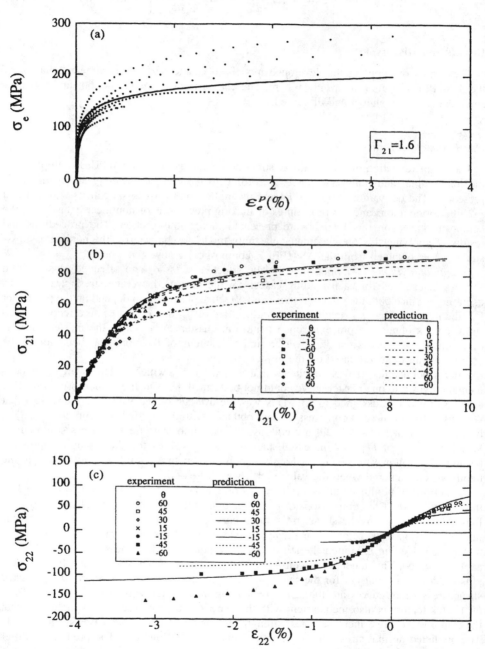

FIG. 17—*One-parameter plasticity model for 2-1 specimens.* (a) *The effective stress-strain curves and master curve,* (b) *the predicted shear response, and* (c) *normal response.*

are shown in Fig. 18a. A master curve was determined by a best curve-fit. It was noticed that, even for the 3-2 specimen, which lies on the isotropic plane, the effective stress-strain data points obtained from different experiments did not collapse onto one curve.

Predictions of nonlinear stress-strain curves in the transverse plane were made (Eq 6) based on the obtained master curve. The comparisons of the predictions and experimental results are shown in Figs. 18b and c. Since the 3-2 specimen is much weaker than the 2-1 specimen, the nonlinear range of the shear stress-strain curves are quite short for most of the experiments conducted. Only the specimens loaded in the −45° and −60° directions exhibited a substantial range of nonlinear behavior. As seen in Figs. 18b and c, the predicted curves roughly followed the experimental curves. Therefore, the performance of the model for predicting the behavior of the 3-2 specimens was considered to be better than it was for the 2-1 specimens. Even so, the predicted shear response for the loading in the −60° direction differed considerably from the measurements. The predictions of the shear response for 3-2 specimens under shear loading was the stiffest and the shear behavior became softer as the loading angle increased. Again, the shear response did not differ between the shear plus tension and shear plus compression. This was inconsistent with the experimental results. The predicted normal stress-strain curves are shown in Fig. 18c. Although the model did not predict the normal stress-strain curves perfectly, the predictions for the 3-2 specimen were much better than those of the 2-1 specimens.

Modified One-Parameter Plasticity Model

It was seen in the experiments that the normal stress has a strong influence on shear yield strength, such that the shear stress-strain curves are different for specimens loaded in the θ and $-\theta$ directions. However, the one-parameter plasticity model makes predictions of shear responses without any difference between specimens loaded in the θ and $-\theta$ directions. Motivated by this fact, a modification of the model to incorporate the effect of normal stress on shear yield strength into the model is needed. Furthermore, it has been suggested that for many polymers, the shear yield strength increases when they are subjected to a state of hydrostatic pressure [26,27] which leads to a modification of yield function for polymers. A similar modification of the yield function used for the composite was then made and is given in Eq 7. The derivation of this modified model and the procedures of making predictions for the nonlinear behavior based on this model for the 2-1 and 3-2 specimens were already given in the previous section. The results of applying the modified model are now presented.

Longitudinal Plane (2-1 Specimens)—The shear yield strength (τ_{21}^y) versus normal stress (σ_{22}) for the 2-1 specimen was shown in Fig. 7a. The influence of normal stress on shear yield strength was expressed in the form $\tau = \tau_0 + \mu p$. For the stress states applied to 2-1 specimens, the tensile stress had a strong influence on the shear strength, although no influence was noticed for the compressive stresses. We now define k_{21} as the slope of τ_{21}^y versus σ_{22} curve. Slopes were determined separately for the portion of shear plus tension and the portion of shear plus compression. Two constants k_{21}^t and k_{21}^c are used to represent the slopes of tensile and compressive sides of the curve, respectively. They were found to be −0.3 and 0, respectively. The values of constant μ are found to be 0.9 and 0 for μ_{21}^t and μ_{21}^c, respectively, where the subscript and superscript associated with constant μ are added to indicate the specimen type and the sign of the normal stress.

Since the effect of tensile stress on shear yield strength is different from that of compressive stress on shear yield strength, the yield function will be different for each case. Thus, as shown by Sun, two master (effective stress-strain) curves are needed in order to characterize the nonlinear composite behavior. One is for specimens under shear and tension, the other one is for specimens under shear and compression.

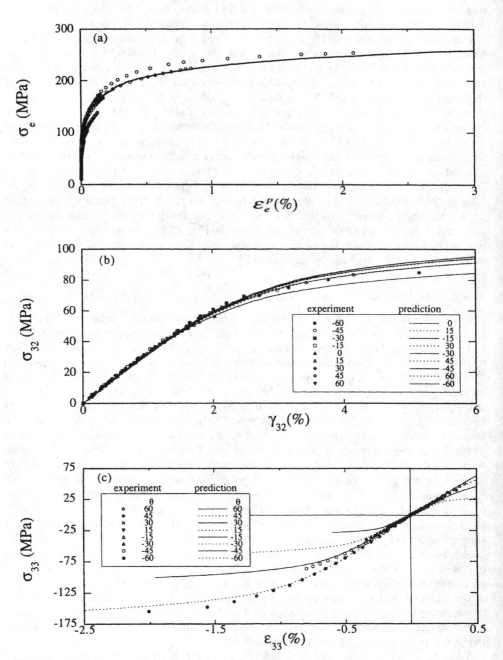

FIG. 18—*One-parameter plasticity model for 3-2 specimens.* (a) *The effective stress-strain curves and master curve,* (b) *the predicted shear response, and* (c) *normal response.*

For the 2-1 specimens, the introduction of the mean stress effect results in a modified plasticity model which contains the parameter $\hat{\Gamma}_{21}$ and a constant $\hat{\alpha}_{21}$. Equation 8 shows the forms of modified yield function, effective stress and strain that were used to determine the master curves. In the case of shear plus tension, when 1.6 was chosen for the parameter $\hat{\Gamma}_{21}$, the experimental data points all collapsed into one curve (Fig. 19a). A master curve was then determined by fitting through the data points and is shown in Fig. 19a. For the experiments conducted under shear plus compression, 5.0 was found to be adequate for the parameter $\hat{\Gamma}_{21}$. The experimental data points again collapsed into one curve, so that the master curve was complete (Fig. 19a). For each case, the modified one-parameter plasticity model was able to collapse all the experimental data points into one curve which was not possible (Fig. 17a) with the original one-parameter plasticity model. Therefore, the inclusion of the mean stress effect into this plasticity model does improve its performance.

Once master curves were determined, Eqs 9 were used to predict the nonlinear behavior of 2-1 specimens under various biaxial loadings. The predicted shear and normal stress-strain curves for specimens with a 2-1 fiber orientation under various biaxial loadings based on the master curves shown in Fig. 19a are now presented in Figs. 19b and c. The predictions for the shear response match with the experimental results reasonably well for each loading case (Fig. 19b). The predictions of the corresponding normal stress-strain curves are shown in Fig. 19c. For normal stress-strain curves, the predictions were very close to the experimental results. Therefore, it can be concluded that the incorporation of the mean stress effect into the one-parameter plasticity model improves performance of the model significantly for this material. The nonlinear behavior of the AS4/PEEK composite in the longitudinal plane is well characterized by the modified model.

Transverse Plane (3-2 Specimens)—The shear yield strength (τ_{32}^y) versus normal stress (σ_{33}) for the 3-2 specimen was shown in Fig. 12a. The tensile stresses had a strong influence on the shear yield strength under biaxial loading, just as had been observed from the 2-1 specimens. However, little influence was observed in the case of compressive stresses. As before, we define k_{32} as the slope of τ_{32}^y versus σ_{33} curve and slopes were determined separately for tensile and compressive stresses. The symbols k_{32}^t and k_{32}^c are used to represent the two slopes where the superscript is used to indicate the sign of the normal stress. The values of k_{32}^t and k_{32}^c were found to be -0.11 and 0, respectively. The values of constant μ were then 0.25 and 0 for μ_{32}^t and μ_{32}^c, respectively. Two master curves were determined due to the different effects between tensile and compressive stresses on shear yield strength.

When considering the application of the modified model to 3-2 specimens, there is only one fitting parameter $\tilde{\Gamma}_{32}$ and a constant $\tilde{\mu}_{32}$ along with $\tilde{\alpha}_{32}$ which was obtained from the τ_{32}^y versus σ_{33} curve (Fig. 12a). This mirrors the development that was followed for the 2-1 specimens. Recall that this was not the case for the original one parameter plasticity model where there was no Γ_{32} term. Equation 10 shows the modified yield function and the effective stress and strain expressions that were used to determine the master curves. In the case of shear plus tension, a value of 1.75 was found to be adequate for $\tilde{\Gamma}_{32}$. All experimental curves collapsed into one curve when this value was used. A master curve was then determined by fitting through all experimental data points which is shown in Fig. 20a. For the experiments conducted under shear plus compression, $\tilde{\Gamma}_{32} = 2.7$ was the best choice. Figure 20b shows the collapsed experimental data points and the master curve. For each case, the use of parameter $\tilde{\Gamma}_{32}$ collapses all experimental data points into one curve, a condition which the original one-parameter plasticity model (Fig. 18a) did not achieve. Therefore, the inclusion of the mean stress effect into this plasticity model provides a significant improvement over the original one-parameter plasticity model for specimens with a 3-2 fiber orientation.

Once the master curve was determined, Eqs 11 were used to predict the nonlinear behavior of 3-2 specimens. The predictions of shear and normal stress-strain curves for specimens

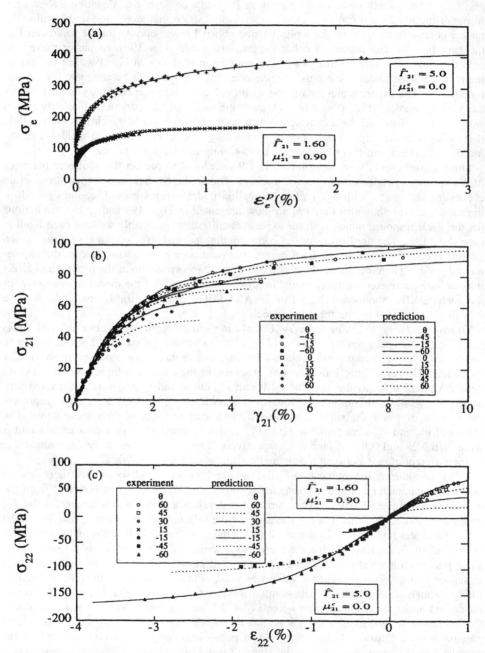

FIG. 19—*Modified one-parameter plasticity model for 2-1 specimens. (a) The effective stress-strain curves and master curves, (b) the predicted shear response, and (c) normal response.*

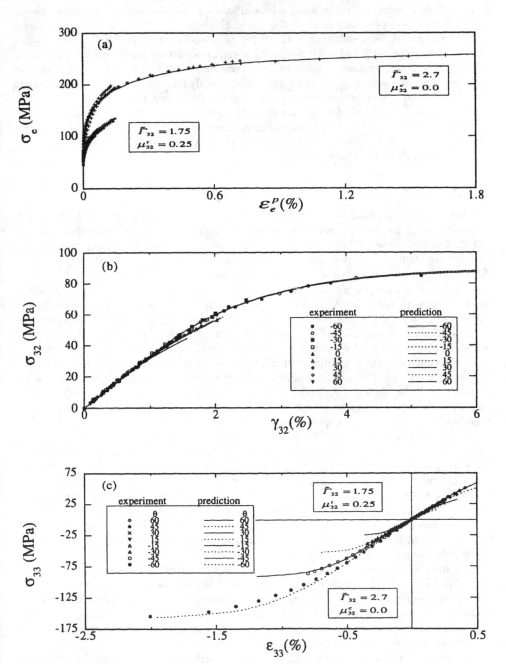

FIG. 20—*Modified one-parameter plasticity model for 3-2 specimens.* (a) *The effective stress-strain curves and master curves,* (b) *the predicted shear response, and* (c) *normal response.*

with a 3-2 fiber orientation under various biaxial loadings based on the master curves shown in Fig. 20a are now presented in Figs. 20b and c. For shear response, the predictions match experimental results very well [Fig. 4.41(a)] for each loading case, especially, for the specimens loaded in the $-45°$ and $-60°$ directions which exhibited the most significant nonlinear behavior. The predicted normal stress-strain curves are shown in Fig. 20c. The predictions and measurements were in very close agreement for all loading except for the specimen loaded in the $-60°$ direction. However, even there the difference was small.

It was again shown that the incorporation of hydrostatic pressure effect into the one-parameter plasticity model provides significant improvement of the model. The nonlinear behavior in both the longitudinal and transverse planes of the AS4/PEEK composite were well characterized by the modified model.

Work Potential Model

In this section, the characterization of the nonlinear stress-strain curves of the AS4/PEEK composite in the longitudinal and transverse planes using the work potential model is presented. The derivation of this model and its application to Arcan specimens with 2-1 and 3-2 fiber orientations were presented in previous section. The results of using this model to characterize the nonlinear composite behavior are now shown in the following sections.

Longitudinal Plane (2-1 Specimens)—As shown previously, the work potential model chooses the work for structural change S ($S_r = S^{1/3}$) as an internal state variable which represents the changes in microstructure. Then the engineering constants of the composite are generalized to be functions of S_r. For Arcan specimens with a 2-1 fiber orientation, only four compliances are needed since it is a plane stress case. They are S_{11}, S_{12}, S_{22}, and S_{66}. Since the composite exhibits little nonlinearity in the fiber direction, the Young's modulus E_{11} and Poisson's ratio v_{12} are assumed to be independent of S_r, which is reasonable if the fibers do not break, therefore S_{11} and S_{12} are independent of S_r. As a result, only S_{22} and S_{66} vary with the internal state variable S_r which means that the nonlinearity exhibited by unidirectional fiber reinforced composites results from changes in S_{22} and S_{66}.

For any off-axis loading ($\theta \neq 0$) of an Arcan specimen, a pair of shear (σ_{21}, γ_{21}) and normal (σ_{22}, ε_{22}) stress-strain curves were obtained. At each strain level, the secant compliances S_{22} and S_{66} and the corresponding value of S_r are determined by Eq 12. Figure 21a shows experimental data points of S_{22} and S_{66} versus S_r obtained from all experiments conducted. For the shear compliance S_{66}, it is obvious that the experimental data points did not collapse onto one curve, especially for shear plus compression. For the compliance S_{22}, the spread of data points was not as severe as was the case with S_{66}. Fourth-order polynomials in S_r were obtained for S_{22} and S_{66} by curve fitting through the data points. The coefficients of the polynomials are listed in Table 1.

After the master curves, $S_{66}(S_r)$ and $S_{22}(S_r)$, were determined, stress states were then prescribed in such a way that they represent the biaxial stress state in the Arcan specimen for each of the loading cases. For each single stress state under a single loading case, the values of stresses σ_{22} and σ_{12} were substituted in Eq 13 to solve for the corresponding S_r. Once S_r was determined, the corresponding compliances were obtained through the master curves. The strains for that stress state were then determined.

The predictions had stiffer responses than their experimental counterparts for all loading cases except $-60°$. Furthermore, the results from the predictions were the same for specimens loaded in the $+\theta$ and $-\theta$ directions, which is inconsistent with the experimental results. The predicted compressive stress-strain curves are in acceptable agreement with experimental results, although the degree of agreement becomes worse for shear plus tension. Overall, the

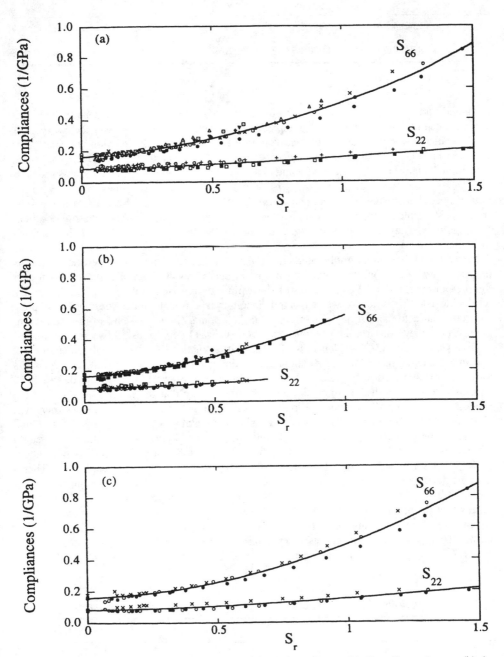

FIG. 21—*Compliances versus corresponding S_r for 2-1 specimens.* (a) *For all experiments,* (b) *for shear plus tension, and* (c) *for shear plus compression.*

TABLE 1—*Coefficients of polynomials* $S_{66}(S_r)$ *and* $S_{22}(S_r)$ *for 2-1 specimens obtained from shear plus tension and shear plus compression.*

$S_{66}(S_r)$	m_0	m_1	m_2	m_3	m_4
	1.5294E-4	2.9752E-4	−2.6037E-4	4.8294E-4	−1.5524E-4
$S_{22}(S_r)$	n_0	n_1	n_2	n_3	n_4
	8.7662E-5	1.8080E-5	6.5814E-5	−7.5919E-5	4.6325E-5

predicted normal stress-strain curves are better than the shear stress-strain curves. This fact was revealed in Fig. 21*a* by the small spread of experimental data points in the S_{22} section.

Motivated by the differences in shear response between shear plus tension and shear plus compression that was observed in the experiments, two sets of master curves were then determined. One set of master $S_{66}(S_r)$ and $S_{22}(S_r)$ curves was obtained through the shear plus tension experimental data points and was only used to predict the behavior of specimens under shear plus tension. They are shown in Fig. 21*b*. The other set of curves were obtained from the shear plus compression data and were used to predict material behavior under those loadings only. They are shown in Fig. 21*c*. The coefficients of these polynomials are listed in Tables 2 and 3, respectively.

It was found that, although the predictions did not perfectly match the experimental results under shear plus tension, using two sets of master curves resulted in a big improvement over the previous predictions. For the cases of shear plus compression, only a small improvement was achieved. This outcome was expected since the experiments under shear plus compression dominated the spreading of experimental data points (Figs. 21*b* and *c*). Overall, compared to the previous case, an obvious improvement was made by using two master curves.

Transverse Plane (3-2 Specimens)—The derivation of the work potential model for 3-2 specimens and the procedure for its application were given in previous section. The Poisson's ratio v_{23} was assumed to be independent of S_r. In order to justify this assumption, a tensile specimen was cut from the composite block in such a way that the longitudinal axis of the specimen was aligned with the 2-direction and the transverse axis of the specimen was aligned with the 3-direction. The measured transverse strain was plotted against longitudinal strain and the slope ($v_{23} = 0.48$) was constant [*18*] all the way to failure. Since the strength of this (3-2) tensile specimen was so low, the maximum longitudinal strain was only about 0.7%. However, this was enough to cover all the biaxial loading cases except −60°. Therefore, neglecting the dependence of v_{23} on S_r is acceptable here. Thus, for Arcan specimens with a 3-2 fiber orientation, only S_{33} and S_{44} vary with the internal state variable S_r which means that the nonlinearity exhibited by 3-2 specimens is mainly attributable to S_{33} and S_{44}.

For any off-axis loading ($\theta \neq 0$) of an Arcan specimen, a pair of shear (σ_{32}, γ_{32}) and normal (σ_{33}, ε_{33}) stress-strain curves were obtained. At each strain level, the secant compliances S_{33} and S_{44} and the corresponding value of S_r is determined through Eq 14. Figure

TABLE 2—*Coefficients of polynomials* $S_{66}(S_r)$ *and* $S_{22}(S_r)$ *for 2-1 specimens under shear plus tension.*

$S_{66}(S_r)$	m_0	m_1	m_2	m_3	m_4
	1.7212E-4	1.3265E-4	2.5079E-4	7.7265E-5	0.0
$S_{22}(S_r)$	n_0	n_1	n_2	n_3	n_4
	9.2800E-5	−5.5128E-6	1.5111E-4	−6.8439E-5	0.0

TABLE 3—*Coefficients of polynomials $S_{66}(S_r)$ and $S_{22}(S_r)$ for 2-1 specimens under shear plus compression.*

$S_{66}(S_r)$	m_0	m_1	m_2	m_3	m_4
	1.5256E-4	1.9658E-4	−6.6980E-5	3.8370E-4	−1.4665E-4
$S_{22}(S_r)$	n_0	n_1	n_2	n_3	n_4
	8.3212E-5	−3.0116E-5	1.5405E-5	−1.0363E-4	3.9404E-5

$22a$ shows the data points of S_{33} and S_{44} versus S_r that were obtained from all the experiments that were conducted. By fitting through the data points, two polynomials, $S_{44}(S_r)$ and $S_{33}(S_r)$, were obtained (Fig. 22a). The coefficients of the polynomials are listed in Table 4. Once master $S_{44}(S_r)$ and $S_{33}(S_r)$ curves were determined, the stress state was prescribed and the corresponding S_r was determined by solving Eq 15. Once S_r was determined, the corresponding compliances and strains for the prescribed stress state were then obtained.

For 3-2 specimens, only the specimens loaded in the −45° and −60° directions exhibited extensive nonlinear behavior. For these two cases, the predictions matched the experimental results quite well. Again, no differences were observed in the predictions between specimens loaded in the +θ and −θ directions. The corresponding normal stress-strain curves did not match with experimental results well.

Similar to the procedure applied in the previous section (2-1 specimens), two sets of master curves were determined. The set of master $S_{44}(S_r)$ and $S_{33}(S_r)$ curves for shear plus tension loading cases are shown in Fig. 22b. Figure 22c shows the other set of curves which were used for shear plus compression. The coefficients of these polynomials are listed in Tables 5 and 6, respectively. The predicted shear and normal responses matched the experimental results quite well. A big improvement was achieved by using two sets of master curves when they were compared to the predictions made by one set of master curves.

To summarize, for using this model, some degree of scatter of experimental data points was noticed in the plots of compliance versus S_r for both specimens. The degree of scattering was more severe in the 2-1 specimen that the 3-2 specimen. The model failed to predict the differences in shear responses of specimens loaded in +θ and −θ directions. This was inconsistent with the experimental results. Motivated by that, two sets of master curves were then used for each specimen. One for the loadings under shear plus tension, the other for the loadings under shear plus compression. Using this procedure, the predicted shear and normal stress-strain curves were in close agreement with the experimental results. Overall, it can be concluded that, for the AS4/PEEK composite, the work potential model works well in the transverse plane, however, it did not characterize the nonlinear behavior in the longitudinal plane very well. However, given the success that including the mean stress effect had in the one-parameter plasticity model, it seems likely that this could be also included in the work potential model approach. In this case, the mean stress effect could be included as a second internal state variable.

Validation

In the previous sections, the nonlinear behavior of AS4/PEEK has been characterized by the one-parameter plasticity model and work potential model by making use of Arcan specimens. However, in Refs *4–8*, the one-parameter plasticity model was used to characterize composite behavior using off-axis specimens. In Refs *12–15* (work potential model), both the off-axis and angle-ply specimens were used. Motivated by their work, 15° off-axis spec-

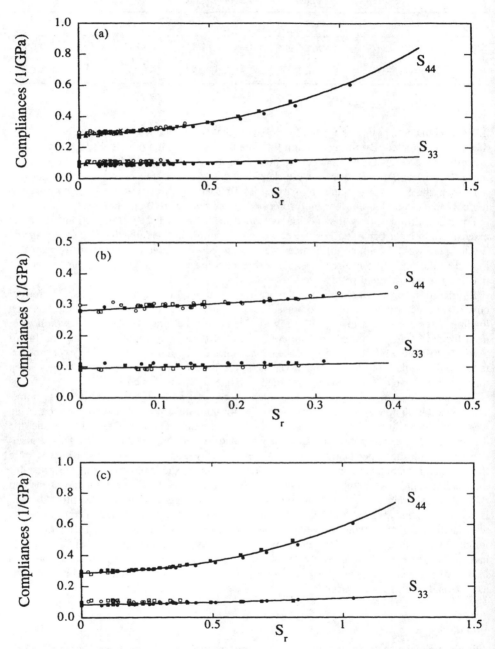

FIG. 22—*Compliances versus corresponding S_r for 3-2 specimens.* (a) *For all experiments,* (b) *for shear plus tension, and* (c) *for shear plus compression.*

TABLE 4—*Coefficients of polynomials $S_{44}(S_r)$ and $S_{33}(S_r)$ for 3-2 specimens obtained from shear plus tension and shear plus compression.*

$S_{44}(S_r)$	m_0	m_1	m_2	m_3	m_4
	2.8186E-4	7.4367E-5	1.3917E-4	9.2217E-5	9.0965E-6
$S_{33}(S_r)$	n_0	n_1	n_2	n_3	n_4
	9.7315E-5	7.2004E-6	2.1726E-5	0.0	0.0

imens were conducted and the experimental results were used to validate the developed models.

Experimental Results—The 15° off-axis specimens were cut from the same AS4/PEEK composite plate that was used for Arcan specimens. These specimens were 203.2 mm long and 12.7 mm wide with an aspect ratio of 16 to ensure a uniform strain field. The thickness of the specimens was 2.54 mm. A special angle tab [28,29] with a 27° tab angle and a 30° taper angle perpendicular to the tab angle was used to minimize nonuniformities in the strain distributions. The strains were measured by two three-arm strain gage rosettes (45°/0°/−45°) bonded to the specimen center on both the front and back faces of the specimen. The strain gages were mounted in such a way that the three arms were 45°, 90° and 135° clockwise from the lengthwise ($+x$) axis. The strains measured from both sides were then averaged to remove any bending effects.

One-Parameter Plasticity Model—The original one-parameter plasticity model was unable to bring the effective stress-strain curves together from all biaxial Arcan tests. However, a master curve based on $\Gamma_{21} = 1.6$ was obtained as a best fit to all the data. In Fig. 23a, the effective stress-strain curve for the off-axis specimen with $\Gamma_{21} = 1.6$ was found to be very close to the master curve for Arcan specimens (originally shown in Fig. 17a). The predicted shear and normal stress-strain curves were also very close to their experimental counterparts (Fig. 23b). Since the master curve for Arcan specimens was obtained through curve-fitting of all Arcan tests under different biaxialities (includes shear + compression), it went to higher strains than the one from the tensile off-axis specimen. It can be concluded that, although this one-parameter plasticity model is unable to collapse all effective stress-strain curves from Arcan specimens into one master curve, the curve based on $\Gamma_{21} = 1.6$ still predicts the nonlinear behavior of the off-axis specimen well.

After including the mean stress effect, the model contained one parameter $\hat{\Gamma}_{21}$ and a constant $\hat{\alpha}_{21}$. This model was then able to collapse all the effective stress-strain curves (either on the shear + tension or shear + compression side) into one master curve. The effective stress-strain curve from the off-axis experiment was obtained using 1.6 and 0.9 for $\hat{\Gamma}_{21}$ and u'_{21}, respectively, and is shown in Fig. 24a. In the same figure, the master curve for the Arcan specimens under shear plus tension (Fig. 19a) is also shown. These two curves were very close. The prediction of the nonlinear response of the 15° off-axis specimen based on the

TABLE 5—*Coefficients of polynomials $S_{44}(S_r)$ and $S_{33}(S_r)$ for 3-2 specimens under shear plus tension.*

$S_{44}(S_r)$	m_0	m_1	m_2	m_3	m_4
	2.8145E-4	1.2199E-4	5.1041E-5	6.0336E-7	0.0
$S_{33}(S_r)$	n_0	n_1	n_2	n_3	n_4
	9.5079E-5	4.4621E-5	0.0	0.0	0.0

TABLE 6—*Coefficients of polynomials $S_{44}(S_r)$ and $S_{33}(S_r)$ for 3-2 specimens under shear plus compression.*

$S_{44}(S_r)$	m_0	m_1	m_2	m_3	m_4
	2.8810E-4	4.3118E-5	1.6304E-4	9.6039E-5	0.0
$S_{33}(S_r)$	n_0	n_1	n_2	n_3	n_4
	8.4658E-5	1.7272E-5	2.2797E-5	0.0	0.0

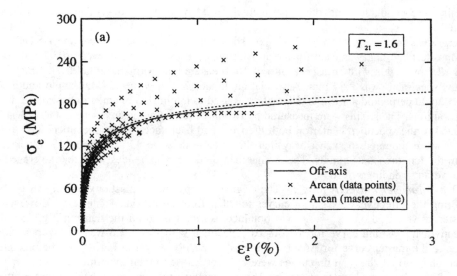

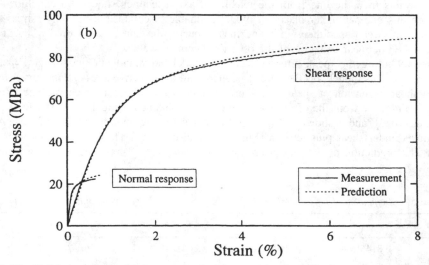

FIG. 23—*Validation of the one-parameter plasticity model using off-axis specimens.* (a) *Effective stress-strain curves, and* (b) *the measured and predicted material responses.*

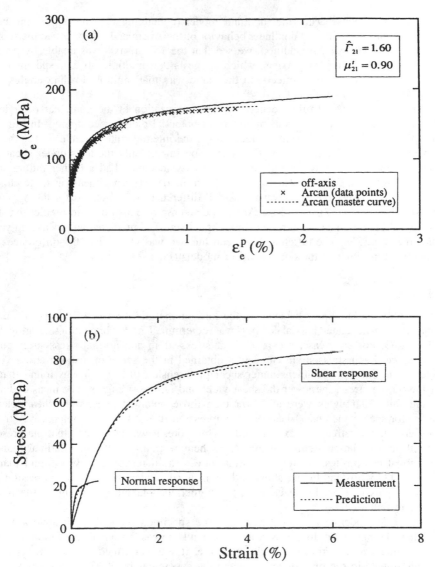

FIG. 24—*Validation of the modified one-parameter plasticity model using off-axis specimens.* (a) *Effective stress-strain curves, and* (b) *the measured and predicted material responses.*

master curve for Arcan specimens is shown in Fig. 24*b*. Good agreement between the measurements and the predictions was obtained. For this case, the master curve for the Arcan specimens is shorter because it was determined solely from Arcan specimens under shear plus tension.

To summarize, the master curve based on the one parameter model and obtained from a best fit of all the data from the Arcan specimens was able to predict the response of a 15° off-axis specimen. However, incorporating the mean stress effect collapsed all the biaxial data from the Arcan specimens onto one curve. This curve also provided excellent predictions of 15° off-axis response. Thus the modified model (containing $\hat{\Gamma}_{21}$ and $\hat{\alpha}_{21}$) is the better approach for characterizing the nonlinear behavior of the Arcan and off-axis specimens made by AS4/PEEK composite. In addition, we see that the 15° off-axis was unable by itself to discriminate between the two models, which suggests that multiple off-axis specimens are required to detect mean stress effects, in the same way that multiple loading angles were required for the Arcan specimens.

Work Potential Model—In this section, the characterization of the nonlinear behavior in the Arcan and off-axis specimens using the work potential model is compared. In Fig. 25a, the compliances, S_{66} and S_{22}, were plotted against the internal state variable S_r, for the off-axis specimen and Arcan specimens. For Arcan specimens, only the tests under shear plus tension are shown. It was found that the 15° off-axis specimen had a higher failure strain than had the Arcan specimens. The S_{22} curve from the off-axis specimen was quite close to those of the Arcan specimens. However, a small difference was observed for the S_{66} curve. The master curves (Fig. 21b) used for Arcan specimens were also used to predict the shear and normal stress-strain curves for the off-axis specimen. The predicted and measured values are shown in Fig. 25b. The difference between the two was larger than the differences that were observed in either of the one-parameter models (Figs. 23 and 24).

Conclusions

In this study, the nonlinear behavior in the longitudinal and transverse planes of a unidirectional composite under biaxial loading was determined under biaxialities ranging from -1.73 to 1.73. For each case, a pair of shear stress-strain and normal stress-strain curves were obtained. Each stress-strain curve was obtained in the sense of average stress versus average strain. For 2-1 specimens under shear plus tension ($\theta > 0$), it was found that the larger the tensile stress, the lower the shear yield and failure strengths. The measured shear yield and failure strengths were all lower than those obtained under only shear loading. However, for specimens under shear plus compression ($\theta < 0$), larger compressive stresses led to higher failure strengths (except $-60°$). Therefore, it was concluded that the presence of normal stresses has a significant effect on shear response in the longitudinal plane of unidirectional composites. As for the normal stress-strain response, it was found that the specimen could withstand higher stress and strain levels under shear plus compression than under shear plus tension. For each case, the higher the biaxiality, the higher the normal strength and failure strain.

Failure of 2-1 specimens was catastrophic. The specimen suddenly split apart along the fibers when the maximum load was reached. For all the cases studied, the failure occurred within the notch roots. From the mode of failure, it was confirmed that the cracks in 2-1 specimens initiate in the fiber direction where the specimens break. Since the cracks propagated in the y-direction, the normal stress σ_{xx} (σ_{22}) plays an important role in the initiation of cracks. The tensile stress σ_{xx} apparently promotes the initiation of cracks, thereby lowering the strength while compressive stress delays the initiation of cracks and therefore raises the strength. Furthermore, the existence of tensile stress not only affects the shear failure strength but also affects the shear yield strength as was mentioned previously.

An examination of fracture surfaces of each 2-1 specimen conducted here showed that hackle markings existed in all 2-1 specimens. The hackle markings are seen in the matrix between fibers and appear as a more or less regular stacking of platelets. The tilt angle of the hackles found in the specimen under shear loading is about 45° into the matrix which is the normal to the direction of maximum principal tensile stress. It was found that, under shear plus compression, the sharp edges of hackles were no longer visible. Instead, the

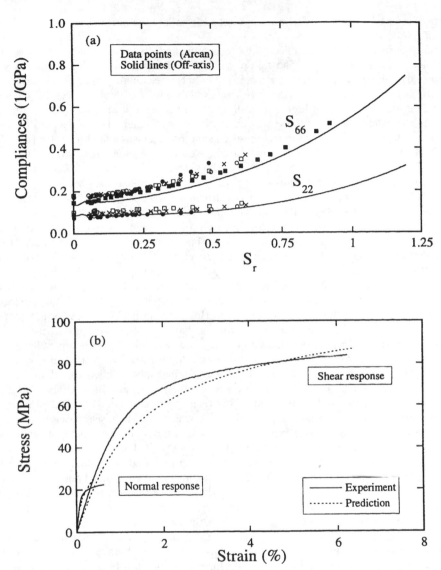

FIG. 25—*Validation of the work potential model using off-axis specimens.* (a) *Compliances versus internal state variable, and* (b) *the measured and predicted material responses.*

hackles appeared to be flattened. It was believed that this feature is due to the existence of compressive stress. Apparently, the sharp edges of hackles seen in the shear loading case were erased due to the friction between the two matching fracture surfaces. It was also noticed that the degree of hackle formation decreased with increasing compressive stress levels. Under shear plus tension, hackles were much more easily identified on the failure surfaces. As the biaxiality changed, the inclination of the principal tensile stress changed and the tilt angles of the hackles varied with load angle. Since tensile stresses tend to separate the crack faces, the hackles exist in a more intact form and the hackle size increased as the level of tensile stresses increased.

When 3-2 specimens were subjected to shear plus tension ($\theta > 0$), larger tensile stress values led to lower shear yield and failure strengths. The yield and failure strengths were all lower than those under pure shear loading. For specimens under shear plus compression ($\theta < 0$), larger compressive stresses gave rise to higher failure strengths. Again, the yield and failure strengths were all higher than those obtained under shear loading.

The normal stress-strain response was characterized by higher stresses and strains under shear plus compression than under shear plus tension. For each case, the higher the biaxiality, the larger the normal strength and failure strain. Again, the tensile stress promotes the initiation of cracks while the compressive stress delays the initiation of cracks. Therefore higher strength was found under shear plus compression than under shear plus tension.

The failure of 3-2 specimens was also catastrophic. Each specimen broke when it maximum load was reached. The failure of 3-2 specimens was due to cracks initiating in the notch roots and propagating into the specimen with an angle ϕ relative to the y-axis. The failure direction of each specimen was different. For all the cases considered here, it was found that the cleavage plane is normal to the direction of the maximum principal tensile stress. It was concluded that the failure of 3-2 specimens (transverse plane) is due to tension. An examination of the fracture surfaces of the 3-2 specimens showed that, in most cases, the fracture path did not exactly follow the fiber matrix interface. There was much evidence of matrix cracking and the surface were rough. However, no hackles were formed perpendicular to the fibers indicating an absence of σ_{32} as has been suggested previously.

Once the nonlinear behavior of the AS4/PEEK was determined, the one-parameter plasticity model and the work potential model were used to characterize the measured nonlinear behavior of the composite. It was found that the original model failed to collapse all the experimental data points into one master curve. Furthermore, the obtained shear response based on the model showed no differences between specimens loaded in the $+\theta$ (shear + tension) and $-\theta$ (shear + compression) directions, which was inconsistent with the experimental results. Motivated by this fact, a modification of the model was made in such a way that the effect of normal stress on shear yield strength was included.

It was found that the inclusion of mean stress effect provided significant improvement in the performance of the model. Based on the modified model, the predicted nonlinear behavior in the 2-1 and 3-2 specimens agreed well with the experimental results. Therefore, the nonlinear behavior of the AS4/PEEK composite in both the longitudinal and transverse planes was well characterized by the modified one-parameter plasticity model.

In using the work potential model with one ISV, some degree of scatter of experimental data points was noticed in the plots of compliance versus S_r for both specimens. The degree of scattering was more severe in the 2-1 specimen than the case of the 3-2 specimen. The model failed to predict the differences in shear response for specimens loaded in the $+\theta$ and $-\theta$ directions. As indicated above, this was inconsistent with the experimental results. Therefore, two sets of master curves were used for each specimen. One was for shear plus tension, while the other one was for shear plus compression. Using this procedure, the predicted shear and normal stress-strain curves were in better agreement with the experimental results. Overall, it can be concluded that, for the AS4/PEEK composite, work potential model works well for the 3-2 specimen, however, some more improvement is needed in order to characterize the nonlinear behavior in the longitudinal plane of the composite. Thus, the success of including the mean stress effect in the one-parameter plasticity model suggests that a second internal state variable is needed to account for the mean stress effect in the work potential model.

A validation of the various models using off-axis specimens was conducted. The 15° off-axis specimen had a slightly higher value of failure strength than the Arcan specimen loaded in the 15° direction. The difference might be due to the small normal stress concentration near the notch roots of the Arcan specimen. The use of angle tabs has reduced the nonuniformity in strains in the off-axis specimen and the occurrence of failure at the tabs, although it has yet to be shown that previously observed bending and twisting effects have been eliminated. On the other hand, the Arcan specimen is much simpler to use for shear plus compression tests and the small size allowed the three dimensional behavior to be examined. The procedure for determining lengthwise averaged strains can be simplified by using strain gages long enough to cover the gage section (from notch to notch).

The master curve based on the one-parameter model and obtained from a best fit of all the data from the Arcan specimens was able to predict the response of a 15° off-axis specimen. However, incorporating the mean stress effect collapsed all the biaxial data from the Arcan specimens onto one curve. This curve also provided excellent predictions of the 15° off-axis response. Thus the modified model is the better approach for characterizing the nonlinear behavior of the Arcan and off-axis specimens made of AS4/PEEK composite. The one-parameter work potential model performed reasonably well for both specimens.

Acknowledgments

The financial support provided by the Office of Naval Research under contract N00014-91-J-4091 is gratefully acknowledged. Discussions with R. A. Schapery on the work potential model were extremely helpful.

References

[1] Hahn, H. T. and Tsai, S. W., "Nonlinear Elastic Behavior of Unidirectional Composite Laminae," *Journal of Composite Materials,* Vol. 7, 1973, pp. 102–118.
[2] Hahn, H. T., "Nonlinear Behavior of Laminated Composites," *Journal of Composite Materials,* Vol. 7, 1973, pp. 257–271.
[3] Sun, C. T., Feng, W. H., and Koh, S. L., "A Theory for Physically Nonlinear Elastic Fiber-Reinforced Composites," *International Journal of Engineering Sciences,* Vol. 12, 1974, pp. 919–935.
[4] Kenaga, D., Doyle, J. F., and Sun, C. T., "The Characterization of Boron/Aluminum Composite in the Nonlinear Range as an Orthotropic Elastic-Plastic Material," *Journal of Composite Materials,* Vol. 21, 1987, pp. 516–531.
[5] Sun, C. T. and Chen, J. L., "A Simple Flow Rule for Characterizing Nonlinear Behavior of Fiber Composites," *Journal of Composite Materials,* Vol. 23, 1989, pp. 1009–1020.
[6] Yoon, K. J. and Sun, C. T., "Characterization of Elastic-Viscoplastic Properites of an AS4/PEEK Thermoplastic Composite," *Journal of Composite Materials,* Vol. 25, 1991, pp. 1277–1296.
[7] Sun, C. T. and Yoon, K. J., "Characterization of Elastic-Plastic Behavior of AS4/PEEK Thermoplastic Composite for Temperature Variation," *Journal of Composite Materials,* Vol. 25, 1991, pp. 1297–1313.
[8] Sun, C. T. and Rui, Y., "Orthotropic Elasto-Plastic Behavior of AS4/PEEK Thermoplastic Composite in Compression," *Mechanics of Materials,* Vol. 10, 1990, pp. 117–125.
[9] Dvorak, G. J. and Bahei-El-Din, Y. A., "Elastic-Plastic Behavior of Fibrous Composites," *Journal of the Mechanics and Physics of Solids,* Vol. 27, 1979, pp. 51–72.
[10] Dvorak, G. J. and Bahei-El-Din, Y. A., "Plasticity Analysis of Fibrous Composites," *Journal of Applied Mechanics,* Vol. 49, 1982, pp. 327–335.
[11] Hill, R., "A Theory of the Yielding and Plastic Flow of Anisotropic Metals," *Proceedings, Royal Society of London,* Vol. A193, 1948, pp. 281–297.

[12] Schapery, R. A., "Mechanical Characterization and Analysis of Inelastic Composite Laminates with Growing Damage," *Mechanics of Composite Materials and Structures,* J. N. Reddy and J. L. Tapley, Eds., ASME AMD-100, American Society of Mechanical Engineers, 1989, pp. 1–9.

[13] Schapery, R. A., "A Theory of Mechanical Behavior of Elastic Media with Growing Damage and Other Changes in Structure," *Journal of the Mechanics and Physics of Solids,* Vol. 38, 1990, pp. 215–253.

[14] Mignery, L. A. and Schapery, R. A., "Viscoelastic and Nonlinear Adherend Effects in Bonded Composite Joints," *Journal of Adhesion,* Vol. 34, 1991, pp. 17–40.

[15] Schapery, R. A. and Sicking, D. L., "On Nonlinear Constitutive Equations for Elastic and Viscoelastic Composites with Growing Damage," *Mechanical Behavior of Materials,* A. Bakker, Ed., Delft University Press, Delft, The Netherlands, 1995, pp. 45–76.

[16] Schapery, R. A., "Prediction of Compressive Strength and Kink Bands in Composites Using A Work Potential," *International Journal of Solids and Structures,* Vol. 32, 1995, pp. 739–765.

[17] Hung, S.-C. and Liechti, K. M., "Finite Element Analysis of the Arcan Specimen for Fiber Reinforced Composites under Shear and Biaxial Loading," EMRL #97-1, Research Center for Mechanics of Solids, Structures and Materials, The University of Texas at Austin, 1997.

[18] Hung, S.-C., "Nonlinear Multiaxial Behavior and Failure of Fiber Reinforced Composites," Dissertation, Department of Aerospace Engineering and Engineering Mechanics, The University of Texas at Austin, 1997, Engineering Mechanics Research Laboratory Report EMRL 97–22.

[19] Kmiec, K. J., "Determination of the Fracture Parameters Associated with Mixed Mode Displacement Fields and Applications of High Density Geometric Moiré," Masters thesis, Dept. Mechanical Engineering, Texas A&M University, 1994.

[20] Hung, S.-C. and Liechti, K. M., "An Evaluation of the Arcan Specimen for Determining the Shear Moduli of Fiber Reinforced Composites," *Experimental Mechanics,* Vol. 37, 1997, pp. 460–468.

[21] Bascom, W. D. and Gweon, S. Y., "Fractography and Failure Mechanisms of Carbon Fiber-Reinforced Composite Materials," *Fractography and Failure Mechanisms of Polymers and Composites,* Anne C. Roulin-Moloney, Ed., Elsevier Applied Science, 1989, pp. 351–382.

[22] Joannesson, T., Sjoblom, P., and Selden, R., "The Detailed Structure of Delamination Fracture Surfaces in Graphite-Epoxy Laminates," *Journal of Materials Science,* Vol. 19, 1984, pp. 1171–1177.

[23] Arcan, L., Arcan, M., and Daniel, I. M., "SEM Fractography of Pure and Mixed-Mode Interlaminar Fractures in Graphite/Epoxy Composites," *Fractography of Modern Engineering Materials: Composites and Metals, ASTM STP 948,* J. E. Masters and J. J. Au, Eds., American Society for Testing and Materials, 1987, pp. 41–67.

[24] Hibbs, M. F. and Bradley, W. L., "Correlations Between Micromechanical Failure Processes and the Delamination Toughness of Graphite/Epoxy Systems," *Fractography of Modern Engineering Materials: Composites and Metals, ASTM STP 948,* J. E. Masters and J. J. Au, Eds., American Society for Testing and Materials, 1987, pp. 68–97.

[25] Daniel, W. K., Maris, J. L., and van Siden, R. C., "Aluminum Metal Matrix Concepts for Missile Airframes," Final Report No. AFWAL-TR-84-3065, LTV Aerospace and Defense Company, Dallas, TX, 1984.

[26] Ward, I. M., *Mechanical Properties of Solid Polymers,* Wiley-Interscience, New York, 1971, pp. 313–314.

[27] Bowden, P. B., "The Yield Behavior of Glassy Polymers," *The Physics of Glassy Polymers,* R. N. Haward, Ed., Applied Science Publisher, 1973, pp. 279–337.

[28] Sun, C. T. and Chung, I., "An Oblique End-Tab Design for Testing Off-axis Composite Specimens," *Journal of Composites,* Vol. 24, 1993, pp. 619–623.

[29] Bocchieri, R. T., "A Baseline Nonlinear Material Characterization for Predicting the Long-Term Durability of Composite Structures," Masters thesis, Dept. of Aerospace Eng. and Eng. Mech., The University of Texas at Austin, 1997.

[30] Bowden, P. B. and Jukes, J. A., "The Plastic Flow of Isotropic Polymers," *Journal of Materials Science,* Vol. 7, 1972, pp. 52–63.

[31] Schofield, A. N. and Wroth, C. P., *Critical State Soil Mechanics,* McGraw-Hill, New York, 1968.

[32] Schapery, R. A., "Prediction of Compressive Strength and Kink Bands in Composites Using a Work Potential," *International Journal of Solids Structures,* Vol. 32, 1995, pp. 739–765.

H. M. Hsiao[1] and I. M. Daniel[2]

Nonlinear and Dynamic Compressive Behavior of Composites with Fiber Waviness

REFERENCE: Hsiao, H. M. and Daniel, I. M., **"Nonlinear and Dynamic Compressive Behavior of Composites with Fiber Waviness,"** *Time Dependent and Nonlinear Effects in Polymers and Composites, ASTM STP 1357,* R. A. Schapery and C. T. Sun, Eds., American Society for Testing and Materials, West Conshohocken, PA, 2000, pp. 223–237.

ABSTRACT: The nonlinear and dynamic behavior of unidirectional composites with fiber waviness under compressive loading was investigated theoretically and experimentally. Unidirectional carbon/epoxy composites with uniform fiber waviness were studied. Complementary strain energy was used to derive the material nonlinear stress-strain relations for the quasistatic case. Nonlinear material properties obtained from shear and longitudinal and transverse compression tests were incorporated into the analysis. An incremental analysis was used to predict the static and dynamic behavior of wavy composites using the basic strain rate characterization data. It is shown that under uniaxial compressive loading, strong nonlinearities occur in the stress-strain curves due to fiber waviness with significant stiffening as the strain rate increases. Stress-strain curves are affected less by fiber waviness under other loading conditions. The major Young's modulus degrades seriously as the fiber waviness increases. It increases moderately as the strain rate increases for the same degree of waviness. Unidirectional composites with uniform waviness across the thickness were prepared by a tape winding method. Compression tests of specimens with known fiber waviness were conducted. Experimental results were in good agreement with predictions based on the complementary strain energy approach and incremental analysis.

KEYWORDS: fiber waviness, nonlinear behavior, dynamic response, strain rate effects, complementary strain energy density, incremental analysis, compression testing of composites

Fiber waviness is a type of manufacturing defect occurring during processing. In the case of filament winding, fiber waviness may result from compressive strains developed in the hoop layers during consolidation. It occurs also in the manufacture of fiber tows, in the prepreg tape impregnation process, or in the subsequent layup and curing using an autoclave process. Fiber waviness has been shown to affect significantly the compressive behavior of composite materials [1–3]. A better understanding of this problem is necessary for more accurate prediction of material strength.

Nonlinear behavior of composites was first studied by Petit and Waddoups [4] using an incremental approach. They employed a piecewise linear representation of lamina stress-strain curves obtained from simple tests to determine the stress-strain response of a laminate. Hashin, Bagchi, and Rosen [5] applied a Ramberg-Osgood power law model to approximate the nonlinearities. Both methods require that the material be quasi-linear, i.e., without cou-

[1] Senior Materials Scientist, Research and Technology, Hexcel Composites, Dublin, CA 94568.
[2] Professor and Director, Center for Intelligent Processing of Composites, Northwestern University, Evanston, IL 60208.

pling among various stress components in the nonlinear range so that superposition of stresses is permissible, and thus, multiaxial nonlinear stress-strain behavior can be uncoupled into a series of uniaxial models. Hahn and Tsai [6] proposed a model based on the complementary strain energy density. It contains the generality necessary for the formulation of stress-strain relations which is capable of describing the nonlinearity inherent in composites. This method can be extended to more complex states of stress and higher degrees of nonlinearity. Luo and Chou [7] applied similar concepts to predict the behavior of flexible composites under finite deformations by considering both material and geometric nonlinearities.

To date, only a limited number of studies have been conducted on the prediction of quasi-static stress-strain behavior of wavy composites and none on the dynamic behavior. Chou et al. [8,9] and Rai et al. [10] developed an incremental loading scheme to predict the general stress-strain response for a curved fiber composite lamina. Bogetti et al. [2] also employed the incremental loading strategy to study the effect of layer waviness on composite laminates based on three-dimensional laminated media analysis. Hsiao and Daniel [11] used the complementary strain energy to derive constitutive relations for three different fiber waviness patterns.

This paper describes an investigation of the effect of fiber waviness on the stress-strain behavior of unidirectional composites under quasi-static and dynamic loadings. Complementary strain energy was used to derive the mathematical form of the constitutive relations under static loading. Incremental analysis was used to predict the stress-strain behavior under both static and dynamic loadings. Stress-strain curves were also obtained experimentally for composites with known fiber waviness and results were compared with theoretical predictions.

Analysis

Analytical Model

In order to quantify the effect of fiber waviness on unidirectional composites under compression, a mathematical description of the wave geometry is required. The fiber waviness is assumed to be planar sinusoidal in the xz plane defined as

$$v = A \sin \frac{2\pi x}{L} \tag{1}$$

where A and L are the amplitude and wavelength of the wavy fiber, respectively.

For the case of uniform fiber waviness, it is sufficient to consider a representative volume element encompassing one period of the waviness (Fig. 1). This volume is divided into infinitesimally thin slices dx, x being the axis of the sinusoidal wave and loading direction. Each slice is treated as an off-axis unidirectional lamina and its stress-strain relations are obtained from the transformation relations. The average stress-strain relations for the representative volume are then obtained by integrating the strains over one wavelength in the loading (x) direction.

Complementary Strain Energy Approach

Complementary strain energy was used to derive the constitutive relations of unidirectional composites with uniform fiber waviness under quasi-static loading. Given the complementary strain energy density, W^*, the stress-strain relations can be derived from

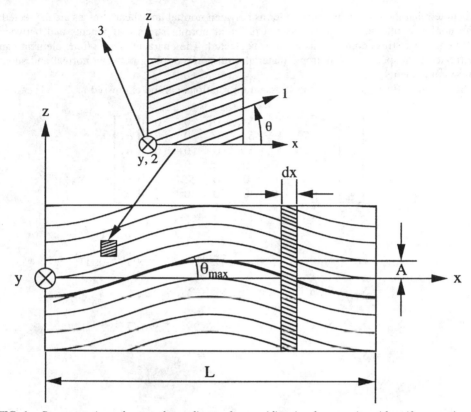

FIG. 1—*Representative volume and coordinates for a unidirectional composite with uniform waviness.*

$$\varepsilon_i = \partial W^* / \partial \sigma_i \tag{2}$$

This relation which has been used by Hahn and Tsai [6] is an approximation justified when strain rate effects are relatively weak, see Schapery [12].

If the material is orthotropic, the polynomial basis of the function W^* referred to the principal material axes is [6,13]

$$W^* = \{\sigma_1, \sigma_2, \sigma_3, \tau_4^2, \tau_5^2, \tau_6^2, \tau_4\tau_5\tau_6\} \tag{3}$$

In the linear elasticity theory W^* is taken to be quadratic in its arguments. However, higher order terms are needed when the nonlinear stress-strain behavior is considered. A polynomial expansion of W^* up to fourth-order terms is

$$W^* = \frac{1}{2} S_{11}\sigma_1^2 + \frac{1}{2} S_{22}\sigma_2^2 + \frac{1}{2} S_{33}\sigma_3^2 + \frac{1}{2} S_{44}\tau_4^2 + \frac{1}{2} S_{55}\tau_5^2 + \frac{1}{2} S_{66}\tau_6^2 + S_{12}\sigma_1\sigma_2$$

$$+ S_{13}\sigma_1\sigma_3 + S_{23}\sigma_2\sigma_3 + \frac{1}{3} S_{111}\sigma_1^3 + \frac{1}{3} S_{222}\sigma_2^3 + \frac{1}{3} S_{333}\sigma_3^3 + \frac{1}{4} S_{1111}\sigma_1^4 + \frac{1}{4} S_{2222}\sigma_2^4$$

$$+ \frac{1}{4} S_{3333}\sigma_3^4 + \frac{1}{4} S_{4444}\tau_4^4 + \frac{1}{4} S_{5555}\tau_5^4 + \frac{1}{4} S_{6666}\tau_6^4 \tag{4}$$

It is noted that the nonlinear coupling terms between normal and shear stresses are neglected. The nonlinear interaction terms between different normal stress components and between different shear stress components are also neglected. The characteristic volume element can be treated as a specially orthotropic material with zero coupling between normal and shear stress components.

Substituting Eq 4 into Eq 2, the stress-strain relations can be expressed as

$$
\begin{bmatrix} \varepsilon_1 \\ \varepsilon_2 \\ \varepsilon_3 \\ \gamma_4 \\ \gamma_5 \\ \gamma_6 \end{bmatrix} = \begin{bmatrix} S_{11}^* & S_{12}^* & S_{13}^* & 0 & 0 & 0 \\ S_{12}^* & S_{22}^* & S_{23}^* & 0 & 0 & 0 \\ S_{13}^* & S_{23}^* & S_{33}^* & 0 & 0 & 0 \\ 0 & 0 & 0 & S_{44}^* & 0 & 0 \\ 0 & 0 & 0 & 0 & S_{55}^* & 0 \\ 0 & 0 & 0 & 0 & 0 & S_{66}^* \end{bmatrix} \begin{bmatrix} \sigma_1 \\ \sigma_2 \\ \sigma_3 \\ \tau_4 \\ \tau_5 \\ \tau_6 \end{bmatrix} \tag{5}
$$

where

$$S_{11}^* = S_{11} + S_{111}\sigma_1 + S_{1111}\sigma_1^2 \qquad S_{22}^* = S_{22} + S_{222}\sigma_2 + S_{2222}\sigma_2^2$$

$$S_{33}^* = S_{33} + S_{333}\sigma_3 + S_{3333}\sigma_3^2 \qquad S_{44}^* = S_{44} + S_{4444}\tau_4^2$$

$$S_{55}^* = S_{55} + S_{5555}\tau_5^2 \qquad S_{66}^* = S_{66} + S_{6666}\tau_6^2$$

$$S_{12}^* = S_{12} \qquad S_{13}^* = S_{13}$$

$$S_{23}^* = S_{23}$$

For composites with out-of-plane fiber orientation, the stress-strain relations referred to an arbitrary xyz coordinate system can be obtained by using transformation relations [14]

$$
\begin{bmatrix} \varepsilon_x \\ \varepsilon_y \\ \varepsilon_z \\ \gamma_q \\ \gamma_r \\ \gamma_s \end{bmatrix} = [R_{ij}] \, [T_{ij}]^{-1} [R_{ij}]^{-1} [S_{ij}^*]_{1,2,3} [T_{ij}] \begin{bmatrix} \sigma_x \\ \sigma_y \\ \sigma_z \\ \tau_q \\ \tau_r \\ \tau_s \end{bmatrix} = [S_{ij}^*]_{x,y,z} \begin{bmatrix} \sigma_x \\ \sigma_y \\ \sigma_z \\ \tau_q \\ \tau_r \\ \tau_s \end{bmatrix} \tag{6}
$$

where

$$
[T_{ij}] = \begin{bmatrix} m^2 & 0 & n^2 & 0 & 2mn & 0 \\ 0 & 1 & 0 & 0 & 0 & 0 \\ n^2 & 0 & m^2 & 0 & -2mn & 0 \\ 0 & 0 & 0 & m & 0 & -n \\ -mn & 0 & mn & 0 & m^2 - n^2 & 0 \\ 0 & 0 & 0 & n & 0 & m \end{bmatrix}
$$

$m = \cos\theta \qquad n = \sin\theta$

$[R_{ij}]$: Reuter matrix

$[S_{ij}^*]_{1,2,3}$: compliance matrix in Eq 5

q, r, s subscripts stand for yz, zx and xy, respectively

Under uniaxial loading, σ_x, Eq 6 can be written as

$$\varepsilon_x = S_{xx}^* \sigma_x = \{[S_{11}m^4 + (2S_{13} + S_{55})m^2n^2 + S_{33}n^4] + (S_{111}\sigma_1 + S_{1111}\sigma_1^2)m^4$$

$$+ (S_{333}\sigma_3 + S_{3333}\sigma_3^2) \, n^4 + S_{5555} \, \tau_5^2 \, m^2n^2\}\sigma_x \tag{7}$$

$$\varepsilon_y = S_{xy}^* \sigma_x = [S_{12}m^2 + S_{23}n^2]\sigma_x$$

$$\varepsilon_z = S_{xz}^* \sigma_x = \{[(S_{11} + S_{33} - S_{55})m^2n^2 + S_{13}(m^4 + n^4)] + (S_{111}\sigma_1 + S_{1111}\sigma_1^2)m^2n^2$$

$$+ (S_{333}\sigma_3 + S_{3333}\sigma_3^2)m^2n^2 - S_{5555} \, \tau_5^2 \, m^2n^2\}\sigma_x$$

Recalling that

$$\sigma_1 = m^2\sigma_x \qquad \sigma_3 = n^2\sigma_x \qquad \tau_5 = -mn \, \sigma_x \tag{8}$$

and applying Eq 8 to Eq 7, we obtain the following stress-strain relations for a composite with out-of-plane (z-direction) fiber orientation θ:

$$\varepsilon_x = \{S_{11}m^4 + (2S_{13} + S_{55})m^2n^2 + S_{33}n^4]\sigma_x + (S_{111}m^6 + S_{333}n^6)\sigma_x^2$$

$$+ (S_{1111}m^8 + S_{3333}n^8 + S_{5555}m^4n^4)\sigma_x^3$$

$$\varepsilon_y = [S_{12}m^2 + S_{23}n^2]\sigma_x \tag{9}$$

$$\varepsilon_z = [(S_{11} + S_{33} - S_{55})m^2n^2 + S_{13}(m^4 + n^4)]\sigma_x + (S_{111}m^4n^2 + S_{333}m^2n^4)\sigma_x^2$$

$$+ (S_{1111}m^6n^2 + S_{3333}m^2n^6 - S_{5555}m^4n^4)\sigma_x^3$$

To average the material response over one wavelength of fiber waviness, it is necessary to relate the fiber orientation θ to the wave parameters A and L

$$\tan \theta = \frac{dv}{dx} = 2\pi \frac{A}{L} \cos \frac{2\pi x}{L} \tag{10}$$

therefore

$$m = \cos \theta = \left[1 + \left(2\pi \frac{A}{L} \cos \frac{2\pi x}{L}\right)^2\right]^{-1/2}$$

$$n = \sin \theta = 2\pi \frac{A}{L} \cos \frac{2\pi x}{L} \left[1 + \left(2\pi \frac{A}{L} \cos \frac{2\pi x}{L}\right)^2\right]^{-1/2} \tag{11}$$

Substituting Eq 11 into 9 and integrating over one period of waviness, L, the average stress-strain relations for the representative volume element under uniaxial loading $\bar{\sigma}_x$ are obtained:

$$\bar{\varepsilon}_x = [S_{11}I_1 + (2S_{13} + S_{55})I_3 + S_{33}I_5]\bar{\sigma}_x + (S_{111}I_9 + S_{333}I_{13})\bar{\sigma}_x^2$$

$$+ (S_{1111}I_{14} + S_{3333}I_{18} + S_{5555}I_{16})\bar{\sigma}_x^3$$

$$\bar{\varepsilon}_y = [S_{12}I_6 + S_{23}I_8]\bar{\sigma}_x \tag{12}$$

$$\bar{\varepsilon}_z = [(S_{11} + S_{33} - S_{55})I_3 + S_{13}(I_1 + I_5)]\bar{\sigma}_x + (S_{111}I_{10} + S_{333}I_{12})\bar{\sigma}_x^2$$

$$+ (S_{1111}I_{15} + S_{3333}I_{17} - S_{5555}I_{16})\bar{\sigma}_x^3$$

where

$$I_1 = \frac{2 + \alpha^2}{2(1 + \alpha^2)^{3/2}} \qquad\qquad I_3 = \frac{\alpha^2}{2(1 + \alpha^2)^{3/2}}$$

$$I_5 = 1 - \frac{2 + 3\alpha^2}{2(1 + \alpha^2)^{3/2}} \qquad\qquad I_6 = \frac{1}{(1 + \alpha^2)^{1/2}}$$

$$I_8 = 1 - \frac{1}{(1 + \alpha^2)^{1/2}} \qquad\qquad I_9 = \frac{8 + 8\alpha^2 + 3\alpha^4}{8(1 + \alpha^2)^{5/2}}$$

$$I_{10} = \frac{4\alpha^2 + \alpha^4}{8(1 + \alpha^2)^{5/2}} \qquad\qquad I_{12} = \frac{3\alpha^4}{8(1 + \alpha^2)^{5/2}}$$

$$I_{13} = 1 - \frac{8 + 20\alpha^2 + 15\alpha^4}{8(1 + \alpha^2)^{5/2}} \qquad\qquad I_{14} = \frac{16 + 24\alpha^2 + 18\alpha^4 + 5\alpha^6}{16(1 + \alpha^2)^{7/2}} \tag{13}$$

$$I_{15} = \frac{8\alpha^2 + 4\alpha^4 + \alpha^6}{16(1 + \alpha^2)^{7/2}} \qquad\qquad I_{16} = \frac{6\alpha^4 + \alpha^6}{16(1 + \alpha^2)^{7/2}}$$

$$I_{17} = \frac{5\alpha^6}{16(1 + \alpha^2)^{7/2}} \qquad\qquad I_{18} = 1 - \frac{16 + 56\alpha^2 + 70\alpha^4 + 35\alpha^6}{16(1 + \alpha^2)^{7/2}}$$

$$\alpha = 2\pi\frac{A}{L}$$

Overbarred symbols denote average values over the representative volume element. Stress-strain relations for other loading conditions can be derived in a similar manner. For example, under shear loading $\bar{\tau}_{xz}$

$$\bar{\gamma}_{xz} = [4(S_{11} + S_{33} - 2S_{13})I_3 + S_{55}(I_1 - 2I_3 + I_5)]\bar{\tau}_{xz}$$

$$+ [16(S_{1111} + S_{3333})I_{16} + S_{5555}(I_{14} - 4I_{15} + 6I_{16} - 4I_{17} + I_{18})]\bar{\tau}_{xz}^3 \tag{14}$$

The compliances are determined experimentally. S_{11} is obtained from the initial slope of the longitudinal stress-strain curve (i.e., $S_{11} = 1/\text{major Young's modulus}$). S_{111} and S_{1111} are determined by curve fitting the same curve. Similarly, S_{33}, S_{333}, and S_{3333} can be obtained from uniaxial testing in the transverse direction. S_{55} and S_{5555} can be obtained from shear testing.

Incremental Analysis

An incremental analysis was used to predict the static and dynamic behavior of wavy composites using the basic strain rate characterization data for the unidirectional material without fiber waviness. Successive stress and time increments were applied with continuous monitoring of the resulting stress-strain behavior. The stress increment can be controlled in such a way that either constant strain rate or constant stress rate can be achieved. The incremental analysis algorithm used consists of the following steps:

(1) Increase the applied stress and time by small increments $\Delta \bar{\sigma}_j$ and Δt, respectively ($j = x, y, z, q, r, s$, in Eq 6)

(2) Calculate stress increments $\{\Delta \sigma_j\}_n^1$ and corresponding stress rates $\{\dot{\sigma}_j\}_n^1$ along the principal material directions of segment dx ($j = 1, 2, 3, 4, 5, 6$) where 1 is the element number in the x direction and n is the current incremental step

(3) Calculate strain increments for each segment dx

$$\{\Delta \varepsilon_i\}_n^1 = [S_{ij}]_{n-1}^1 \{\Delta \bar{\sigma}_j\}_n \qquad (i,j = x, y, z, q, r, s) \qquad (15)$$

(4) Update effective (average) strain increments for entire representative volume

$$\{\Delta \bar{\varepsilon}_i\}_n = \frac{1}{m} \sum_{1=1}^{m} \{\Delta \varepsilon_i\}_n^1 \qquad (i,j = x, y, z, q, r, s) \qquad (16)$$

(5) Update the effective stress and strain at every step to obtain the cumulative stress-strain behavior of the wavy composites

$$\{\bar{\varepsilon}_i\}_n = \{\bar{\varepsilon}_i\}_{n-1} + \{\Delta \bar{\varepsilon}_i\}_n$$

$$\{\bar{\sigma}_i\}_n = \{\bar{\sigma}_i\}_{n-1} + \{\Delta \bar{\sigma}_i\}_n \qquad (i,j = x, y, z, q, r, s) \qquad (17)$$

(6) For each segment use the cumulative stresses along the local principal material directions and stress rates obtained in step 2 to find the corresponding instantaneous tangential stiffnesses from the basic strain rate characterization curves by interpolation/extrapolation on the $\log S_{ij}(\sigma_j) - \log \dot{\sigma}_j(\sigma_j)$ scale ($i,j = 1, 2, 3, 4, 5, 6$) to update the material properties $[S_{ij}]_n^1$ for the next increment.

(7) Change $\Delta \bar{\sigma}_j$ for the next increment if constant strain rate is desired; otherwise keep the applied stress increment the same to maintain constant stress rate.

Experimentation

Material and Specimen Fabrication

The material investigated in this study was IM6G/3501-6 carbon/epoxy composite (Hexcel Corporation). Techniques were developed for fabrication of thick composite specimens with and without controlled waviness. Unidirectional composites with uniform waviness across the thickness were prepared by a tape winding method [15]. The prepreg tape was wound around a rectangular mandrel with rounded ends and the entire layup was enclosed in a mold and cured in an autoclave following a curing cycle developed for thick composites. This resulted in composites with plies of sine wave shape of uniform amplitude through the thickness.

Test Procedure

A servohydraulic testing machine and a falling weight impact system were used for static and dynamic characterization of the composite material in compression at strain rates up to several hundred per second [16]. Strain rates below 10 s^{-1} were generated using the servohydraulic testing machine. Strain rates above 10 s^{-1} were generated using the drop tower apparatus developed. Detailed description of the specimen configuration and test set-up can be found in [16].

Seventy-two-ply unidirectional laminates with aligned fibers loaded in the longitudinal and transverse directions were characterized under quasi-static and high strain rate loadings. Forty-five degree off-axis compression tests of the same laminates were also conducted to obtain the static and dynamic in-plane shear stress-strain behavior. These basic characterization data were generated as the input information for theoretical predictions. Special 150-ply unidirectional specimens with known degree of uniform fiber waviness were tested to compare with these predictions.

Results and Discussion

Material Characterization

Figures 2–4 show the longitudinal and transverse compressive and in-plane shear stress-strain curves of aligned IM6G/3501-6 composites under quasi-static and high strain rates, respectively. The 90-degree properties, which are governed by the matrix, show an increase in modulus and strength over the static values but no significant change in ultimate strain. The in-plane shear behavior, which is also matrix-dominated, shows high nonlinearity with a plateau region at a stress level that increases significantly with increasing strain rate. The 0-degree laminates show higher strength and strain values as the strain rate increases, whereas the modulus increases only slightly over the static value. In all cases, the dynamic stress-strain curves stiffen as the strain rate increases. The stiffening is lowest in the longitudinal case and highest in the transverse and shear cases. The experimentally determined higher-order compliances (e.g., S_{5555}) for this material under quasi-static compression can be found in [11].

Unidirectional Composite with Uniform Fiber Waviness

Unidirectional specimens with uniform fiber waviness were fabricated and tested. The degree of waviness was characterized by the amplitude to period ratio, A/L, or the maximum angle θ_{max} of fiber misalignment. The specimens were 150-plies thick (19 mm; 0.75 in.), 1.27 cm (0.50 in.) wide, and 14.6 cm (5.75 in.) long. They were tabbed with glass/epoxy tabs leaving a 3.81 cm (1.50 in.) gage length. The specimens had the following average waviness parameters:

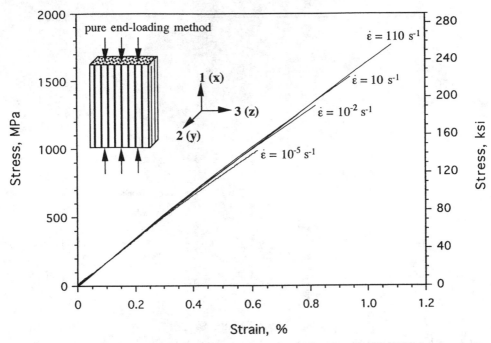

FIG. 2—*Longitudinal compressive stress-strain curves for unidirectional IM6G/3501-6 carbon/epoxy under quasi-static and high strain rate loading.*

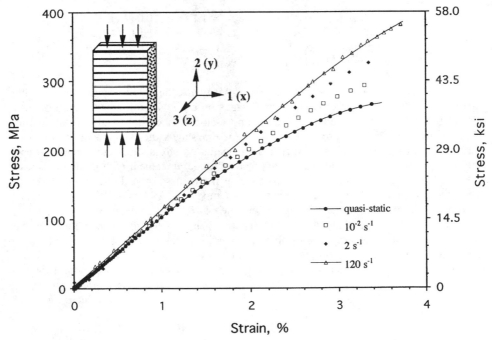

FIG. 3—*Transverse compressive stress-strain curves for unidirectional IM6G/3501-6 carbon/epoxy under quasi-static and high strain rate loading.*

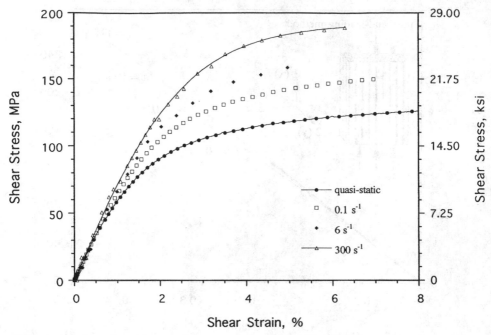

FIG. 4—*In-plane shear stress-strain curves obtained from 45° off-axis tests under quasi-static and high strain rate compressive loading.*

Amplitude:	A = 1.19 mm (0.047 in.)
Period:	L = 27.9 mm (1.10 in.)
Amplitude/Period Ratio:	A/L = 0.043
Maximum Misalignment:	θ_{max} = 15°

The specimens were instrumented with two back-to-back longitudinal gages spanning half a period of the waviness, and two transverse gages on a line spanning half a period on the 1-2 and 1-3 planes. They were tested under compression at quasi-static and high strain rates.

Complementary Strain Energy Approach—Figure 5 shows the comparison between the analytical constitutive model and experimental results on stress-strain curves under quasi-static compression $\overline{\sigma}_x$. Experimental results were in good agreement with predictions based on Eqs 12 and 13. These results show that uniform waviness of A/L = 0.043 causes appreciable reductions (42%) in the major Young's modulus. As far as Poisson's ratios are concerned, the in-plane ratio v'_{12} is higher and the out-of-plane ratio v'_{13} is lower than the value of $v_{12} = v_{13} = 0.31$ for the aligned specimen. It is noted that strong nonlinearities occur in the x and z directions due to fiber waviness, especially in the z direction. Figure 6 shows the predicted stress-strain curves under quasi-static compression $\overline{\sigma}_x$ as a function of waviness parameter A/L. It shows again how material nonlinearity increases with fiber waviness. Figure 7 illustrates the predicted stress-strain curves under quasi-static shear loading τ_{xz} obtained from Eq 14. Stress-strain curves are affected less by fiber waviness under this loading condition.

Incremental Analysis—Figures 8a and 8b illustrate the predicted stress-strain curves (A/L = 0.075) under quasi-static and high strain rate compression $\overline{\sigma}_x$ for constant strain rate and constant stress rate, respectively. The nonlinear stress-strain curve stiffens significantly

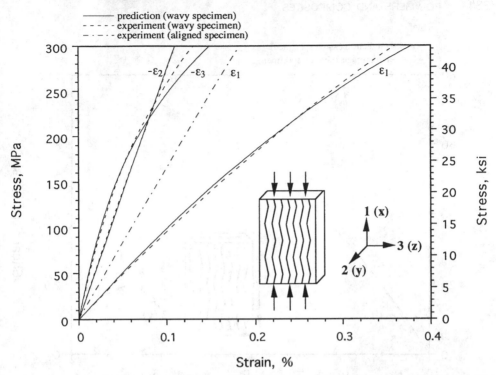

FIG. 5—*Comparison between predictions and experiments on stress-strain curves for 150-ply unidirectional IM6G/3501-6 carbon/epoxy with uniform fiber waviness under quasi-static compressive loading in the* x *direction* ($A/L = 0.0425$).

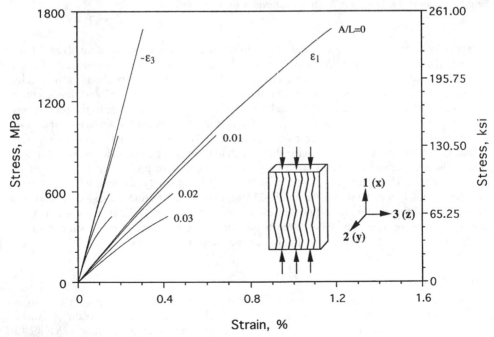

FIG. 6—*Predicted stress-strain curves as a function of waviness parameter A/L for unidirectional IM6G/3501-6 carbon/epoxy with uniform fiber waviness under quasi-static compressive loading in the* x *direction.*

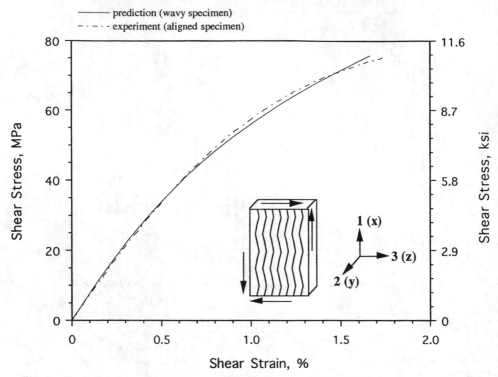

FIG. 7—*Predicted shear stress-strain curves for unidirectional IM6G/3501-6 carbon/epoxy with uniform fiber waviness under shear in the* x-z *directions* (A/L = 0.0425).

at higher loads as the strain rate increases due to the larger shear component involved. Figure 9 shows the comparison between predictions and experiments on stress-strain curves under quasi-static and high strain rate compression $\bar{\sigma}_x$ (A/L = 0.043). The specimens tested failed prematurely due to fiber discontinuities on the specimen surfaces (Fig. 1) and thus comparison can only be made up to approximately 350 MPa (51 ksi). According to the predictions, the strain rate effect is not significant up to this load level; however, as the load is increased above this level, a strong strain rate effect is expected. The experimental stress-strain curves indeed prove this point and show only slight stiffening over the static ones within this load range. These experimental results were in good agreement with predictions based on incremental analysis. The predicted z-direction behavior is surprisingly accurate when compared to experiments. Figure 10 shows the predicted stress-strain curves under quasi-static and high strain rate compression $\bar{\sigma}_x$ as a function of waviness parameter A/L. It is shown that, for this type of severe fiber waviness, strong nonlinearities occur in the stress-strain curves due to fiber waviness with significant stiffening as the strain rate increases.

Conclusions

A systematic investigation was conducted of the effect of fiber waviness on the stress-strain behavior of unidirectional laminates under quasi-static and dynamic loadings. An analytical constitutive model was developed to predict the nonlinear quasi-static behavior for wavy composites based on the complementary strain energy. Incremental analysis was also

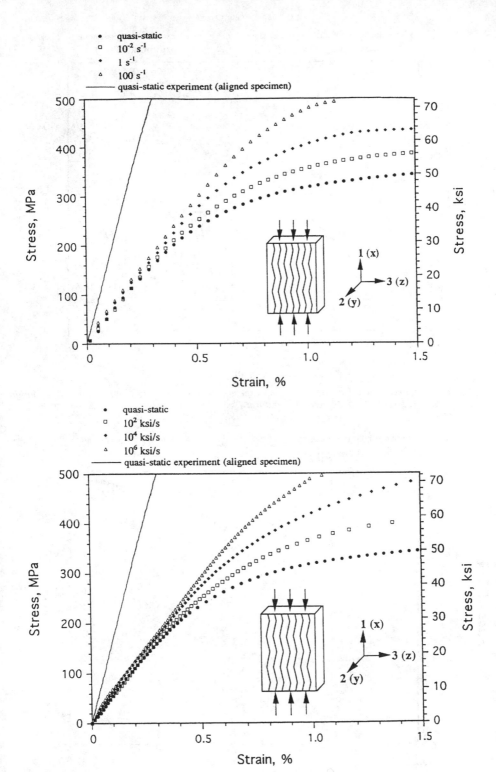

FIG. 8—*Predicted stress-strain curves for unidirectional IM6G/3501-6 carbon/epoxy with uniform fiber waviness under quasi-static and high strain rate compressive loading in the* x *direction: (a) Constant strain rate and (b) Constant stress rate.*

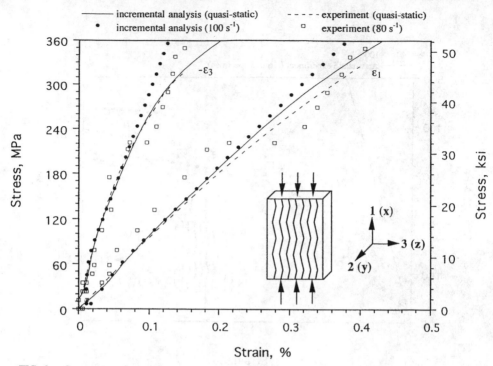

FIG. 9—*Comparison between predictions and experiments on stress-strain curves for 150-ply unidirectional IM6G/3501-6 carbon/epoxy with uniform fiber waviness under quasi-static and high strain rate compressive loading in the* x *direction* ($A/L = 0.0425$).

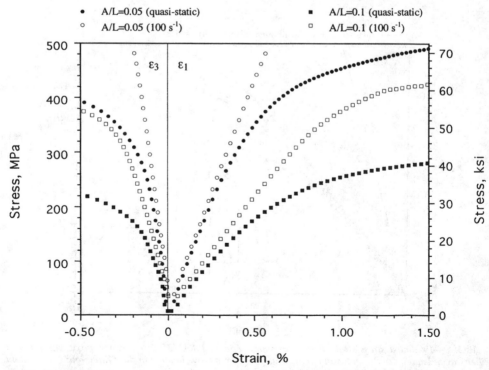

FIG. 10—*Predicted stress-strain curves as a function of waviness parameter A/L for unidirectional IM6G/3501-6 carbon/epoxy with uniform fiber waviness under quasi-static and high strain rate compressive loading in the* x *direction.*

used to predict the stress-strain behavior under both static and dynamic loadings. It is shown that under uniaxial compressive loading $\overline{\sigma}_x$, strong nonlinearities occur in the stress-strain curves due to fiber waviness with significant stiffening as the strain rate increases. Stress-strain curves are affected less by fiber waviness under other loading conditions which relate to matrix-dominated behavior under transverse or shear loading. Experiments were conducted to verify the theoretical predictions. Unidirectional carbon/epoxy specimens were fabricated with uniform fiber waviness. Experimental results were in good agreement with predictions.

Acknowledgments

The work described in this paper was sponsored by the Office of Naval Research. We are grateful to Dr. Y. D. S. Rajapakse of ONR for his encouragement and cooperation and to Mrs. Yolande Mallian for typing the manuscript.

References

[1] Adams, D. O. and Hyer, M. W., "Effects of Layer Waviness on the Compression Strength of Thermoplastic Composite Laminates," *Journal of Reinforced Plastics and Composites*, Vol. 12, 1993, pp. 414–429.

[2] Bogetti, T. A., Gillespie, J. W., and Lamontia, M. A., "Influence of Ply Waviness with Nonlinear Shear on the Stiffness and Strength Reduction of Composite," *Journal of Thermoplastic Composite Materials*, Vol. 7, No. 2, 1994, pp. 76–90.

[3] Hsiao, H. M. and Daniel, I. M., "Effect of Fiber Waviness on Stiffness and Strength Reduction of Unidirectional Composites under Compressive Loading," *Composites Science and Technology*, Vol. 56, No. 5, 1996, pp. 581–593.

[4] Petit, P. H. and Waddoups, M. E., "A Method of Predicting the Nonlinear Behavior of Laminated Composites," *Journal of Composite Materials*, Vol. 3, 1969, pp. 2–19.

[5] Hashin, Z., Bagchi, D., and Rosen, B. W., "Non-Linear Behavior of Fiber Composite Laminates," NASA CR-2313, Materials Sciences Corp., Blue Bell, PA, 1974.

[6] Hahn, H. T. and Tsai, S. W., "Nonlinear Elastic Behavior of Unidirectional Composite Laminae," *Journal of Composite Materials*, Vol. 7, 1973, pp. 102–118.

[7] Luo, S.-Y. and Chou, T.-W., "Finite Deformation and Nonlinear Elastic Behavior of Flexible Composites," *Journal of Applied Mechanics*, Trans. ASME, Vol. 55, 1988, pp. 149–155.

[8] Chou, T.-W. and Takahashi, K., "Non-linear Elastic Behavior of Flexible Fiber Composites," *Composites*, Vol. 18, No. 1, 1987, pp. 25–34.

[9] Kuo, C.-H., Takahashi, K., and Chou, T.-W., "Effect of Fiber Waviness on the Nonlinear Elastic Behavior of Flexible Composites," *Journal of Composite Materials*, Vol. 22, 1988, pp. 1004–1025.

[10] Rai, H. G., Rogers, C. W., and Crane, D. A., "Mechanics of Curved Fiber Composites," *Journal of Reinforced Plastics and Composites*, Vol. 11, 1992, pp. 552–566.

[11] Hsiao, H. M. and Daniel, I. M., "Nonlinear Elastic Behavior of Unidirectional Composites with Fiber Waviness under Compressive Loading," *Journal of Engineering Materials and Technology*, Trans ASME, Vol. 118, No. 4, 1996, pp. 561–570.

[12] Schapery, R. A., "Nonlinear Viscoelastic and Viscoplastic Constitutive Equations Based on Thermodynamics," *Mechanics of Time-Dependent Materials*, Vol. 1, 1997, pp. 209–240.

[13] Green, A. E. and Adkins, J. E., *Large Elastic Deformations*, Oxford University Press, London, 1970.

[14] Daniel, I. M. and Ishai, O., *Engineering Mechanics of Composite Materials*, Oxford University Press, New York, 1994.

[15] Hsiao, H. M., Wooh, S. C., and Daniel, I. M., "Fabrication Methods for Unidirectional and Crossply Composites with Fiber Waviness," *Journal of Advanced Materials*, Vol. 26, No. 2, 1995, pp. 19–26.

[16] Hsiao, H. M., Daniel, I. M., and Cordes, R. D., "Dynamic Compressive Behavior of Thick Composite Materials," *Experimental Mechanics*, Vol. 38, No. 3, 1998, pp. 172–180.

Robert T. Bocchieri[1] and Richard A. Schapery[1]

Nonlinear Viscoelastic Behavior of Rubber-Toughened Carbon and Glass/Epoxy Composites

REFERENCE: Bocchieri, R. T. and Schapery, R. A., **"Nonlinear Viscoelastic Behavior of Rubber-Toughened Carbon and Glass/Epoxy Composites,"** *Time Dependent and Nonlinear Effects in Polymers and Composites, ASTM STP 1357,* R. A. Schapery and C. T. Sun, Eds., American Society for Testing and Materials, West Conshohocken, PA, 2000, pp. 238–265.

ABSTRACT: A previously-developed constitutive equation for nonlinear viscoelastic materials is first used to characterize mathematically the ply-level stress-strain behavior of a rubber-toughened carbon/epoxy composite. Emphasized are some practical aspects of carrying out such a characterization. Constant stress rate tests are used to derive first-loading material response in the dry state, at room temperature. It is found for a large portion of the loading curve that the shear and transverse compliances can be described by a quasi-elastic model expressed in terms of a single scalar function of stress state, a ratio of compliances, a time or rate exponent, and two elastic terms. This model offers a simple way to incorporate nonlinearity and time dependence in a material model for limited loading conditions. Results are then compared to those for a glass/epoxy composite with the same rubber-toughened resin system. Special considerations for using off-axis coupons are examined and unique tabbing is employed to derive material properties out to high stress levels. The consistency of the carbon/epoxy characterization is checked by comparing experimental stress-strain response to results not used in the characterization. Behavior of the two composite systems is compared and checked for consistency using a micromechanical model.

KEYWORDS: polymer matrix composites, nonlinear viscoelasticity, constitutive equation, off-axis angle-tabs, damage, rubber-toughened

Glass and carbon fiber-reinforced plastics offer important structural advantages over conventional metals in many applications as a result of their high strength and stiffness, low density, tailorability of their properties and corrosion resistance, among others. However, questions on their long-term durability have often inhibited engineers in their use. Here, we are concerned with one aspect of the durability issue—how the deformation characteristics change due to the nonlinear viscoelastic behavior of the polymer matrix.

In most textbooks and in the preliminary design for most structural applications, unidirectional fiber composites (the building block of multidirectional fiber laminates) are modeled as anisotropic linear elastic solids. However, their behavior is, in reality, more complex than this. These materials have been shown to exhibit significant amounts of time dependence and nonlinearity in those properties dominated by the polymer matrix due to viscoelasticity and damage growth [1,2]. Under adverse environmental conditions, such as elevated tem-

[1] Graduate Research Assistant and Professor, respectively, The University of Texas at Austin, Austin, TX 78712-1085; and The Offshore Technology Research Center, 2901 North IH 35, Suite 101, Austin, TX 78722.

perature, moisture, or stress, the time dependence of the matrix properties is greatly enhanced. Even when the overall deformation of a fiber-dominated laminated composite structure is essentially immune to matrix properties prior to significant levels of cracking such as delamination, the ply-level stresses are not. These stresses affect matrix cracking, softening and fiber-matrix debonding which can lead to premature failure of the fibers, and shortening the life of the structure. In some circumstances, "structural failure" may not mean fiber breakage but matrix cracking leading to leakage of some fluid, such as in pressurized composite tubulars. With a theoretical basis previously established for such predictions [3,4], good experimental techniques and accelerated testing methodologies, such as use of elevated temperatures, need to be developed to predict long-term response from short-term experimental studies and to validate the theory. One important issue is whether or not the temperature dependence of the damage growth rate and intrinsic viscoelastic behavior are the same.

Presented first are some experimental techniques used to derive the initial constant stress and constant stress-rate tensile response of the rubber-toughened carbon/epoxy AS4C/E719LT in the dry state, at room temperature. In the first loading of the material, significant nonlinear rate-dependent behavior occurs, some of which may be due to damage, thereby concealing the intrinsic viscoelasticity of the polymer matrix [5,6]. The present characterization establishes only a baseline that will be used in the future with other data for isolating the effect of damage growth from intrinsic viscoelasticity and for determining property dependence on temperature and moisture. We emphasize that experimental results from this study are not sufficient to separate damage and viscoelastic effects.

Extracting the nonlinear time-dependent properties of these materials requires some special considerations experimentally. Specifically, one needs to carefully choose the fiber angles and layups which provide the most useful information in isolating stress and time dependence. Some of these considerations are reviewed here. A new configuration of tapered-fiberglass angle tabs used for testing off-axis specimens is also examined. After demonstrating the time-dependence and nonlinearity of AS4C/E719LT, a quasi-elastic form of a constitutive equation, used successfully in the past for constant stress and stress-rate conditions, is introduced. A method for reducing the data to extract the ply-level properties is also described. Material behavior derived from unidirectional and angle-ply laminates are compared and the accuracy of the material characterization is checked by comparing experimental stress-strain response to a predicted response based on averaged properties of all material samples.

Many foreseeable applications are structures comprised of a hybrid of glass and carbon fibers. From the standpoint that much of the material dependence on time, stress, temperature, and moisture is intrinsic to the polymer matrix, many similarities in the mechanical response of these two materials are anticipated, given the same resin system. A complete characterization of one of the two materials with limited testing of the second may be sufficient. Presented here is a look at what similarities exist. Different fiber properties are taken into account using the semi-empirical Halpin-Tsai equations. Response of the glass/epoxy 158B/E719LT is then compared to a predicted response based on the carbon/epoxy characterization.

Experimental Program—Carbon/Epoxy

Material and Processing Information

The first material under study is comprised of continuous AS4C carbon fibers with a rubber-toughened epoxy resin E719LT manufactured by BP (now Fiberite). Fiber volume fracture, V_f, was spot-checked on specimens of different layups and thickness by the method of matrix digestion [7]. The fractions fell between 56.5 to 57.8% with an average of approximately 57%. For details on the manufacture and physical characteristics of this material, the reader is referred to Ref 8.

Material Model and Physical Mechanisms of Behavior

It should be mentioned briefly for the sake of clarity and notation that the materials under consideration are modeled as having transversely isotropic, homogenized unidirectional layers in a state of plane stress. Unidirectional lamina stress-strain behavior can then be described as [9]

$$\begin{Bmatrix} \varepsilon_1 \\ \varepsilon_2 \\ \varepsilon_6 \end{Bmatrix} = \begin{bmatrix} S_{11} & S_{12} & 0 \\ S_{12} & S_{22} & 0 \\ 0 & 0 & S_{66} \end{bmatrix} \begin{Bmatrix} \sigma_1 \\ \sigma_2 \\ \sigma_6 \end{Bmatrix} \tag{1}$$

where S_{ij} are compliances, ε are strains and σ are stresses referred to the material coordinates x_1, x_2, in Fig. 1; with nonlinear and/or time-dependent behavior, following from appendix A, the S_{ij} are symmetric and should be interpreted as secant compliances. The compliance matrix may be expressed in terms of engineering properties as

$$[S_{ij}] = \begin{bmatrix} \dfrac{1}{E_1} & \dfrac{-\upsilon_{21}}{E_2} & 0 \\ \dfrac{-\upsilon_{12}}{E_1} & \dfrac{1}{E_2} & 0 \\ 0 & 0 & \dfrac{1}{G_{12}} \end{bmatrix} \tag{2}$$

Single index notation is used for stresses and strains referred to material coordinates, as shown in Fig. 1, where x_1 is the fiber direction. That is

$$\begin{Bmatrix} \varepsilon_1 \\ \varepsilon_2 \\ \varepsilon_6 \end{Bmatrix} = \begin{Bmatrix} \varepsilon_{11} \\ \varepsilon_{22} \\ \gamma_{12} \end{Bmatrix} \tag{3}$$

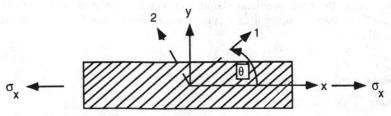

FIG. 1—*Unidirectional specimen with arbitrary fiber angle loaded in the axial direction.*

The objective of this characterization is to determine the in-plane lamina compliances shown in Eq 1.

Fiber-dominated properties, S_{11} and S_{12}, may show nonlinearity due to fiber straightening [6]. This behavior is not time-dependent, but is elastic, provided that a significant amount of fiber breakage does not occur. Shear and transverse components, S_{66} and S_{22}, are dominated by the polymer matrix. Consequently, these properties are time-dependent. Nonlinearity in the matrix stems from intrinsic viscoelasticity and possible changes in the microstructure including microcracking, shear banding, and crazing. These mechanisms will be referred to as damage since they normally degrade the mechanical properties of the material. Matrix dominated properties are also affected by temperature and moisture as both increase the free volume of the polymer and thus accelerate viscoelastic motion. The material under study is further complicated by rubber-toughening which enhances some of these mechanisms and causes time-dependent rubber particle cavitation [10]. Time dependence of the material is increased even at room temperature conditions due to the rubber-toughening [11].

Experimental Procedure for Obtaining Nonlinear, Time-dependent Mechanical Properties

Fiber-dominated properties S_{11} and S_{12}, which show negligible time dependence, can be determined in a straight-forward manner from uniaxial testing of 0° coupons as specified in ASTM Standard D3039M-95a. Matrix-dominated properties, however, require a collection of experiments which isolate the stress, time, temperature, and moisture-dependence of the material. Presently, however, we are only concerned with stress and time. Time dependence can be isolated by testing the material with various loading rates, as is the case in this study, or by performing creep/recovery tests. Determining stress dependence, however, warrants some further explanation.

Under uniaxial loading, stress-dependent mechanical properties are only functions of the uniaxial stress. However, in a real composite structure, any combination of multiaxial stress states can exist. Strong evidence suggests that the matrix-dominated compliances, S_{22} and S_{66}, vary with some quadratic scalar function of this stress field. Several invariants or effective stresses have been proposed [3,5,12–14]. For plane stress, a scalar function of the form

$$\sigma_0 = \sqrt{\sigma_2^2 + c_3 \sigma_6^2} \qquad (4)$$

has been successfully used in earlier studies, and we use it here. There is one free coefficient, c_3, to be determined experimentally. Stresses σ_2 and σ_6 are the transverse and shear stresses in primary material directions and σ_0 is called the effective stress.

Unidirectional off-axis, 90°, and $(\pm 45)_{ns}$ angle-ply laminates were employed in the present study to obtain four different combinations of stress in the matrix. The $(\pm 45)_{ns}$ and 90° unidirectional specimens are simple, but important cases where a ply-level single-stress component dominates (there are, of course, stress concentrations around the fibers at the microstructural scale); only shear stress is significant in the $(\pm 45)_{ns}$ specimen provided the fibers are much stiffer than the matrix. Unidirectional, off-axis specimens were selected to achieve multiaxial stresses in the material coordinates. Loading for this configuration is a special case of proportional loading (i.e., the stress components are a constant geometric function of one another and therefore increase proportionally under applied uniaxial stress – σ_x). Off-axis samples offer an economical method of obtaining information about both the shear and transverse compliances simultaneously. The compliance S_{22} can also be found to much higher levels of effective stress than in 90° samples (which have a low ultimate strength). As will be shown, two stress combinations are sufficient for this type of characterization; however, additional combinations provide a check on the model. One remaining question is which fiber angles relative to the loading axis will yield the most useful information.

Selection of Fiber Angle for Off-axis Specimens

There is a tradeoff in selecting the fiber angle of an off-axis specimen. One should consider the degree of nonlinearity observed prior to failure, sensitivity of material properties to measurable laminate properties, sensitivity to fiber angle, and strain gage misalignment. As there are many details to consider in this process, the reader is referred to Ref 8 for a detailed description of all the factors considered in selecting the best fiber angles for this characterization.

Two off-axis configurations were chosen so that the effect of different combinations of stress could be correlated independently of the potentially different shear response in angle-ply $(\pm 45)_{ns}$ samples (mismatch of the fiber and matrix thermal expansion coefficients in angle-ply laminates induce residual stresses from cure and may promote damage in the matrix [15,16]). Angles of 15° and 30° were chosen because both offer sufficient sensitivity to the shear and transverse modulus (Fig. 2) and represent two significantly different stress ratios without compromising the degree of nonlinearity observed (as will be seen in the results). As both angles chosen provide results that are moderately sensitive to fiber misalignment, the average off-axis fiber angle was determined from measurements under a microscope and this average fiber angle was used for data analysis.

A 10° off-axis specimen has traditionally been used to characterize the shear strength and modulus of fibrous composites, as originally proposed by Chamis and Sinclair [17]. How-

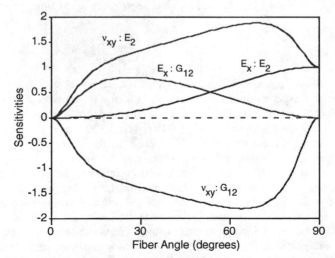

FIG. 2—*Sensitivity of* G_{12} *and* E_{12} *to* E_x *and* v_{xy} *for a unidirectional laminate. Preliminary results based on testing 0°, 90°, and* $(\pm 45)_{ns}$ *specimens of AS4C/E719LT were used for this sensitivity study.*

ever, an angle of 10° was not the best choice in this characterization as, among other reasons, the measurable properties of E_x and v_{xy} do not become appreciably sensitive to E_2 until an angle of approximately 15°, as shown in Fig. 2. Sensitivities are logarithmic derivatives [6]; for example, $E_x : G_{12}$ is the plot of

$$\frac{\partial \log E_x}{\partial \log G_{12}} = \frac{G_{12}}{E_x} \frac{\partial E_x}{\partial G_{12}}. \tag{5}$$

In addition, the 10° specimen was originally suggested with the intent of only obtaining shear properties as the normalized shear strain, $\gamma_{12}/\varepsilon_x$, reached a maximum at approximately 10° for the material they studied. Carbon/epoxy AS4C/E719LT reaches this maximum at approximately 12° and would be the best choice based on that criterion alone. In this study, however, it is desired to obtain both shear and transverse properties out to high levels of stress and damage.

Complications in Testing Off-axis Specimens

Off-axis specimens impose certain experimental complications. Rigid end constraints typically used for tensile testing resist the rotational tendency of the material due to the shear coupling effect. A non-uniform stress distribution results in the specimen. Many researchers have investigated possible experimental methods to solve this problem as discussed in Ref 8. For instance, Chang et al. [18] found that a sufficiently large aspect ratio (length/width = 17) can be used to obtain accurate shear moduli. However, our need for shorter

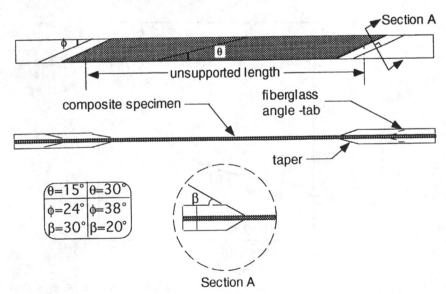

FIG. 3—*Off-axis specimen configuration with a tapered-fiberglass/angle-tab.*

specimens to conserve material led to a tabbing option similar to that proposed by Sun and Chung [19]. Use of a correctly angled tab results in a more uniform strain field than the standard rectangular tab. This angle ϕ (Fig. 3) is a function of the material compliances as,

$$\cot \phi = \frac{\bar{S}_{xy}}{\bar{S}_{xx}} \tag{6}$$

where $\bar{S}_{xy}$ and $\bar{S}_{xx}$ are the shear coupling and axial terms of the material compliance matrix in coordinates coincident with the loading axis. More uniform strain fields can be obtained in significantly shorter specimens (aspect ratios as small as 5) by this tab design. However, the aluminum sheet tabs used in their study may cause premature failure at the tabs due to the stress concentration caused by an abrupt change in geometry [20]. Use of fiberglass tabs with a taper reduces the stress concentration. Therefore, a design which incorporates the desirable aspects of both configurations was used. In this way, greater degrees of nonlinearity and accurate strength information can be obtained from a relatively short specimen. The final tab design is shown in Fig. 3 where θ represents the fiber angle, ϕ the angle of the tab, and β the taper angle in a plane perpendicular to the tab edge.

Design of the Off-axis Specimen: A Linear Elastic Finite Element Analysis

During loading, the transverse and shear compliances vary due to nonlinear viscoelasticity and damage growth. Consequently, the tab angle, ϕ, necessary to maintain a nearly uniform stress/strain field changes as the ratio of $\bar{S}_{xy}/\bar{S}_{xx}$ changes (due to degradation of matrix-dominated properties E_2 and G_{12}). Variation of tab angle, ϕ, with degree of material degra-

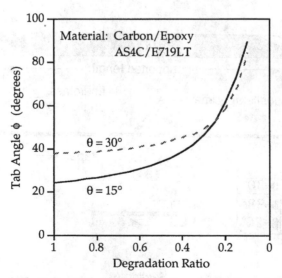

FIG. 4—Variation of tab angle with level of material degradation. Degradation Ratio = Degraded E_2/Initial E_2 = Degraded G_{12}/Initial G_{12}.

dation is demonstrated in Fig. 4. For this reason, a finite element study was performed to determine which aspect ratio is best. A two-dimensional, plane stress model with rigid tabs was developed using ABAQUS for the off-axis specimen. Rigid tabs are conservative as any compliance in the tabs will allow the specimen to deform in shear. Full details about the model can be found in Ref 8.

The most realistic model would have nonlinear viscoelastic material properties as inputs. As these properties are stress and time-dependent, they degrade nonuniformly throughout the specimen due to stress concentrations. The model would need to be stepped through time to allow a redistribution of stresses and a corresponding redistribution of material properties to occur. However, without a priori knowledge of these properties, we simply degraded matrix dominated secant moduli, E_2 and G_{12} uniformly to 20% of their original values to simulate the effect of nonlinear viscoelasticity and damage.

Model Results

An appropriate load was applied to the rigid tab such that a 0.1% axial strain is predicted by transformation of the elastic properties to the axes coincident with the loading direction, x. The resulting transverse strain fields in the 15° off-axis specimen for the initial and degraded conditions are shown in Fig. 5, where each contour interval represents 1% difference in strain from the uniform value predicted by transforming the material properties. As the largest difference in strains is in the transverse direction of the 15° configuration, only these results are shown. For a more complete picture of the strain fields in 15° and 30° off-axis models, the reader is referred to Ref 8. To aid in interpreting the transverse strain contours,

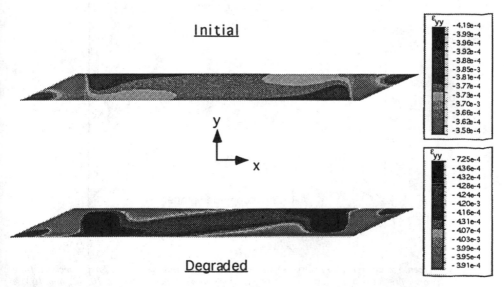

FIG. 5—*Variation of transverse strain in a 15° off-axis specimen of aspect ratio 12 with rigid angle—tabs.*

note that the in-plane Poisson's ratios predicted by transformation of the material properties are 0.375 initially and 0.409 in the degraded condition.

Special attention should be placed on the center section of each specimen as strain gages were attached in this region. There is a dramatic redistribution of strains when the material properties are degraded. For all levels of degradation, axial strain in the angle-tab configurations are within 1.5% of the uniform value in the area of the strain gages (3% for rectangular tabs with aspect ratio 18). Therefore, transverse strain at the centerline of the 15° model was used as a guide for determining an adequate specimen length in the degraded condition.

Figure 6 shows the percentage difference in transverse strain at the centerline across the specimen width for various aspect ratio models. In the degraded condition, the angle-tab design offers a more uniform strain field than the rectangular tab specimen with aspect ratio 18 (largest possible with the available composite plates). An aspect ratio of 12 (15.24 cm (6 in.) length and 1.27 cm (0.5 in.) width) significantly reduces the error in transverse strain over that offered by the rectangular tab configuration, while optimizing use of the available plates. This value was chosen to perform all testing of off-axis samples.

Performance of the Off-axis Specimen Design

Two adjacent 15° specimens were cut from the same plate of material and affixed with angled tabs, one with aspect ratio 17 and the other with aspect ratio 12. Figure 7 shows a comparison of the transverse strains, ε_y, for these two configurations. At the highest stress level prior to failure, and of course the highest degree of material degradation achieved,

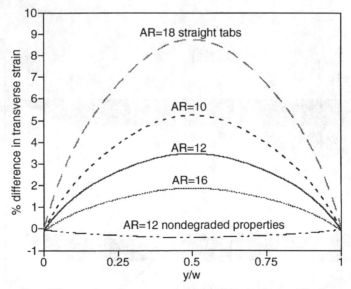

FIG. 6—*Percentage difference in transverse strains across the width of a 15° off-axis specimen from the uniform value predicted by material property transformation. Aspect ratios of 10, 12, and 16 are shown for the angle-tab model in the degraded condition. Aspect ratio of 12 for the non-degraded condition is shown for comparison.*

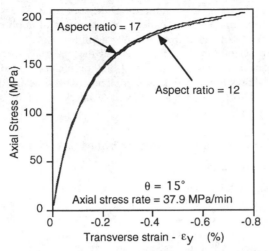

FIG. 7—*Axial stress versus transverse strain in two 15° off-axis specimens of aspect ratios 12 and 17.*

transverse strains are in agreement to within 9%. As some of this difference may be attributed to specimen variation (seen especially at high stresses), the lower aspect ratio design was felt to yield sufficiently accurate strain information.

Use of a taper on the angle-tabs allowed higher stresses to be reached prior to failure. Failures predominantly occurred in the unsupported length once taper angles were reduced to those given in Fig. 3. However, this configuration was primarily successful for the 6 and 12-ply samples, the few 24-ply off-axis samples tested failed at the tabs. Stress concentration at the tabs is greatest for the thicker samples due to the delivery of higher loads. Recognizing that a reduction in taper angle, β, reduces the stress concentration [20], further reductions in taper angle may yield better results for the thickest specimens.

Specimen Preparation and Test Conditions

Desiccated 0°, 90°, $(\pm 45)_{ns}$, and off-axis specimens were prepared with bonded fiberglass tabs and tested at three constant stress rates at room temperature. Each sample was equipped with micromeasurements strain gages and transverse strains were corrected according to the manufacturers specifications [21]. Gages were placed on both faces of each specimen to correct for out-of-plane bending. A dummy specimen of identical fiber orientation and gage configuration was used to compensate for temperature fluctuations. Further details on specimen preparation and testing can be found in Ref 8.

Demonstration of Time Dependence and Nonlinearity in Material Response

The rate or time-dependent nature of this rubber-toughened composite at room temperature is plainly demonstrated in the axial stress-strain behavior of three 30° off-axis specimens, as shown in Fig. 8*a*. An increase in compliance with decreasing stress rate is seen as the material is allowed more time to creep under each increment of applied load.

To separate this time dependence from possible nonlinearity, the axial compliance versus time can be examined at constant levels of stress. If the material is linear, but viscoelastic, there will be no dependence on stress level and the compliance will only be a function of time [22]. Figure 8*b* shows the variation of compliance with time for the same 30° off-axis samples at various levels of stress. Clearly, this material exhibits strong nonlinear behavior.

Theoretical Approach and Material Characterization

Constitutive Equation—A Quasi-elastic Model

A constitutive theory which incorporates effects of nonlinear viscoelasticity and time-dependent damage growth is needed to make long-term durability predictions for polymeric composites (e.g., that proposed by Schapery [3,4]). However, our purpose currently is to first characterize the dry, room temperature response for the initial loading of the material and compare the glass and carbon material systems. To that end, a simple constitutive equation was used which has far fewer material parameters. This model offers a simple way to incorporate nonlinearity and time dependence in a material model for limited loading conditions and to quickly compare the glass and carbon systems. We define ΔS_{ij} through,

$$S_{ij} = S_{ij}^0 + \Delta S_{ij}(\sigma_o, t) \tag{7}$$

where S_{ij}^0 are constant elastic compliances, the time under load is t and the effective stress σ_0 is given in Eq 4. The ΔS_{ij} obey a power-law in t and σ_0

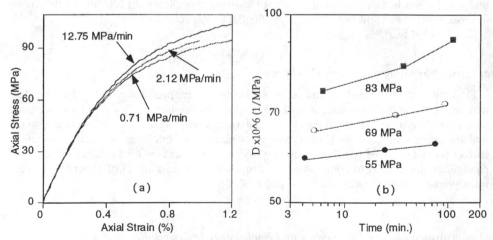

FIG. 8—(a) *Axial stress-strain response of three 30° off-axis specimens loaded with three different stress rates.* (b) *Axial secant compliance versus time for constant levels of stress as taken from three 30° off-axis specimens tested at different stress rates.*

$$\begin{Bmatrix} \varepsilon_1 \\ \varepsilon_2 \\ \varepsilon_6 \end{Bmatrix} = \begin{bmatrix} S_{11}^0 & S_{12}^0 & 0 \\ S_{12}^0 & S_{22}^0 + \dfrac{A}{C}\sigma_0^p t^n & 0 \\ 0 & 0 & S_{66}^0 + A\sigma_0^p t^n \end{bmatrix} \begin{Bmatrix} \sigma_1 \\ \sigma_2 \\ \sigma_6 \end{Bmatrix} \qquad (8)$$

where A, C, p, and n are constants. This form has been used by several researchers in the past to model this type of material [e.g., Ref 5] and allows the effect of loading rate to be isolated in the time exponent, n. It is demonstrated in Appendix A that this model can be derived from the more general model proposed by Schapery if certain simplifications are made.

Equation 8 is convenient for several reasons. First, strain behavior can be easily viewed as a function of stress level by plotting $\Delta S_{66}^0 \sigma_x$ and $\Delta S_{22}^0 \sigma_x$ against axial stress, as shown for some sample unidirectional specimens in Fig. 9. Appropriate elastic strains, $S_{12}^0 \sigma_x$, $S_{22}^0 \sigma_x$, and $S_{66}^0 \sigma_x$, with S_{ij}^0 assumed constant, have been subtracted from the data (S_{12}^0 as seen in Eq 8 is known from testing 0° unidirectional samples). Figure 9 has been plotted with logarithmic scales to clearly show that these compliances behave approximately as a power law in stress over a significant portion of the curve, as indicated by the straight regions. Beyond some threshold level of stress, there is a deviation from this power law behavior as indicated by the upward swing of the curves. Second, data on other polymer

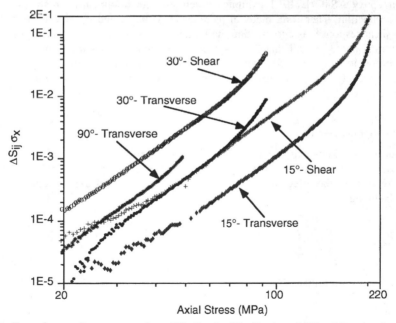

FIG. 9—*Example compliance curves from 30° off-axis, 15° off-axis and 90° unidirectional specimens. Compliances were multiplied by axial stress in order to avoid amplifying scatter in the strain data prior to curve fitting.*

composites shows that the time-dependent components of S_{22} and S_{66} are approximately proportional to one another [5,14,23,24]. That is

$$\Delta S_{66} = C \Delta S_{22} \tag{9}$$

where C is the constant in Eq 8. As shown in Fig. 9, this relation was found to hold over a significant range of the loading curve. For both off-axis angles shown, transverse compliances are essentially a vertical shift from their shear counterparts. In order for this relation to hold true, the initial secant shear compliance, S_{66}^0, and a value approximately 1.5% lower than the estimated initial value of S_{22}^0 (due to slight nonlinearity in this term) was subtracted from the transverse data.

In light of this behavior, and recalling that the data are for different stress rates, material compliance can be formulated over an appropriate stress range as

$$\Delta S_{66} = C \Delta S_{22} = \frac{A}{\dot{\sigma}_0^n} \sigma_0^m \tag{10}$$

where A and n are constants from Eq 8 and $m = p - n$. Thus, to characterize this material for constant stress and constant stress-rate conditions, the elastic terms and parameters A, C, m, n, and c_3 (cf. Eq 4) must be extracted from experimental data.

Experimental Results and Data Reduction

Carbon/epoxy AS4C/E719LT exhibited much stronger nonlinearity in shear than in the transverse direction. Therefore, material parameters A, m, n, and c_3 were evaluated strictly from the shear response, as curve-fitting the larger strain readings is less susceptible to error from electrical noise (Recall that $\Delta S_{66} = C \Delta S_{22}$). Separate values of A and m were found for the unidirectional and $(\pm 45)_{ns}$ samples to check for possible residual stress effects. These effects imply the $(\pm 45)_{ns}$ samples start at an elevated stress state prior to any mechanical loading.

Power-law exponent—m

Without a priori knowledge of the value of c_3 in the effective stress, we must work with axial stress and axial stress rate. However, since all stresses are proportional in load control, they are related by the effective stress parameter c_3, and for off-axis specimens, the fiber angle θ. Defining

$$h(\theta, c_3) = \frac{\sigma_0}{\sigma_x}, \tag{11}$$

using Eq 4 and noting that

$$\sigma_2 = \sigma_x \sin^2 \theta \tag{12}$$

$$\sigma_6 = -\sigma_x \cos\theta \sin\theta \tag{13}$$

TABLE 1—*Relation between the axial and effective stress for each specimen type. Note only the shear stresses are significant in the* $(\pm 45)_{ns}$ *samples.*

Specimen Type	$h(\theta,c_3)$
Off-axis	$\sin\theta\sqrt{\sin^2\theta + c_3\cos^2\theta}$
$(\pm 45)_{ns}$	$\sqrt{c_3}/2$
90°	1

we get the relations given in Table 1 for $h(\theta,c_3)$. Since C is typically greater than 3 and σ_2 is less than 17% of σ_6 in the $(\pm 45)_{ns}$ samples, σ_2 has been neglected. Its inclusion would only affect the results by 0.3%. The time-dependent properties in Eq 10 may then be written as

$$\Delta S_{66} = C\Delta S_{22} = \frac{Ah^p(\theta,c_3)}{\overline{\sigma}_x^n}\,\sigma_x^m = A_1\sigma_x^m \qquad (14)$$

The power law coefficient, A_1, and exponent, m, can be found from the change in transverse and shear compliance of each specimen type on the logarithmic scale of Fig. 9. Stress exponents extracted from off-axis and $(\pm 45)_{ns}$ samples are shown in Table 2. Note that a much broader spread occurs for the off-axis layups. A power law exponent of $m = 2.5$ was used to evaluate all other material properties from unidirectional specimens. Notice that the values of m determined from the $(\pm 45)_{ns}$ and off-axis specimens are different, as anticipated from residual stress effects.

Ratio of Nonlinear Compliances—C

With an exponent of $m = 2.5$ for off-axis samples, a linear elastic transverse compliance, S_{22}^0, and shear-to-transverse ratio, C, were then found to match the slope, m, and intercept, A_1 in Fig. 9. An average value of $C = 5.4$ was found from seven 30° and six 15° off-axis specimens. This value is higher than any reported previously implying there is an unusually large shear compliance relative to the transverse compliance; the largest reported is $C = 4.2$ for another rubber-toughened system [5].

TABLE 2—*Range of power law exponents extracted from off-axis and angle-ply samples.*

Specimen Type	Minimum m	Maximum m
Off-axis	2.27	2.65
$(\pm 45)_{ns}$	2.23	2.35

Normalization for the Effect of Stress Rate—n

Normalization for the rate effect was performed by dividing the change in compliance ΔS_{ij} for each specimen type by the time under load raised to some exponent, n. That is, with the correct n

$$\frac{\Delta S_{66}}{t^n} = \frac{C\Delta S_{22}}{t^n} = A[h(\theta,c_3)\sigma_x]^p \qquad (15)$$

should be the same for each specimen type regardless of load rate; $h\sigma_x$ is the effective stress which has not yet been determined. The exponent n was varied and the mean square error between compliance curves for different stress rates was calculated. Due to plate-to-plate variability, only comparisons between specimens cut out of the same plate were considered in determining an average value of n inherent to the material. Table 3 shows the average values of n determined from the off-axis and $(\pm 45)_{ns}$ configurations. A value of $n = 0.15$ was chosen as the best time exponent (subsequent creep testing for the time-scale of the constant stress rate tests yielded the same exponent), however, specimen-to-specimen variations still complicated this procedure. Figures 10 and 11 show the result of this normalization for $(\pm 45)_{ns}$ and 30° off-axis specimens. The top of Fig. 10 shows ΔS_{66} derived from $(\pm 45)_{ns}$ samples prior to normalization for the effect of stress rate. The bottom figure shows normalized curves.

Correlation of Stress States—c_3 and the Power Law Coefficient—A

Having determined the effect of stress rate, stress level, and relationship between the transverse and shear compliances, we need only to correlate the material response for different combinations of tension and shear stresses and thus find c_3. Only the unidirectional samples were used for this purpose as the $(\pm 45)_{ns}$ laminates showed a slightly different response. After c_3 has been found, agreement of the $(\pm 45)_{ns}$ specimens with this correlation using σ_0 will be examined.

Returning to Eq 15, and since all off-axis data were curve fit with $m = 2.5$ (forcing all variation into the power law coefficient—A_1), c_3 can be found by introducing the rate-normalized compliance coefficient, A_2, where

$$A = \frac{A_2(\theta)}{h^p(\theta,c_3)} = CA_2(90°). \qquad (16)$$

TABLE 3—*Average values of the time exponent,* n, *as derived from each specimen configuration.*

Specimen Configuration	Average n
15° Off-axis	0.146
30° Off-axis	0.138
$(\pm 45)_{ns}$	0.157

The constant A is the power law coefficient introduced in Eq 8. Notice that extracting the effective stress parameter through the 15° and 30° off-axis specimens is desirable as the result is influenced by only the time exponent, n, and not C. However, the effect of the different stress states in the two off-axis angles chosen was not significant enough to overcome the large specimen-variation seen for this material. A correlation may still be made between the two off-axis angles and 90° data making use of the ratio, C. Using Eq 16, and Table 1

$$c_3 = \frac{1}{\cos^2 \theta} \left\{ \left[\frac{1}{\sin\theta} \left(\frac{A_2(\theta)}{CA_2(90°)} \right)^{1/p} \right]^2 - \sin^2 \theta \right\}. \tag{17}$$

A value of $c_3 = 4.4$ best correlated the off-axis to the 90° data. This c_3 is just outside of the

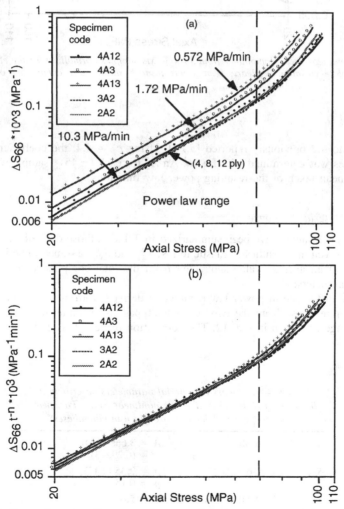

FIG. 10—*Nonlinear shear compliance as a function of axial stress curve fit from* $(\pm 45)_{ns}$ *laminates. Both the* (a) *actual response and* (b) *curves normalized for the effect of stress rate are shown.*

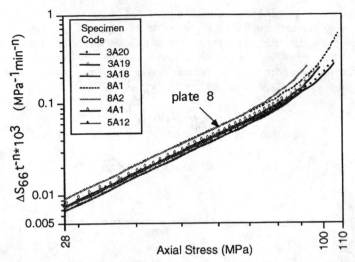

FIG. 11—*Nonlinear shear compliance from 30° off-axis samples normalized for the effect of loading rate. Plate-to-plate variation is clearly seen, where plate 8 shows more compliance over the others tested.*

range of 2.5 to 4.2 previously reported [5,14]. Using $c_3 = 4.4$, the coefficient A for both specimen types was determined with Eq 16. An A_2 for each $(\pm 45)_{ns}$ sample was found and a geometric mean taken of the resulting power law functions.

Summary of Material Response

Final parameter values have been summarized in Table 4; those derived from the $(\pm 45)_{ns}$ laminates are given in parentheses. Elastic terms, S_{22}^0 and S_{66}^0, assumed to be independent of stress, represent an average value subtracted from the total compliances in evaluating the power law parameters.

Despite the difference in power law parameters derived from unidirectional and $(\pm 45)_{ns}$ samples, shear response from the two specimen types are in quite good agreement in the power law range as shown in Fig. 12. The correlation made between stress states, using c_3

TABLE 4—*Summary of material parameters describing the behavior of AS4C/E719LT in the nonlinear range. Parameters for the $(\pm 45)_{ns}$ laminates are given in parentheses.*

$S_{11}^0 = 7.90\mathrm{e}{-}6 \ \mathrm{MPa}^{-1}$	$A = 3.66\mathrm{e}{-}3 \ \mathrm{MPa}^{-(p+1)}\mathrm{min}^{-n}$
	$\quad\ = (6.29\mathrm{e}{-}3)$
$S_{12}^0 = -2.61\mathrm{e}{-}6 \ \mathrm{MPa}^{-1}$	$p = 2.35 \ (2.13)$
	$n = 0.15$
$S_{22}^0 = 1.23\mathrm{e}{-}4 \ \mathrm{MPa}^{-1}$	$C = 5.4$
$S_{66}^0 = 2.19\mathrm{e}{-}4 \ \mathrm{MPa}^{-1} \ (2.21\mathrm{e}{-}4)$	$c_3 = 4.4$

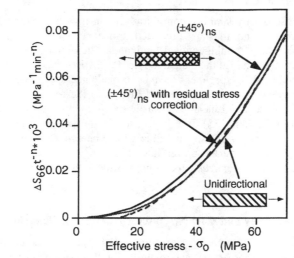

FIG. 12—*Comparison of the time normalized nonlinear shear compliance as found from* $(\pm 45)_{ns}$ *and unidirectional off-axis laminates versus effective stress. The unidirectional curve with residual stress has had 2.07 MPa (300 psi) of residual stress added.*

derived from unidirectional samples, is, therefore, applicable to the $(\pm 45)_{ns}$ samples. Qualitatively, the addition of 13.8 MPa (2 ksi) residual stress, σ_2, into σ_0 for the $(\pm 45)_{ns}$ laminate shows better agreement between the two curves. However, the amount of residual stress is significantly less than the 21 to 28 MPa which would be predicted immediately after cool down from the glass transition temperature, apparently indicating a significant amount of stress relaxation has occurred in these five-year-old samples.

Stress Range for Power Law Behavior

Table 5 shows the average stress ranges for each specimen type over which the nonlinear shear and transverse compliances exhibited power law behavior to within 5%. Apparently, the effective stress chosen is adequate for indicating the onset of the power law region, as shown in the third column of Table 5. However, it does a poor job in correlating the maximum stress for which a power law is applicable, as indicated in the fifth column.

TABLE 5—*Stress ranges in which nonlinear power law behavior exists.*

Specimen Configuration	Min σ_x (MPa)	Min σ_0 (MPa)	Max σ_x (MPa)	Max σ_0 (MPa)	Failure σ_x (MPa)
15°	51.0	25.5	154.5	78.6	217
30°	27.6	25.5	75.2	71.0	105
90°	23.1	23.1	45.5	45.5	48
$(\pm 45)_{ns}$	20.0	20.7	69.0	72.4	—

Prediction of Stress-strain Curves

Stress-strain response of any laminate configuration can now be predicted and compared to experiment. Figures 13 and 14 show predicted and experimental axial stress-strain response for 30° off-axis and 90° unidirectional laminates. The dotted lines indicate the range of responses from samples tested at the indicated rate. Predictions based on properties derived from unidirectional and (± 45)$_{ns}$ specimens have been made and are in good agreement. Each figure has been segregated by horizontal dashed lines into the regions in which linear and nonlinear behavior was observed. Linear viscoelastic creep functions from preliminary testing of the same material by Soriano [25] have been used in the linear portions of each stress-strain curve. Material parameters as listed in Table 4 have been used in the power law region. Predicted strain at stresses above the power law range are simply an average of the data from samples tested at that rate; recall the time normalization used here was ineffective beyond the power-law range.

Comparison of Glass/epoxy and Carbon/epoxy Deformation Behavior

A glass/epoxy composite with Owens-Corning 158B E-glass fibers and presumably identical resin to that in the carbon/epoxy was manufactured, desiccated and tested under the

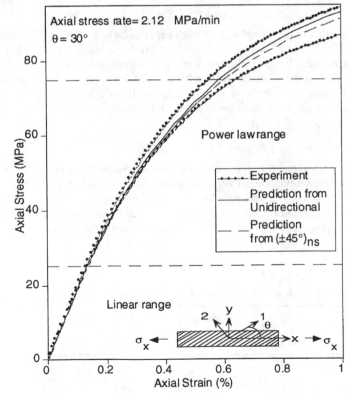

FIG. 13—*Experimental and predicted axial stress-strain response of a unidirectional off-axis laminate with a fiber angle of 30°.*

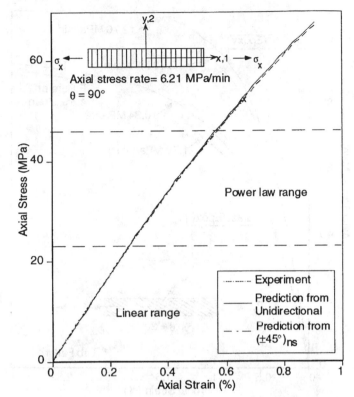

FIG. 14—*Experimental and predicted axial stress-strain response of a unidirectional off-axis laminate with a fiber angle of 90°.*

same conditions as the carbon/epoxy. Figures 15 and 16 show the stress-strain response of 30° off-axis and 90° unidirectional samples. Qualitatively, the glass/epoxy demonstrates a greater degree of ductility and nonlinearity prior to failure, despite having the same resin. An important question is whether or not this difference can be attributed to the different elastic response of the fibers or whether this implies different matrix behavior.

Prediction of the Glass/epoxy Response Through Micromechanics

An expedient way to examine the effect of the fibers on overall stiffness is through the semi-empirical Halpin-Tsai equations [9]

$$E_2 = E_m \frac{1 + \xi_1 \eta_1 V_f}{1 - \eta_1 V_f} \tag{18}$$

$$G_{12} = G_m \frac{1 + \xi_2 \eta_2 V_f}{1 - \eta_2 V_f} \tag{19}$$

where

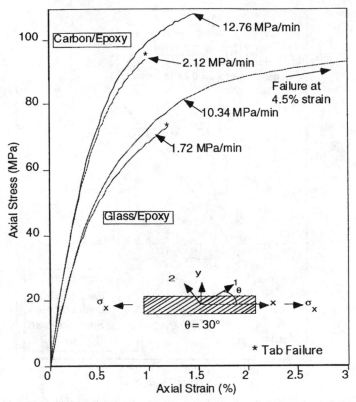

FIG. 15—*Comparison of 30° off-axis carbon/epoxy and glass/epoxy stress-strain response. Both materials have the same matrix.*

$$\eta_1 = \frac{E_{2f} - E_m}{E_{2f} + \xi_1 E_m} \tag{20}$$

$$\eta_2 = \frac{G_{12f} - G_m}{G_{12f} + \xi_2 G_m}. \tag{21}$$

Longitudinal modulus is E and shear modulus is G. Subscripts 'f' and 'm' stand for the fiber and matrix respectively. Fiber volume fraction is V_f. The ξ_1 and ξ_2 are transverse and shear reinforcing efficiency factors. Initial elastic properties derived from mechanical testing of both materials, along with fiber and matrix elastic properties [26,27], were used to estimate the reinforcing factors (Carbon/Epoxy: $\xi_1 = 2$, $\xi_2 = 1.4$; Glass/Epoxy: $\xi_1 = 1.15$, $\xi_2 = 1$). Although there is no explicit reason for us to assume these equations are accurate in the nonlinear range, they do serve at least to normalize the response. They are fairly accurate for linear elastic response and appropriately become proportional to the matrix properties when significant softening occurs (i.e., η_1 and η_2 go to 1).

Predicting the stress and time-dependence of the glass/epoxy requires that we first extract the in-situ matrix properties (representing smeared-out matrix properties between the fibers)

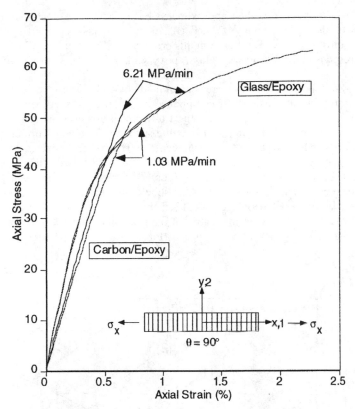

FIG. 16—*Comparison of 90° unidirectional carbon/epoxy and glass/epoxy stress-strain response. Both materials have the same matrix.*

from viscoelastic properties derived for the carbon/epoxy. These matrix properties are then substituted in the Halpin-Tsai equations along with properties of the glass fibers assuming the same reinforcing factors can be extended into the nonlinear range. Solving Eqs 18 and 19 for the matrix properties E_m and G_m

$$E_m = \frac{\beta_1 + \sqrt{\beta_1^2 + 4\xi_1 V_m^2 E_{2f} E_2}}{2\xi_1 V_m} \tag{22}$$

$$G_m = \frac{\beta_2 + \sqrt{\beta_2^2 + 4\xi_2 V_m^2 G_{12f} G_{12}}}{2\xi_2 V_m} \tag{23}$$

where

$$\beta_1 = E_2(\xi_1 + V_f) - E_{2f}(1 + \xi_1 V_f) \tag{24}$$

$$\beta_2 = G_{12}(\xi_2 + V_f) - G_{12f}(1 + \xi_2 V_f). \tag{25}$$

Time and stress-dependence is then approximated by replacing the elastic shear and transverse moduli of the carbon/epoxy, E_2 and G_{12}, with viscoelastic properties for the same composite effective stress, σ_0. Results from this process are shown in Fig. 17 where predicted stress-strain response of a 30° off-axis glass/epoxy sample is compared with micromechanical predictions.

Average carbon/epoxy properties were used in prediction {1}. The second prediction, {2}, allows nonlinearity in the elastic shear compliance, S_{66}^0, in the glass/epoxy; the initial value was sufficient for the carbon-epoxy. The largest difference, however, is in the magnitude of the nonlinear compliance, A. Prediction {3} is the same as {2} along with an increase in A by a factor of 2.2, which offers excellent agreement. That is, the same c_3 in the effective stress, stress exponent, p, and time exponent, n, are used to describe the behavior of the glass/epoxy. Thus, the softening in the glass/epoxy composite is more than predicted by simply replacing carbon with glass fibers. One should also notice that the applicable stress range of power-law type behavior is different (c.f. Figs. 13 and 17). For angles greater than 15°, the average effective stress in the matrix of the glass-fiber composite is actually slightly *smaller* than that in the carbon-fiber composite (for the same applied stress) [28], and thus

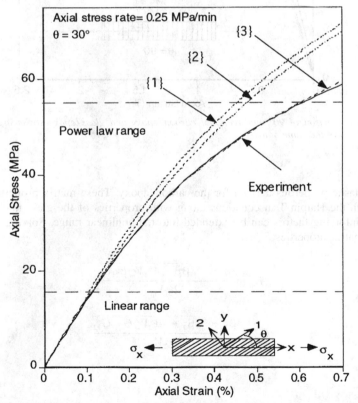

FIG. 17—*Experimental axial stress-strain response of a 30° off-axis glass/epoxy laminate compared with micromechanical predictions. (1) Average carbon/epoxy properties. (2) Allows nonlinearity in the elastic shear compliance in the glass/epoxy. (3) Is the same as (2) along with an increase in A by a factor of 2.2.*

the effect of fiber modulus difference cannot explain the higher degree of softening in the former composite.

These differences imply that the in-situ matrix behavior is different in the two materials. Differences in fiber-matrix adhesion [e.g., *29,30*] and the interphase regions may cause this behavior both directly (i.e., different mechanical properties locally) and indirectly by affecting the growth of damage in the microstructure. Acoustic emission and microscopy studies are currently underway to help determine the source of these differences.

Concluding Remarks

Nonlinear time-dependent mechanical properties of the carbon/epoxy AS4C/E719LT were characterized for the initial loading of the material for constant stress and stress-rate conditions. A large portion of the loading curve can be described by a quasi-elastic model of the shear and transverse compliances in terms of a single power-law scalar function of stress state, a ratio of compliances, a time or rate exponent, and two elastic terms. This model offers a simple way to incorporate nonlinearity and time dependence in a material model for limited loading conditions. Different material systems can also be quickly compared. Beyond the power law range, further analysis is needed to characterize behavior. Intrinsic variability in this material complicated the isolation of time and stress effects. In general, to avoid such a problem, creep testing is recommended to supplement this type of testing so time dependence can be isolated separately.

The off-axis specimen configuration with tapered-fiberglass angle tabs enabled us to derive nonlinear properties out to high levels of stress with relatively short samples. This is an important sample in a nonlinear study as it allows transverse and shear compliances to be related directly as both evolve concurrently during loading. In this way, their relationship is not influenced by specimen to specimen variation. Although a 10° off-axis specimen traditionally has been used in characterizing shear response, 15° and 30° specimens proved to be a better choice for this study as both shear and transverse properties can be assessed with the same specimen. Careful selection of off-axis angles, specimen length/width and tab-angle taper, as mentioned herein, is recommended prior to carrying out a full characterization.

The glass/epoxy displayed many commonalities but some differences in behavior from the carbon composite; for instance, magnitude of the nonlinearity and the stress range of power-law behavior differed. There may be different interface and interphase regions with different mechanical properties locally which in turn affect the growth of damage. A study of local deformation and damage growth in the two materials is needed to determine what similarities may exist on the microscale.

Acknowledgment

Sponsorship of this research by the National Science Foundation through the Offshore Technology Research Center is gratefully acknowledged.

References

[*1*] Hahn, H. T. and Tsai, S. W., "Nonlinear Elastic Behavior of Unidirectional Composite Laminae," *Journal of Composite Materials,* Vol. 7, 1973, pp. 102–118.
[2] Schapery, R. A. and Sicking, D. L., "On Nonlinear Constitutive Equations for Elastic and Visco-elastic Composites with Growing Damage," In: A. Bakker, Ed., *Mechanical Behavior of Materials,* Delft University Press, Delft, The Netherlands, 1995, pp. 45–76.

[3] Schapery, R. A., "Nonlinear Viscoelastic and Viscoplastic Constitutive Equations Based on Thermodynamics," *Mechanics of Time-Dependent Materials 1,* Kluwer Academic Publishers, Netherlands, 1997, pp. 209–240.

[4] Schapery, R. A., "Nonlinear Viscoelastic and Viscoplastic Constitutive Equations with Growing Damage," *International Journal of Fracture,* 1999, Vol. 97, in press.

[5] Mignery, L. A. and Schapery, R. A., "Viscoelastic and Nonlinear Adherend Effects in Bonded Composite Joints," *Journal of Adhesion,* 1991, Vol. 34, pp. 17–40.

[6] Schapery, R. A., "Mechanical Characterization and Analysis of Inelastic Composite Laminates with Growing Damage," *Mechanics of Composite Materials and Structures,* ASME AMD, Vol. 100, 1989, pp. 1–9.

[7] Carlsson, L. A. and Pipes, R. B., *Experimental Characterization of Advanced Composite Materials,* Prentice-Hall, Inc., 1987.

[8] Bocchieri, R. T., "A Baseline Nonlinear Material Characterization for Determining the Long-term Durability of Composite Structures," Master's Thesis, University of Texas at Austin, 1996.

[9] Daniel, I. M. and Ishai, O., *Engineering Mechanics of Composite Materials,* Oxford University Press, New York, 1994.

[10] Huang, Y., Hunston, D. L., Kinloch, A. J., and Riew, C. K., "Mechanisms of Toughening Thermoset Resins," In: C. K. Riew and A. J. Kinloch, Eds., *Toughened Plastics I—Science and Engineering,* American Chemical Society, Washington, 1993.

[11] Bucknall, C. B., *Toughened Plastics,* Applied Science Publishers, Ltd, London, 1977.

[12] Peretz, D. and Weitsman, Y., "Nonlinear Viscoelastic Characterization of FM-73 Adhesive," *Journal of Rheology,* Vol. 26, 1982, p. 245.

[13] Hiel, C., Cardon, A. H., and Brinson, H. F., "The Nonlinear Viscoelastic Response of Resin Matrix Composite Laminates," Technical Report, *NASA CR 3772,* 1984.

[14] Sun, C. T. and Chen, I. L., "A Simple Flow Rule for Characterizing Nonlinear Behavior of Fiber Composites. *ICCM+ECCM Second European Conference on Composite Materials,* 1987, p. 1.250.

[15] Friedrich, K., *Application of Fracture Mechanics to Composite Materials,* Elsevier, New York, 1989.

[16] Hahn, H. T., "Residual Stresses in Polymer Matrix Composite Laminates," *Journal of Composite Materials,* Vol. 10, 1976, pp. 266–278.

[17] Chamis, C. C. and Sinclair, J. H., "Ten-Deg Off-Axis Shear Properties in Fiber Composites," *Experimental Mechanics,* September 1977, pp. 339–346.

[18] Chang, B., Huang, P., and Dallas, S., "A Pinned-end Fixture for Off-axis Testing," *Experimental Techniques,* Vol. 8, 1984, pp. 339–346.

[19] Sun, C. T. and Chung, I., "An Oblique End-Tab Design for Testing Off-Axis Composite Specimens," *Composites,* Vol. 24, 1993, pp. 619–623.

[20] Cunningham, M., Schoultz, S., and Toth, J., "Effect of End-Tab Design on Tension Specimen Stress Concentrations," In: Vinson, J. and Taya, M. Eds., *Recent Advances in Composites in the United States and Japan, ASTM STP 864,* American Society for Testing and Materials, West Conshohocken, PA, 1985, pp. 253–262.

[21] Micromeasurements-TN-509, "Errors Due to Transverse Sensitivity in Strain Gages," Micromeasurements Group, Inc., 1982.

[22] McCrum, N., Buckley, C., and Bucknall, C., *Principles of Polymer Engineering,* Oxford University Press, 1988.

[23] Schapery, R. A., "Prediction of Compressive Strength and Kink Bands in Composites Using Work Potential," Technical Report, Report No. *SSM-94-1,* University of Texas, 1994.

[24] Zhang, L., "Time-Dependent Behavior of Polymers and Unidirectional Polymeric Composites," Ph.D. Thesis, Shanghai Jiao Tong University, 1995.

[25] Soriano, E., "Preliminary Test Results for Low Stress Level, First Cycle Creep of AS4C/E719LT in the Dry State at Room Temperature," University of Texas, Dept. of Aero. Eng. and Eng. Mechanics, 1996.

[26] Wood, C. A., "Determining the Effect of Seawater on the Interfacial Strength of an Interlayer E-Glass-Graphite/Epoxy Composite Using Observations of Transverse Cracking Made In-situ in an Environmental SEM," Master's Thesis, Texas A&M University, 1996.

[27] Correspondence with Scott Lewis of Hexcel Corp., 1997.

[28] Lou, Y. C. and Schapery, R. A., "Viscoelastic Characterization of a Nonlinear Fiber-Reinforced Plastic," *Journal of Composite Materials,* 1971, Vol. 5, pp. 208–234.

[29] Drzal, L. T., "Fibre-Matrix Adhesion and its Relationship to Composite Mechanical Properties," *Journal of Materials Science,* Vol. 28, 1993, pp. 569–610.

[30] Drzal, L. T. and Larson, B. K., "Interphase Formed by Glass Fiber Sizings and its Effects on Adhesion and Composite Properties," *Proceedings of the American Society for Composites Pro-*

ceedings of the 8th Technical Conference on the American Society of Composites, 1993, pp. 187–196.

[31] Schapery, R. A., "Further Development of Thermodynamic Constitutive Theory: Stress Formulation," Technical Report, Report No. 69-2, Purdue University, 1969.

[32] Schapery, R. A., "Simplifications in the Behavior of Viscoelastic Composites with Growing Damage," Inelastic Deformation of Composite Materials—IUTAM Symposium, Troy, New York, Edited by G. J. Dvorak, Springer-Verlag, 1990, pp. 695–706.

[33] Schapery, R. A., "On Viscoelastic Deformation and Failure Behavior of Composite Materials with Distributed Flaws," 1981 Advances in Aerospace Structures and Materials, AD-Vol. 1, ASME Edited by S. S. Wang and W. J. Renton, 1981, pp. 5–20.

[34] Fung, Y. C., Foundations of Solid Mechanics, Prentice-Hall, New Jersey, 1965.

Appendix A

The following discussion illustrates how the thermodynamically-based constitutive theory developed by Schapery [3,4] for nonlinear viscoelastic materials with growing damage can be simplified to the quasi-elastic model used in this paper through selection of specific functional dependence of the material functions. This theory has been generalized to account for behavior due to changes in the microstructure, such as are seen in the initial loading of the material, where it originally required an equilibrium level of damage (i.e., a mechanically conditioned material) [31]. By no means do we suggest that this experimental study is sufficient to fully define the parameters in this model. On the contrary, such definition will require extensive mechanical testing and monitoring of the damage mechanisms which occur.

Assuming a state of plane stress, a special version of the model can be written in 2-D form as

$$\varepsilon_i = S_{ij}^0 \sigma_j + \frac{\partial \hat{\sigma}_j}{\partial \sigma_i} \int_0^t \Delta S_{jk}^l (\psi - \psi') \frac{d(g_2 \hat{\sigma}_k)}{d\tau} \, d\tau \quad (i,j = 1,2,6) \tag{A-1}$$

where

$$\psi = \int_0^t \frac{dt}{a_\sigma} \tag{A-2}$$

$$\psi' = \int_0^\tau \frac{dt}{a_\sigma} \tag{A-3}$$

are reduced times. The S_{ij}^0 are elastic compliances and may be a function of stress and damage level. The ΔS_{jk}^l are the linear viscoelastic creep compliances. Parameters $\hat{\sigma}_j$, a_σ, and g_2 are stress, temperature, moisture and damage-dependent material properties. We first assume that $\hat{\sigma}_j = \sigma_j$ to perserve tensorial linearity and cause the term in front of the integral to disappear. Currently, specific dependence on environmental factors, such as temperature and moisture is not shown as this study is focused on establishing a baseline dry material response at room temperature. Guided by viscoelastic fracture mechanics, we assume that the damage, if any, depends on a Lebesgue Norm of stress [32,33]. Also, for simplicity, we assume that the only effect of damage can be accounted for through a scalar power-law function g_2. That is

$$g_2 \sim \left[\int_0^t \sigma_0^\beta dt' \right]^{\alpha/\beta} \tag{A-4}$$

where α and β are constants. Here, for monotone loading, $g_2 \sim t^r$ where $r = \alpha(\beta + 1)/\beta$. Damage effects are therefore combined with other stress effects.

Constant stress rate loading and the proportionality of stresses in all specimens tested allow several simplifications to be made to the material model. Under these conditions, all stresses (i.e., σ_1, σ_2, and σ_6) are related to the time under load by some time-independent stress rate, $\dot{\sigma}_0$ for instance, and a geometric parameter relating them to the effective stress, $h(\theta,c_3)$, from Eq 10. The variable of integration in Eqs A-1 through A-3 can then be changed to σ_0.

Rate or time dependence stems from both intrinsic viscoelasticity and damage growth. For this characterization, several assumptions were therefore made. Power law linear viscoelastic creep compliances,

$$\Delta S_{ij}^l = S_{ij}^l t^q \tag{A-5}$$

were assumed where t is the creep time and q and S_{ij}^l are constants. The damage-dependent material function g_2 and linear viscoelastic creep compliance are then multiplicative and we get a total time-dependence t^n, where $n = q + r$. Elastic compliances S_{ij}^0 have also been assumed to be constant. Inelastic compliances may then be defined as

$$\Delta S_{22} = \frac{\varepsilon_2 - S_{12}^0 \sigma_1 - S_{22}^0 \sigma_2}{\sigma_2} = \frac{1}{\dot{\sigma}_0^n} \left\{ \frac{S_{22}^l}{\sigma_0} \int_0^{\sigma_0} (\eta - \eta')^q \frac{d(g_2(\sigma_0')\sigma_0')}{d\sigma_0'} d\sigma_0' \right\} \tag{A-6}$$

$$\Delta S_{66} = \frac{\varepsilon_6}{\sigma_6} - S_{66}^0 = \frac{1}{\dot{\sigma}_0^n} \left\{ \frac{S_{66}^l}{\sigma_0} \int_0^{\sigma_0} (\eta - \eta')^q \frac{d(g_2(\sigma_0')\sigma_0')}{d\sigma_0'} d\sigma_0' \right\} \tag{A-7}$$

where

$$\eta = \psi\dot{\sigma}_0 = \int_0^{\sigma_0} \frac{d\sigma_0}{a_\sigma(\sigma_0)} \tag{A-8}$$

$$\eta' = \psi\dot{\sigma}_0 = \int_0^{\sigma_0'} \frac{d\sigma_0}{a_\sigma(\sigma_0)} \tag{A-9}$$

The quantity σ_0' is the effective stress at time τ. Therefore, both ΔS_{22} and ΔS_{66} are functions of the effective stress as shown in the braces of Eqs A-6 and A-7 and independent of loading rate. This function was seen in Fig. 9 to be a power-law. Thus we arrive at Eq 10

$$\Delta S_{66} = C\Delta S_{22} = \frac{A}{\dot{\sigma}_0^n} \sigma_0^m \tag{A-10}$$

and the simplified constitutive equation used in this paper, Eq 8.

As a final note, elastic compliances S_{ij}^0 and linear viscoelastic compliances S_{ij}^l are symmetric on the basis of thermodynamics [34]. With $\hat{\sigma}_j = \sigma_j$ and noting that a_σ and g_2 are scalars, the constitutive Eq A-1 reduces to Eq 1 in which the total compliance matrix S_{ij} is also symmetric.

Changming Zhu[1] and Chin-Teh Sun[2]

A Viscoplasticity Model for Characterizing Loading and Unloading Behavior of Polymeric Composites

REFERENCE: Zhu, C. and Sun, C. T., **"A Viscoplasticity Model for Characterizing Loading and Unloading Behavior of Polymeric Composites,"** *Time Dependent and Nonlinear Effects in Polymers and Composites, ASTM STP 1357,* R. A. Schapery and C. T. Sun, Eds., American Society for Testing and Materials, West Conshohocken, PA, 2000, pp. 266–284.

ABSTRACT: An overstress viscoplasticity model is proposed to describe the rate-dependent behavior of a polymeric composite during loading and unloading. In the model, a three-parameter function is used to describe viscoplastic strain rate. In the loading stage, the equilibrium stress is determined using a multi-step relaxation test performed during loading. During the initial unloading stage, owing to the fact that the viscoplastic strain rate is still positive, the material still experiences "loading," and the corresponding equilibrium stress is the equilibrium stress-strain curve for loading. In the second unloading stage, the viscoplastic strain rate becomes negative, and the material is in a true unloading mode for which the equilibrium stress is determined again using the multi-step "relaxation" test. The viscoplasticity model is found to be capable of capturing the characteristics of the rate-dependent loading and unloading behavior.

KEYWORDS: viscoplasticity, equilibrium stress, loading and unloading, multi-step relaxation, polymeric composites, strain rate

Carbon fiber reinforced polymeric composites have emerged as an important class of structural materials. Except for loading in the fiber direction, these materials generally exhibit nonlinear and rate-dependent behavior. Viscoplasticity models that can describe the nonlinear and rate-dependent behavior of composites are highly desirable.

Several models have been proposed to describe the rate-dependent behavior of polymeric composites. Utilizing the one-parameter plastic potential function introduced by Sun and Chen [1] along with the overstress concept introduced by Malvern [2], Gates and Sun [3] proposed an elastic/viscoplastic constitutive model for describing rate-dependent behavior of AS4/PEEK thermoplastic composite at room temperature. Later, this model was extended to characterize the behavior of polymeric composites at both room and elevated temperatures for AS4/PEEK [4], IM7/5260, and IM7/8320 [5]. In the above investigations, a two-parameter function was used to describe the viscoplastic strain rate. Recently, Weeks and Sun [6] conducted experiments on AS4/PEEK using both the MTS machine and a Split Hopkinson Pressure Bar to study low to high strain rate behavior of AS4/PEEK composite. Instead of the overstress viscoplasticity model, a rate-dependent power law model based on

[1] Currently, Research Scientist, Dresser-Rand, Olean, NY.

[2] Neil A. Armstrong Professor, School of Aeronautics and Astronautics, Purdue University, West Lafayette, IN 47907-1282.

the one-parameter plasticity model [1] and a modified Johnson rate-dependent model [7] were employed to describe the viscoplastic behavior of AS4/PEEK.

Until now, most investigations have focused on monotonic loading, while much less effort has been made to examine unloading behavior, which may be highly nonlinear and rate-dependent for polymeric composites. The main purpose of the present research is to investigate the loading and unloading behavior of the IM7/5260 polymeric composite, using off-axis specimens. Because the overstress viscoplasticity model is relatively simple, it was used by choosing a different functional form (i.e., a three-parameter function) to describe the viscoplastic strain rate. A multi-step relaxation procedure was used to establish the equilibrium stress for both loading and unloading. The viscoplasticity model was used to predict the rate-dependent loading and unloading behavior of the IM7/5260 composite.

Experimental Procedure and Results

The material system selected for the research was the carbon fiber reinforced bismaleimide IM7/5260. Coupon specimens with fiber orientations 0°, 15°, 30°, and 45° were cut from a 12-plied unidirectional composite panel, which was 0.168 cm thick. The panel was provided by NASA Langley Research Center. The shape and dimensions of the specimens are given in Fig. 1.

All tests were performed using a closed-loop servo-hydraulic MTS 22 kip machine at room temperature. A stroke control mode was selected for all tests. Three different strain

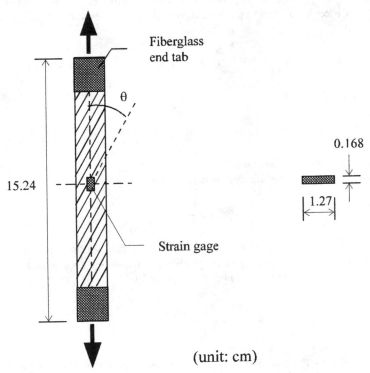

FIG. 1—*Shape and dimensions of specimen.*

rates 0.00001/sec, 0.001/sec and 0.1/sec were chosen to conduct the tests. In a loading-unloading cycle, the loading and unloading rates could be different. Strains were measured by using two Micro-Measurement EA-13-125AC-350 strain gages mounted back-to-back at the center of the specimen. Data were collected using an analog to digital signal converter and were recorded by a Gateway 2000 computer using LabVIEW.

Figure 2 shows typical stress-strain curves for constant loading and unloading at three different strain rates of 0.00001/s, 0.001/s, and 0.1/s, respectively. It is clear that both loading and unloading curves are nonlinear.

Figures 3 and 4 show typical results for tests with different loading and unloading rates. Compared with the case of constant loading-unloading rate, the slow-loading-fast-unloading case results in a straighter unloading curve and a larger strain when unloading stress approaches zero (Fig. 3), while the fast-loading-slow unloading case results in a much more distorted unloading curve during the initial unloading stage (Fig. 4). These phenomena indicate that the unloading behavior of IM7/5260 composite strongly depends on its preceding loading rate history.

The above strain rate-dependent loading-unloading behavior was also observed by Kremple and Kallianpur [8] for metals and by Bordonaro and Kremple [9] for some polymers. In Ref 8, models were proposed to interpret the effect of the jump in loading rate on the subsequent unloading behavior.

Overstress Viscoplasticity Model

The general description of the overstress viscoplasticity model can be found elsewhere [2] and [3]. In this model, the strain rate is decomposed into an elastic part and a viscoplastic part

$$\dot{\varepsilon}_{ij} = \dot{\varepsilon}_{ij}^e + \dot{\varepsilon}_{ij}^p \tag{1}$$

The stress rate is given by

$$\dot{\sigma}_{ij} = c_{ijkl}\dot{\varepsilon}_{kl}^e = c_{ijkl}(\dot{\varepsilon}_{kl} - \dot{\varepsilon}_{kl}^p) \tag{2}$$

where c_{ijkl} is the elastic constants tensor.

The viscoplastic strain rates are assumed to be of the form

$$\dot{\varepsilon}_{ij}^p = F\xi_{ij} \tag{3}$$

where F may be a function of stress, temperature, and equilibrium stress (i.e., a state of stress produced at vanishing loading rates); and ξ_{ij} are quantities used to describe the direction of plastic flow. In the present model, ξ_{ij} is given by

$$\xi_{ij} = \frac{\partial f(\sigma)}{\partial \sigma_{ij}} \tag{4}$$

where $f(\sigma)$ is the plastic potential function. For composite materials in the plane stress condition, this function reduces to (Sun and Chen [1])

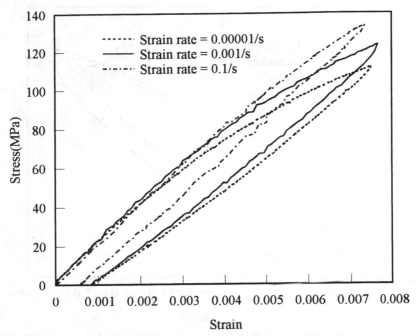

FIG. 2—*Experimental results for 30° specimens.*

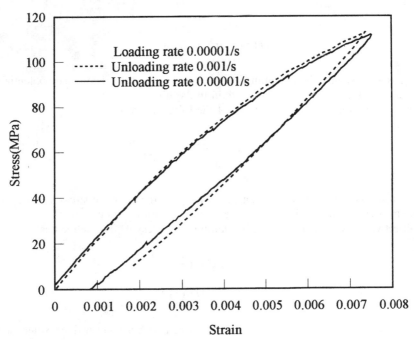

FIG. 3—*Experimental results for constant loading and unloading and slow loading-fast unloading cases (fiber orientation 30°).*

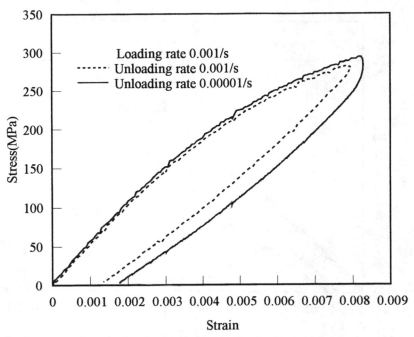

FIG. 4—*Experimental results for constant loading and unloading and fast loading-slow unloading cases (fiber orientation 15°).*

$$2f(\sigma) = \sigma_{22}^2 + 2a_{66}\sigma_{12}^2 \tag{5}$$

where a_{66} is a constant to be determined from experimental data. The plastic potential given by Eq 5 implies that there is no plastic strain in the fiber direction.

The scalar function F is assumed to be of the following form.

$$F = B\,\frac{1}{\bar{\sigma}}\left|\sinh \zeta\,\frac{\bar{\sigma} - \bar{\sigma}^*}{\bar{\sigma}^*}\right|^{1/m}\,\mathrm{sign}(\bar{\sigma} - \bar{\sigma}^*) \tag{6}$$

where, B, ζ and m are material constants, $\bar{\sigma}$ and $\bar{\sigma}^*$ are the effective stress and the effective equilibrium stress, respectively, and sign (x) is the signum function.

The effective stress $\bar{\sigma}$ and the effective equilibrium stress $\bar{\sigma}^*$ are defined as

$$\bar{\sigma} = \sqrt{3f(\sigma)} \tag{7}$$

$$\bar{\sigma}^* = \sqrt{3f(\sigma^*)} \tag{8}$$

where σ_{ij} and σ_{ij}^* are the "dynamic" stress (produced with a nonvanishing strain rate) and the equilibrium stress, respectively.

Combining Eqs 3, 4 and 6, we obtain the plastic strain rates

$$\dot{\varepsilon}_{ij}^{p} = B \, \frac{1}{\bar{\sigma}} \, \left| \sinh \zeta \, \frac{\bar{\sigma} - \bar{\sigma}^{*}}{\bar{\sigma}^{*}} \right|^{1/m} \, \mathrm{sign}(\bar{\sigma} - \bar{\sigma}^{*}) \, \frac{\partial f(\sigma)}{\partial \sigma_{ij}} \qquad (9)$$

The above functional form for the viscoplastic strain rates is different from the power form used by Gates and Sun [3], Yoon and Sun [4], and Gates [5]. The present functional form has one more parameter and allows more accurate fittings between the model and experimental data.

For a state of plane stress, we obtain from Eqs 5 and 9

$$\dot{\varepsilon}_{11}^{p} = 0$$

$$\dot{\varepsilon}_{22}^{p} = B \, \frac{1}{\bar{\sigma}} \, \left| \sinh \zeta \, \frac{\bar{\sigma} - \bar{\sigma}^{*}}{\bar{\sigma}^{*}} \right|^{1/m} \, \mathrm{sign}(\bar{\sigma} - \bar{\sigma}^{*}) \sigma_{22} \qquad (10)$$

$$\dot{\gamma}_{12}^{p} = B \, \frac{1}{\bar{\sigma}} \, \left| \sinh \zeta \, \frac{\bar{\sigma} - \bar{\sigma}^{*}}{\bar{\sigma}^{*}} \right|^{1/m} \, \mathrm{sign}(\bar{\sigma} - \bar{\sigma}^{*}) 2a_{66} \sigma_{12}$$

In the above model, the equilibrium stress needs to be determined. A simple way to establish the equilibrium stress is to perform the loading at a very slow loading rate [4–5]. In the present research, an alternative method [10–11] was used to determine the equilibrium stress. In this method, multi-step relaxation tests are performed during the loading and unloading cycles (see Fig. 5); each relaxation event lasted 20 min. The repeated relaxation tests were

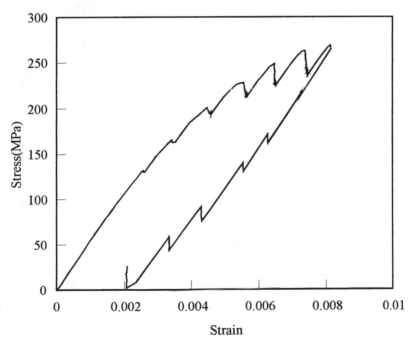

FIG. 5—*Relaxation tests during a tensile test.*

also performed by Bordonaro and Kremple [9] on polymers for a shorter duration (210 s). At the end of the relaxation, the stress is taken to be the time-independent equilibrium stress. For loading, the equilibrium stresses were determined experimentally for different off-axis specimens with the results shown in Fig. 6.

In the present research, the equilibrium stress for loading is described using the one-parameter plasticity model [1]. Following this model, the effective plastic strain increment corresponding to the definition of $\bar{\sigma}^*$ is

$$d\bar{\varepsilon}^{p*} = \sqrt{\frac{2}{3}} [(d\varepsilon_{22}^{p*})^2 + \frac{1}{2a_{66}} (d\gamma_{12}^{p*})^2]^{1/2} \tag{11}$$

The relation between $\bar{\sigma}^*$ and $\bar{\varepsilon}^{p*}$ is assumed to be in the form of a power law

$$\bar{\varepsilon}^{p*} = A_1(\bar{\sigma}^*)^{n_1} \tag{12}$$

where A_1 and n_1 are constants. The plastic strain increments are obtained from the flow rule as

$$d\varepsilon_{ij}^p = \frac{\partial f}{\partial \sigma_{ij}^*} d\lambda \tag{13}$$

where

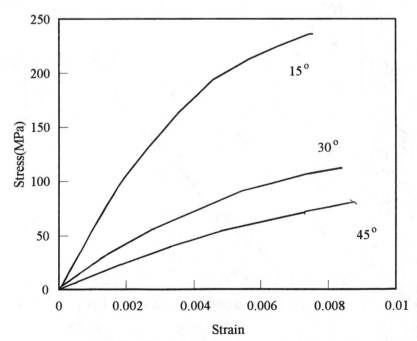

FIG. 6—*Off-axis equilibrium stress-strain curves.*

$$d\lambda = \frac{3}{2} \frac{d\bar{\varepsilon}^{p*}}{\bar{\sigma}*} \tag{14}$$

Once A_1 and n_2 in Eq 12 are determined, the equilibrium stress for loading can be obtained.

The equilibrium stress for unloading is more complicated. Before we defined the equilibrium stress for unloading, we first investigate the unloading behavior and the relaxation behavior during unloading. From Fig. 4, we note that at the initial unloading stage, the viscoplastic strain rate is positive, and from Fig. 7 we note that the equilibrium stress-strain curve is located beneath the dynamic stress-strain curve (because the dynamic stress tends to approach the equilibrium stress in a relaxation test). However, as unloading proceeds to a certain point, the viscoplastic strain rate becomes negative and the equilibrium stress-strain curve lies above the dynamic stress-strain curve.

Based on these observations, two assumptions about equilibrium stress for unloading are made. a) During the initial unloading stage, the viscoplastic strain rate is positive, (i.e., the material is in a loading mode), thus, the equilibrium stress for this stage is the equilibrium stress-strain curve for loading. b) The dynamic unloading stress-strain curve intersects the equilibrium stress-strain curve for loading. Beyond that point, the viscoplastic strain rate becomes negative, and the material is in a true unloading mode. The equilibrium stress for this stage is determined using the multi-step "relaxation" tests performed during unloading.

Based on the above assumptions, the equilibrium stress for unloading is determined. A schematic illustration of the equilibrium stress and the dynamic stress is given in Fig. 8.

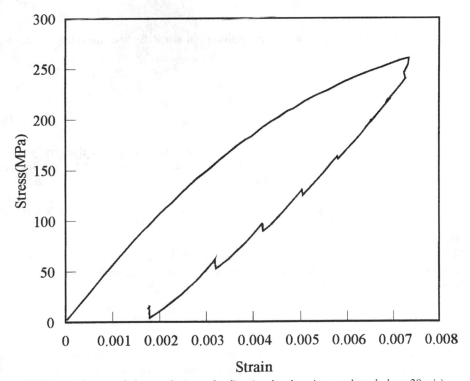

FIG. 7—*Relaxation behavior during unloading (each relaxation test lasted about 20 min).*

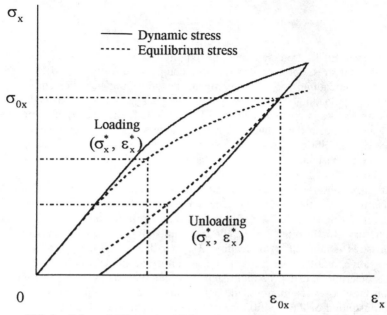

FIG. 8—*Schematic illustration of dynamic stress and equilibrium stress.*

In order to mathematically describe the equilibrium stress for the unloading mode, we define

$$\tilde{\sigma}_x^* = \sigma_{0x} - \sigma_x^* \tag{15}$$

$$\tilde{\varepsilon}_x^* = \varepsilon_{0x} - \varepsilon_x^* \tag{16}$$

where, σ_x^* and ε_x^* are the equilibrium stress and strain, respectively, for unloading; and σ_{0x} and ε_{0x} are the stress and strain, respectively, at the intersection point of the equilibrium stress-strain curve for loading and dynamic stress-strain curve for unloading, as shown in Fig. 8.

The unloading $\tilde{\sigma}_x^* - \tilde{\varepsilon}_x^*$ relation is similar to the $\sigma_x^* - \varepsilon_x^*$ relation for loading, and therefore, the one-parameter plasticity model can also be used to describe the $\tilde{\sigma}_x^* - \tilde{\varepsilon}_x^*$ relation.

The unloading effective stress $\bar{\tilde{\sigma}}_x^*$ and effective plastic strain increment $d\bar{\tilde{\varepsilon}}^{p*}$ are defined in a manner similar to Eq 7 and Eq 11, respectively. The plastic stress-strain relation in unloading is given by the power law as

$$\bar{\tilde{\varepsilon}}^{p*} = A_2(\bar{\tilde{\sigma}}^*)^{n_2} \tag{17}$$

where coefficients A_2 and n_2 are determined using the unloading equilibrium stress-strain curves. The parameter a_{66} in the plastic potential is found to be the same for loading and unloading.

It is found that the power law amplitude A_2 for unloading depends on the effective equilibrium plastic strain at which the unloading mode begins. A simple linear relation

$$A_2 = a\bar{\varepsilon}^{p*} + b \tag{18}$$

was found to fit the experimental data very well. In Eq 18, a and b are constants, and $\bar{\varepsilon}^{p*}$ is the equilibrium effective plastic strain corresponding to the starting point of the unloading mode. Based on the experimental data, we found that the power index n_2 in Eq 17 is a constant and is independent of the unloading point.

Once the equilibrium stress for unloading as given by Eq 17 is determined, the unloading plastic strain rates are obtained from Eq 9. To use Eq 9 for unloading analysis, the equilibrium stress must be converted from $\bar{\sigma}^*$ to $\overline{\sigma}^*$, and the analysis is performed in the original stress-strain space. For IM7/5260, the same function F given by Eq 6 is valid for both loading and unloading. Of course, in the case of unloading, sign $(\bar{\sigma} - \bar{\sigma}^*) = -1$.

Uniaxial Off-Axis Loading

For a state of off-axis tension, the normal and shear stress components in the material principal directions are given by

$$\sigma_{11} = \sigma_x \cos^2 \theta$$

$$\sigma_{22} = \sigma_x \sin^2 \theta \tag{19}$$

$$\sigma_{12} = -\sigma_x \sin \theta \cos \theta$$

where θ is the fiber orientation as shown in Fig. 1.

The effective stress and the effective equilibrium stress are given explicitly by

$$\bar{\sigma} = \sqrt{2f} = \left[\frac{3}{2} (\sigma_x^2 \sin^4 \theta + 2a_{66}\sigma_x^2 \sin^2 \theta \cos^2 \theta) \right]^{1/2} = h(\theta)\sigma_x \tag{20}$$

$$\bar{\sigma}^* = \sqrt{3f} = \left[\frac{3}{2} \left((\sigma_x^*)^2 \sin^4 \theta + 2a_{66}(\sigma_x^*)^2 \sin^2 \theta \cos^2 \theta \right) \right]^{1/2} = h(\theta)\sigma_x^* \tag{21}$$

where

$$h(\theta) = \left[\frac{3}{2} (\sin^4 \theta + 2a_{66} \sin^2 \theta \cos^2 \theta) \right]^{1/2} \tag{22}$$

From the coordinate transformation law, we have

$$\dot{\varepsilon}_x^p = \cos^2 \theta\dot{\varepsilon}_{11}^p + \sin^2 \theta\dot{\varepsilon}_{22}^p - \sin \theta \cos \theta\dot{\gamma}_{12}^p \tag{23}$$

Substituting Eq 19 into Eq 10 and then into Eq 23 results in

$$\dot{\varepsilon}_x^p = \frac{2}{3} B \left| \sinh \zeta \frac{\bar{\sigma} - \bar{\sigma}^*}{\bar{\sigma}^*} \right|^{1/m} \text{sign}(\bar{\sigma} - \bar{\sigma}^*) h(\theta) \qquad (24)$$

in which Eq 20 has been used. Using Eqs 21 and 24 and adding the elastic part of the strain, we obtain the total axial strain rate

$$\dot{\varepsilon}_x = \frac{\dot{\sigma}_x}{E_x} + \dot{\varepsilon}_x^p = \frac{\dot{\sigma}_x}{E_x} + \frac{2}{3} B \left| \sinh \zeta \frac{\sigma_x - \sigma_x^*}{\sigma_x^*} \right|^{1/M} \text{sign}(\sigma_x - \sigma_x^*) h(\theta) \qquad (25)$$

where E_x is the apparent elastic modulus of the off-axis specimen which can be obtained from the transformation equation

$$\frac{1}{E_x} = \frac{1}{E_1} \cos^4 \theta + \left(\frac{1}{G_{12}} - \frac{2v_{12}}{E_1} \right) \sin^2 \theta \cos^2 \theta + \frac{1}{E_2} \sin^4 \theta \qquad (26)$$

Equation 25 is in the form of a first order nonlinear ordinary differential equation (ODE) with respect to time. For a given strain rate, this ODE can be solved for stress using the numerical method, and consequently, the stress-strain curves for different strain rates can be obtained.

Parameter Evaluation

The elastic constants E_1, E_2, v_{12}, and G_{12} can be determined using 0°, 90°, and 45° off-axis test results. To determine material parameters for the equilibrium stress, multi-step relaxation tests were performed during off-axis tests. Each relaxation test lasted about 20 min. From these relaxation tests, the off-axis equilibrium stress-strain curves were determined. Figure 6 shows the equilibrium stress-strain curves for 15°, 30°, and 45° off-axis specimens. Parameters a_{66}, A_1, and n_1 can be determined from these curves. The procedure is exactly the same as that for determining material parameters in the one parameter plasticity model [1].

After a_{66}, A_1, and n_1 are determined, we proceed to determine constants B, ζ and m in the viscoplasticity model. For this purpose, consider Eq 9. During loading, $\bar{\sigma} > \bar{\sigma}^*$ and Eq 9 becomes

$$\dot{\varepsilon}_{ij}^p = B \frac{1}{\bar{\sigma}^*} \left(\sinh \zeta \frac{\bar{\sigma} - \bar{\sigma}^*}{\bar{\sigma}^*} \right)^{1/m} \frac{\partial f(\sigma)}{\partial \sigma_{ij}} \qquad (27)$$

Corresponding to the effective dynamic stress defined in Eq 7, we can define an effective viscoplastic strain rate $\bar{\varepsilon}^p$ using the rate of plastic work

$$\bar{\sigma} \bar{\dot{\varepsilon}}^p = \sigma_{ij} \dot{\varepsilon}_{ij}^p \qquad (28)$$

Using Eqs 5, 7, 27, and 28, we obtain

$$\bar{\dot{\varepsilon}}^p = \frac{2}{3} B \left(\sinh \zeta \frac{\bar{\sigma} - \bar{\sigma}^*}{\bar{\sigma}^*} \right)^{1/m} \qquad (29)$$

Taking the logarithm of Eq 29 we have

$$\log (\bar{\dot{\varepsilon}}^p) = \log \left(\frac{2}{3} B\right) + \frac{1}{m} \log\left(\sinh \zeta \, \frac{\bar{\sigma} - \bar{\sigma}^*}{\bar{\sigma}^*}\right) \qquad (30)$$

In Eq 30, for a given ζ, $\log(\sinh \zeta(\bar{\sigma} - \bar{\sigma}^*/\bar{\sigma}^*))$ against $\log(\bar{\dot{\varepsilon}}^p)$ is a straight line. The value of ζ is chosen to best correlate the experimental data. In this study, the off-axis test results for $\dot{\varepsilon}_x = 0.00001/\text{sec}$ are used to determine parameters B, ζ, and m. The log-log plot with $\zeta = 0.167$ is shown in Fig. 9. From this plot, B and m can be determined.

Following the procedure similar to that in the loading case, the power law constants A_2 and n_2 for unloading can be determined. The only difference is that for the unloading case, at least two sets of test results for different effective equilibrium plastic strain levels are needed, because A_2 is a function of effective equilibrium plastic strain at which unloading mode begins (see Eq 18).

Generally, the viscoplastic constants B, ζ, and m may have different values for loading and unloading. For the IM7/5260, we found that the values determined from loading can also be used for unloading.

Table 1 lists all the material parameters for IM7/5260 at room temperature, which were determined according to the aforementioned procedures.

Model Predictions

For a given strain rate, Eq 25 can be solved to give stress. A FORTRAN program was written to integrate the equation by using Euler's one-step algorithm. The material parameters given in the previous section were used to make the predictions.

As mentioned in the preceding section, the experimental results for $\dot{\varepsilon}_x = 0.00001/\text{sec}$ were used to determine the material parameters. Therefore, only results for other strain rates, i.e.,

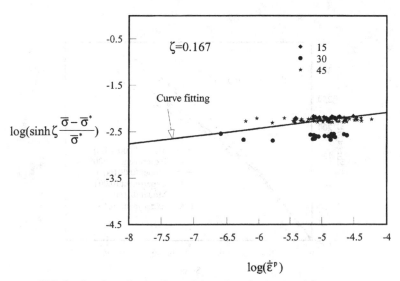

FIG. 9—*Log-log plot for determining viscoplastic material constants.*

TABLE 1—*Material parameters of IM7/5260.*

E_1(GPa)	152.8
E_2(GPa)	8.7
G_{12}(GPa)	5.2
v_{12}	0.30
a_{66}	1.4
A_1(MPa)$^{-n_1}$	3.0×10^{-8}
n_1	2.50
a(MPa)$^{-n_2}$	2.6×10^{-5}
b(MPa)$^{-n_2}$	1.03×10^{-7}
n_2	2.05
B	5.0×10^8
ζ	0.167
m	0.08

$\dot{\varepsilon}_x = 0.001/\text{sec}$ and $\dot{\varepsilon}_x = 0.1/\text{sec}$, were used for verifying the model. Figures 10–12 show both the experimental and model results for loading at the three strain rates. From the results for $\dot{\varepsilon}_x = 0.001/\text{sec}$ and $\dot{\varepsilon}_x = 0.1/\text{sec}$, it is evident that the model predicts the rate-dependent behavior of IM7/5260 quite well.

Figures 13–15 show results for loading and unloading at constant strain rates. It is interesting to note that, although no experimental results for unloading have been used to determine the material parameters, the predicted unloading stress-strain curves for all three strain rates agree with the experimental curves quite well.

Figures 16–19 show results for the cases where the loading rate differs from the unloading rate. It is clear that the model predicts both slow loading-fast unloading and fast loading-

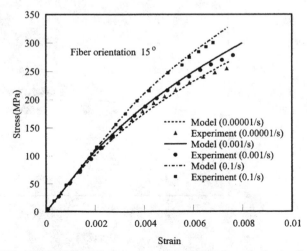

FIG. 10—*Model predictions versus experimental results for loading.*

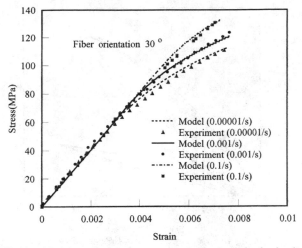

FIG. 11—*Model predictions versus experimental results for loading.*

slow unloading cases very well. For the fast loading-slow unloading case, the model correctly predicts the greatly distorted unloading curve during the initial unloading stage.

Summary

Experimental results indicate that the unloading response of the polymeric composite IM7/5260 is significantly influenced by the strain rate in the preceding loading. Moreover, at the initial stage of the apparent unloading, the material may actually experience loading in the

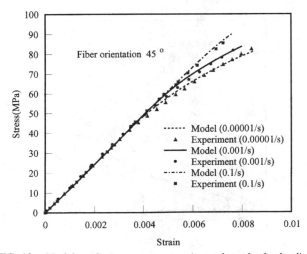

FIG. 12—*Model predictions versus experimental results for loading.*

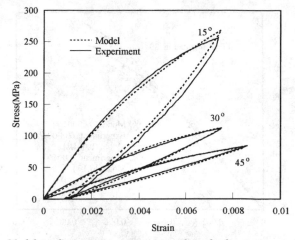

FIG. 13—*Model predictions versus experimental results for strain rate 0.00001/s.*

sense that the plastic strain rate is positive. In this study, a three-parameter viscoplasticity constitutive model has been developed for the rate-dependent loading and unloading behavior of the IM7/5260 composite. The model is based on the overstress concept with the equilibrium stress determined using a multi-step relaxation procedure. The material constants required for the model can easily be obtained from uniaxial tests on off-axis unidirectional composite coupon specimens. This overstress-based viscoplasticity model is shown to predict the loading-unloading behavior of the composite at any combination of loading and unloading rates.

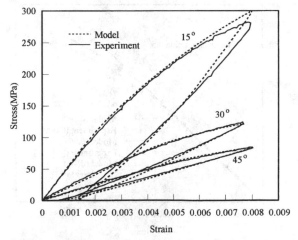

FIG. 14—*Model predictions versus experimental results for strain rate 0.001/s.*

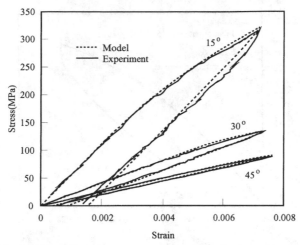

FIG. 15—*Model predictions versus experimental results for strain rate 0.1/s.*

Finally, it should be noted that in this study, the equilibrium stresses were obtained based on a 20-min relaxation time. With longer relaxation times, the equilibrium stress-strain curve may be somewhat different, and consequently, the numerical coefficients in the overstress viscoplasticity model may also be affected.

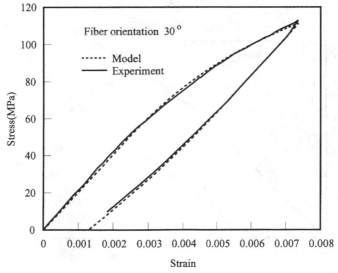

FIG. 16—*Model predictions versus experimental results for loading strain rate 0.00001/s and unloading strain rate 0.001/s case.*

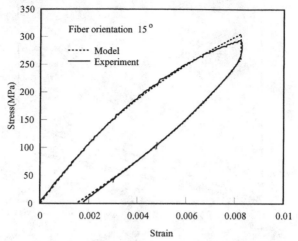

FIG. 17—*Model predictions versus experimental results for loading strain rate 0.001/s and unloading strain rate 0.00001/s case.*

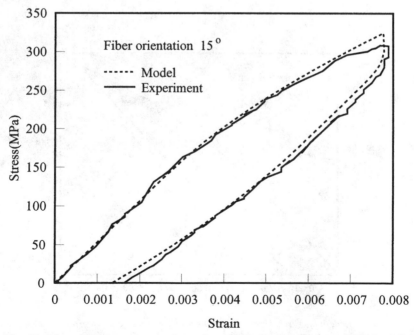

FIG. 18—*Model predictions versus experimental results for loading strain rate 0.01/s and unloading strain rate 0.001/s case.*

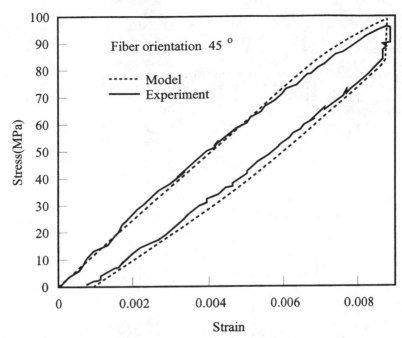

FIG. 19—*Model predictions versus experimental results for loading strain rate 0.1/s and unloading strain rate 0.00001/s case.*

Acknowledgment

This work was supported by NASA Langley Research Center through grant No. NAG-1-1366 to Purdue University. Dr. T. S. Gates was the technical monitor.

References

[1] Sun, C. T. and Chen, J. L., "A Simple Flow Rule for Characterizing Nonlinear Behavior of Fiber Composites," *Journal of Composite Materials,* Vol. 23, 1989, pp. 1009–1020.

[2] Malvern, L. E., "The Propagation of Longitudinal Waves of Plastic Deformation in a Bar of Material Exhibiting a Strain-Rate Effect," *Journal of Applied Mechanics,* Vol. 18, 1951, pp. 203–208.

[3] Gates, T. S. and Sun, C. T., "Elastic/Viscoplastic Constitutive Model for Fiber Reinforced Thermoplastic Composites," *AIAA Journal,* Vol. 29, 1991, pp. 457–463.

[4] Yoon, K. J. and Sun, C. T., "Characterization of Elastic-Viscoplastic Properties of an AS4/PEEK Thermoplastic Composite," *Journal of Composite Materials,* Vol. 25, 1991, pp. 1277–1298.

[5] Gates, T. S., "Effects of Elevated Temperature on the Viscoplastic Modeling of Graphite/Polymeric Composites," *NASA TM 104160,* 1991.

[6] Weeks, C. A. and Sun, C. T., "Nonlinear Rate Dependent Response of Thick-Section Composite Laminates," *Proceedings of the ASME International Mechanical Congress and Exposition,* San Francisco, Nov. 12–17, 1995, pp. 81–95.

[7] Johnson, G. R., Hoegfeldt, J. M., Lindholm, U. S., and Nagy, A., "Response of Various Metals to Large Torsional Strains Over a Large Range of Strain Rates—Part 1: Ductile Metals," *Journal of Engineering Materials and Technology,* Vol. 105, 1983, pp. 42–47.

[8] Kremple, E. and Kallianpur, V. V., "The Uniaxial Unloading Behavior of Two Engineering Alloys at Room Temperature," *Journal of Applied Mechanics,* Vol. 52, 1995, pp. 654–658.

[9] Bordonaro, C. M. and Kremple, E., "The Rate-Dependent Mechanical Behavior of Plastics. A Comparison Between Nylon 6/6, Ultem and PEEK," *ASME MD, Use of Plastics and Plastic Composites: Materials and Mechanics Issues* (Edited by V. K. Stokes). The American Society of Mechanical Engineers, New York, NY, Vol. 46, 1993, pp. 43–56.
[10] Yen, C. F., Wang, K. F., and Hsiao, W. C., "An Experimental Study of the Uniaxial Rate-Dependent Behavior of Type 304 Stainless Steel at Room Temperature," *Spring Conference on Experimental Mechanics,* Society for Engineering Mechanics, New Orleans, LA, 8–13 June 1986.
[11] Krempl, E. and Bordonaro, C. M., "A State Variable Model for High Strength Polymers," *Polymer Engineering and Science,* Vol. 35, No. 4, 1995, pp. 310–316.

Michael G. Castelli,[1] *James K. Sutter,*[2] *and Dianne Benson*[3]

Durability and Damage Tolerance of a Polyimide Chopped Fiber Composite Subjected to Thermomechanical Fatigue Missions and Creep Loadings

REFERENCE: Castelli, M. G., Sutter, J. K., and Benson, D., **"Durability and Damage Tolerance of a Polyimide Chopped Fiber Composite Subjected to Thermomechanical Fatigue Missions and Creep Loadings,"** *Time Dependent and Nonlinear Effects in Polymers and Composites, ASTM STP 1357,* R. A. Schapery and C. T. Sun, Eds., American Society for Testing and Materials, West Conshohocken, PA, 2000, pp. 285–309.

ABSTRACT: Although polyimide based composites have been used for many years in a wide variety of elevated temperature applications, very little work has been done to examine the durability and damage behavior under more prototypical thermomechanical fatigue (TMF) loadings. Synergistic effects resulting from simultaneous temperature and load cycling can potentially lead to enhanced, if not unique, damage modes and contribute to a number of nonlinear deformation responses. The goal of this research was to examine the effects of a TMF loading spectrum, representative of a gas turbine engine compressor application, on a polyimide sheet molding compound (SMC). High performance SMCs present alternatives to prepreg forms with great potential for low cost component production through less labor intensive, more easily automated manufacturing. To examine the issues involved with TMF, a detailed experimental investigation was conducted to characterize the durability of a T650-35/PMR-15 SMC subjected to TMF mission cycle loadings. Fatigue damage progression was tracked through macroscopic deformation and elastic stiffness. Additional properties, such as the glass transition temperature (T_g) and dynamic mechanical properties were examined. The fiber distribution orientation was also characterized through a detailed quantitative image analysis. Damage tolerance was quantified on the basis of residual static tensile properties after a prescribed number of TMF missions. Detailed micro-structural examinations were conducted using optical and scanning electron microscopy to characterize the local damage. The imposed baseline TMF missions had only a modest impact on inducing fatigue damage with no statistically significant degradation occurring in the measured macroscopic properties. Microstructural damage was, however, observed subsequent to 100 h of TMF cycling which consisted primarily of fiber debonding and transverse cracking local to predominantly transverse fiber bundles. The TMF loadings did introduce creep related effects (strain accumulation) which led to rupture in some of the more aggressive stress scenarios examined. In some cases, this creep behavior occurred at temperatures in excess of 150°C below commonly cited values for T_g. Thermomechanical exploratory creep tests revealed that the SMC was subject to time dependent deformation at stress/temperature thresholds of 150 MPa/230°C and 170 MPa/180°C.

KEYWORDS: thermomechanical fatigue, damage tolerance, sheet molding compound, PMR-15, creep, residual properties, micro-structural damage, fiber distribution, graphite fiber, T650-35, polymer matrix composites

[1] Senior Research Engineer, Ohio Aerospace Institute/NASA Lewis Research Center, MS 49-7, Cleveland, OH.

[2] Research Chemist, NASA Lewis Research Center, MS 49-3, Cleveland, OH, 44135.

[3] Engineer, ProTech Lab Corporation, Cincinnati, OH 45241.

High performance polymeric composites (PMCs) continue to be the focus of a number of research efforts aimed at developing cost effective, light weight material alternatives for advanced aerospace and aeropropulsion applications. These materials not only offer significant advantages in specific stiffness and strength over their current metal counterparts, but present the further advantage that structures can be designed and manufactured to eliminate joints and fasteners by combining individual components into integral subassemblies, thus making them extremely attractive for commercial applications. Of particular interest to elevated temperature applications, are polyimide matrix based composite materials which exhibit outstanding thermal stability providing for short and long term uses to 550 and 300°C, respectively [1]. PMR-15 is one such polyimide which has seen considerable use in aeropropulsion applications due to its good thermo-oxidative stability, relatively low cost and availability in a variety of forms [2].

With current emphasis on low cost manufacturing aspects of advanced composite structures, there is heightened interest on high performance sheet molding compounds (SMCs). SMCs effectively serve to reduce the costs associated with component production using prepregs, where variable costs are generally associated with labor, secondary processes, and scrap. Using compression molding, SMCs can be molded into complicated shapes facilitating the use of simple charge patterns, part consolidation and molded-in inserts, which reduce labor, equipment, and operation costs for preparatory and secondary processes [3]. Specific to the present study is a carbon fiber reinforced PMR-15 SMC which has been used in a number of elevated temperature static aero applications, including oil exposed helicopter gearboxes [2] and shrouds for gas turbine engine-inlet housings [4].

The primary objective of the present research was to evaluate the durability of PMR-15 SMC subjected to a thermomechanical fatigue (TMF) loading spectrum. The TMF mission spectrum was representative of conditions found at a mid-stage within a gas turbine engine compressor. Researchers at Allison Advanced Development Company (AADC) and NASA Lewis Research Center investigated the use of PMR-15 SMC as the material comprising a mid-stage inner vane endwall [5]. Such a component resides in the engine flow path and is subjected to not only high airflow rates, but also elevated temperatures and pressures. Thus, the application represents a much more aggressive use of the material than those cited previously and raises obvious concerns related to the fatigue durability and damage tolerance. A survey of the literature on polyimide SMCs and their various applications reveals that very little information is available which details the structural durability of such materials, particularly in the light of more prototypical thermomechanical loading conditions. Therefore, one of the first goals of the current research was to determine reasonable maximum stress and temperature parameters for the representative engine mission cycle, so as to take full advantage of the SMC's capabilities.

Toward this end, a detailed experimental investigation was conducted to characterize the fatigue durability and damage tolerance of a T650-35/PMR-15 SMC subjected to TMF mission cycle loadings. Fatigue damage progression was tracked on the basis of longitudinal stiffness degradation and strain accumulation, while damage tolerance was quantified by residual static tensile properties after a prescribed number of TMF missions. The two parameters of stiffness and static response were selected not only because of application design and performance criteria, but due to their commonplace use in the area of damage mechanics and material life modeling [6–9]. Sufficient tests were conducted for all of the conditions investigated so that the statistical significance of the results could be assessed. Additional properties, such as fiber distribution orientation, glass transition temperature, T_g, and dynamic mechanical properties were examined. Detailed inspections were conducted using optical and scanning electron microscopy to characterize the microstructural damage. As the TMF cycle promoted damage associated with creep deformation, the creep behavior was investigated through a series of unique thermomechanical exploratory tests and compared with the response of neat PMR-15. Emphasis was placed on determining stress/temperature thresholds for time dependent deformation.

Material Details

Composition and Properties

The chopped carbon fiber polyimide based sheet molding compound examined in this study was T650-35/PMR-15 SMC [10] supplied by Quantum Composites, Inc., Midland, MI, (QCI 15C, lot No. 092343; comparable to HyComp 310). The carbon fiber, Amoco's T650-35 (3K tow, UC309 sized), was chopped to 25 mm lengths and sprinkled with a randomized orientation (2-D) onto the matrix layer [3]. The composite panels had a nominal dimension of $10 \times 20 \times 0.22$ (2 ply) cm and were compression molded at AADC, Indianapolis, IN, using the conditions presented in Table 1.

The T_g was measured on selected representative panels with an RMS 800 instrument (Rheometrics Scientific™) deforming the specimen in torsion at a frequency of 1 Hz, a temperature ramp rate of 5°C/min, and a nominal specimen geometry of 40×5 mm (length $\times$ width). T_g was calculated by the intercept method using the storage modulus curve (for more details see Ref 11, p. 245). The results indicated a dry T_g in the range of 284 to 306°C which is generally considered to be unacceptably low for this material and indicative of insufficient postcuring. Therefore, all panels were subjected to a second postcure to raise the T_g. In an effort to optimize the T_g and also prevent extensive thermo-oxidative degradation, the effects of three potential secondary postcure cycles consisting of soaks at 316°C in 1 atm air were evaluated. Specifically, test samples were postcured at 316°C for either 4, 8, or 12 h. Prior to the secondary postcure, all panels were vacuum dried at 140°C/76 cm Hg for 48 h. The average T_g values after the 4, 8, and 12 h postcures were 334, 339, and 345°C, respectively. Given that the anticipated maximum test temperature would be approaching the target temperature of 316°C, and desiring to test at a maximum temperature of at least 28°C (i.e., 50°F) below the T_g, the secondary postcure of 12 h at 316°C was selected and implemented for all of the test panels.

The quality of all the postcured panels was evaluated by nondestructive analysis using ultrasonic C-scan. Past research in this area has successfully established a correlation between signal attenuation and void content [12]. Good consistency was found among the panels used in the study and C-scan results generally indicated void volumes in the range of 0.5 to 1.5%.

TABLE 1—*T650-35/PMR-15 SMC panel processing details.*

Imidization Cycle	Cure Cycle	Postcure Cycle
• Room temperature to 121°C over 30 min	• Preheat mold to 260°C	• Heat from temperature to 249°C over 3 h
• Hold at 121°C for 30 min	• Load imidized part into mold	• Hold at 249°C for 3 h
• Ramp to 204°C over 30 min	• Close mold, apply no pressure and wait 2 min	• Ramp to 288°C over 2 h
• Hold at 204°C for 1 h	• Apply 4.1 MPa and heat mold to 316°C	• Hold at 288°C for 3 h
	• Hold pressure and heat for 60 min	• Ramp to 316°C over 2 h
	• Cool while maintaining 4.1 MPa until mold temp is 204°C and unload part	• Hold at 316°C for 12 h
		• Cool to room temperature over 6 h

Additionally, the void content and fiber volumes were measured by the methods described in the Test Method for Void Content of Reinforced Plastics (ASTM D 2734) and Test Method for Fiber Content of Resin-Matrix Composites by Matrix Digestion (ASTM D 3171). The void content and fiber volume values ranged from 0.9 to 1.7% and 54 to 61%, respectively. It is important to note that the fiber volumes are higher than those recommended for this type of SMC, which is generally desired to be in the range of 50 to 55%. Fiber volume fractions above 55% can lead to poor fiber wetting, reducing the overall mechanical performance of the material (as will be seen in the discussion to follow).[4]

Fiber Orientation Distribution

Carbon fiber orientation distribution has a significant effect on mechanical properties for SMCs. Therefore, a representative sample was subjected to a detailed examination to quantify the fiber orientation distribution at various depths into the thickness of the sample. This analysis was performed at ProTech Lab Corp., Cincinnati, OH. Fiber orientations at three depths (0.7, 1.0, and 1.5 mm) were examined through progressive polishing. The second depth represented the SMC's mid-section. The area scanned was 218 mm^2 (approximately square) and was deemed to be representative of the plane. Fifty-five photographs, each representing approximately 3.95 mm^2, were used to compose the scanned area at each of the three depths. The detailed procedure for measuring the fiber orientation distribution is given in Appendix A.

The image analysis was performed by analyzing the carbon fiber distribution in 10° increments from 0 to 180°, where the 0 and 180° orientations coincide and are parallel to the longitudinal specimen axis (i.e., the specimen loading axis). In theory, the fiber orientation in the plane being examined should be random for sheet molding compound composites. However, achieving a completely random distribution of carbon fibers during the prepreg process is very difficult. The image analysis results for depths 1 to 3 suggest that the carbon fiber distributions are not random, but rather, bi-modal. The fiber distributions through each of the three depths differ slightly, as might be statistically anticipated and depth 2 is only slightly bi-modal. However, depths 1 and 3 suggest a strong bi-modal distribution as illustrated in Fig. 1, where depth 3 is shown in histogram form. The bi-modal distribution tends toward fiber alignment along the longitudinal axis (i.e., 0 or 180 ± 30°), leaving a significantly reduced fiber volume fraction along the 90 ± 30° axis. The tendency for such a trend is not entirely unanticipated by the manufacturers[4] given that the fibers are dropped onto the matrix sheet as it moves along on a conveyor system [3]. That is, the fibers tend to align in the moving direction, though a minimization of this tendency is desired and sought.

In addition to acid digestion, fiber volumes were determined by image analysis. The fiber volume percent at each depth was determined by randomly selecting ten digitized photographs at a magnification of 200× and assuming that the general bundle orientation had no effect on the fiber volume. The results from these fiber volume analyses are given in Table 2. Note that the values (56 to 61%) correspond well to the fiber volumes obtained from acid digestion (54 to 61%).

Testing Details

All coupon specimens were cut using abrasive water-jet machining; each of the panels (10 × 20 cm) yielded three samples. After cutting and prior to testing, specimens were dried

---

[4] Personal communication with Dr. Joseph Reardon, HyComp, Inc., Cleveland, OH, 44130.

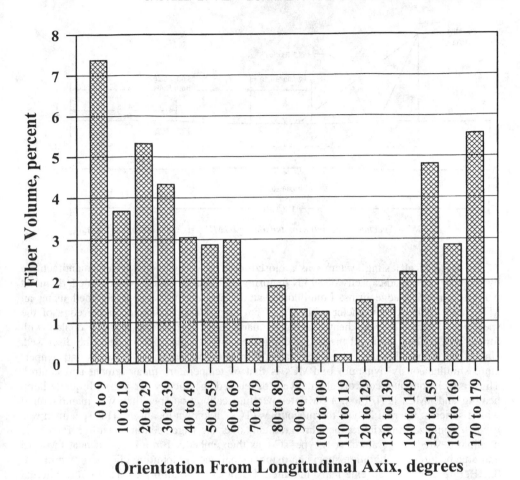

Orientation From Longitudinal Axix, degrees

FIG. 1—*Fiber orientation distribution found at depth 3 (1.5 mm into thickness) revealing bi-modal trend.*

for 48 h at 140°C and 76 cm Hg vacuum and then stored in a desiccator until immediately prior to testing. The specimen geometry was a reduced gage section dogbone geometry, shown in Fig. 2. The relatively large radius of 36.8 cm forming the transition section has been used extensively in advanced metal matrix composite testing for a variety of laminates [13]. This geometry was successfully extended to high temperature PMC testing, facilitating gage section failures while avoiding the use of tabs [14].

TABLE 2—*Fiber volume precentages determined by quantitative image analysis.*

Polishing Depth	Fiber Volume, %
1	60.6
2	56.4
3	55.7

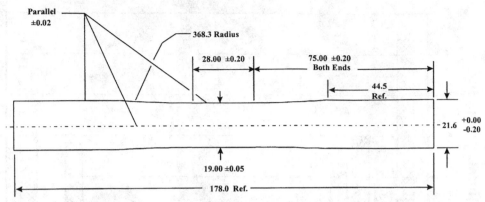

FIG. 2—*Specimen geometry for tensile and TMF tests (dimensions in mm).*

The mechanical testing system was a closed-loop, servo-hydraulic system manufactured by MTS™ with a load capacity of 89 kN featuring hydraulic actuated, water cooled, diamond pattern serrated, wedge grips. Longitudinal strain measurements were obtained using an MTS™ air cooled extensometer with a 1.27 cm gage length mounted on the edge of the specimen. Specimens were heated using a quartz lamp system and actively cooled with forced air enabling the rapid thermal cycling necessary for the TMF cycle to be discussed. One of the formidable difficulties associated with true TMF testing (i.e., stress and temperature simultaneously dynamic) of PMCs is that of temperature measurement and control. Thus, a technique and control scheme were developed specifically for use with quartz lamp heating and PMCs [14]. First, a SMC temperature calibration specimen was manufactured with a series of internal K-type thermocouples (TCs) at known locations. This specimen was used to optimize the axial temperature gradients over the gage section to with ±1% of the nominal desired temperature. A K-type TC was then embedded in a block of neat PMR-15 and attached to the calibration specimen using a small metal mounting clip as shown in Fig. 3a. Before each test, an externally mounted TC block was calibrated against the internal TCs of the SMC calibration specimen as illustrated in Fig. 3b. The relationship between the imbedded and external TCs was found to be linear. The block was then mounted on a test specimen in the precise calibration location and used subsequently to measure and control the temperature. Repeatability of the relationship between the block and the calibration specimen was verified by removing and installing the setup several times and examining consistency; acceptable variations of ±2°C were observed.

All static tensile tests were conducted in accordance with *the Standard Test Method for Tensile Properties of Plastics (ASTM D 638)* in displacement control with a loading rate of 0.5 mm/min. The TMF mission cycle used for this study is shown in Fig. 4 (note that all loads are tensile). This generic cycle was determined by AADC researchers to be representative of the gas turbine compressor mid-stage inner vane endwall application. The mission consists of eight secondary segments, seven of which represent idle to maximum engine conditions, and one which represents a redline engine condition where the stress level is prescribed to be a 5% increase over maximum. Note that the idle, maximum, and redline stresses and temperatures (σ^I, σ^{Max}, σ^{RL}, and T^I, T^{Max}, T^{RL}, respectively) are not specified, since determining these parameters was part of the research objective. Here, the goal was to determine a mission where the material was loaded as aggressively as possible, but survive a minimum of 100 h of mission cycling. As shown, the TMF mission time is approximately 50 min; therefore, 100 h of mission cycling corresponded to approximately 120 missions.

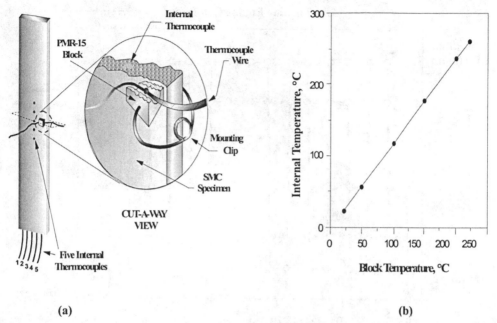

(a)   **(b)**

FIG. 3—*Temperature measurement and control scheme showing:* (a) *the internal and external block thermocouple configuration and* (b) *representative calibration data.*

Once the TMF mission was specified, the test matrix consisted of testing to the two conditioning states of 50 and 100 h, in addition to the 0 h TMF state which represents the unconditioned material. Residual tensile properties were then examined. A minimum of six tests was conducted at each of the TMF conditioning states satisfying issues regarding statistical significance. The experiments were intentionally designed with respect to i) order of tests, and ii) specimen selection for condition. Given that each panel yielded three samples, one of each of these were used to examine the three conditioning states. For example, panel-A yielded samples A1, A2, and A3; A1 was used in a 0 h TMF residual test, A2 in a 50 h TMF residual test, and A3 in a 100 h TMF residual test. Thus, the design of experiments incorporated panel to panel variations for any one TMF state, but not specimen to specimen variation within a panel. The panel to panel variation was considered more significant. Several specimens were also tested to 50 and 100 TMF h states for purposes of destructive examination to detail the state of microstructural damage using optical and scanning electron microscopy.

Results and Discussion

Static Tensile Properties

One of the first issues to resolve on the PMR-15 SMC was the as-manufactured static properties. These properties would in turn assist in formulating the maximum and redline parameters specified for the TMF mission cycle. Shown in Fig. 5 are the room temperature (i.e., 22°C) and 260°C static tensile behaviors. The 22 and 260°C tests yielded an average strength (σ^{ult}) of 259 and 230 MPa and an average strain to failure (ε^f) of 1.0 and 0.47%, respectively. Two points were worthy of noting. First, the σ^{ult} values appeared to fall short

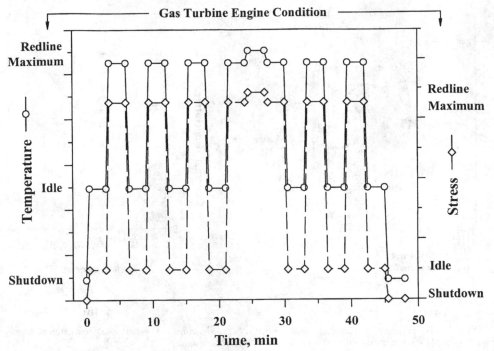

FIG. 4—*TMF mission cycle representative of a gas turbine engine compressor inner vane endwall application.*

of those advertised by the manufacturers of the comparable SMC, HyComp 310, (Dexter/HyComp) [*15*] by approximately 25% at 22°C and 17% at 260°C. This shortfall in tensile properties was likely due to the high fiber volume fraction state noted earlier. Further, the properties given in [*15*] were generated using four and six ply based materials, as opposed to two plies. The thicker materials will tend to show less scatter, as the through thickness characteristics become more homogeneous.

Second, note the distinctive "reverse" curvature (concave upward) of the tests performed at 260°C indicative of a stiffening effect with increased loading. The curvature becomes particularly noticeable at an approximate stress/strain of 170 MPa/0.4%. This effect is likely due to fiber straightening and/or rotation of the fiber segments into the loading direction, enabled by the reduced viscosity and stiffness of the matrix at 260°C. Although this behavior was somewhat surprising, given that 260°C is significantly below the T_g (~337°C), it is not entirely unreasonable. In addition to matrix viscosity issues, one must also consider the micro stress state induced by the mismatch in coefficient of thermal expansion (CTE), where the longitudinal CTE for T650-35 fiber is -0.5×10^{-6} °C^{-1} and that of PMR-15 is 50×10^{-6} °C^{-1}. With increasing temperature, the matrix will tend to literally expand away from the fiber tow segments, enhancing the propensity for localized fiber rotation with the application of a macroscopic load. Thus, to understand the manifestation of this complex behavior, it is not sufficient to examine simply the longitudinal macroscopic stress/strain state induced by the applied load, but rather, there are several mechanisms which must be considered. With regard to the material's tendency to exhibit "viscous" behavior, it will be discussed later in

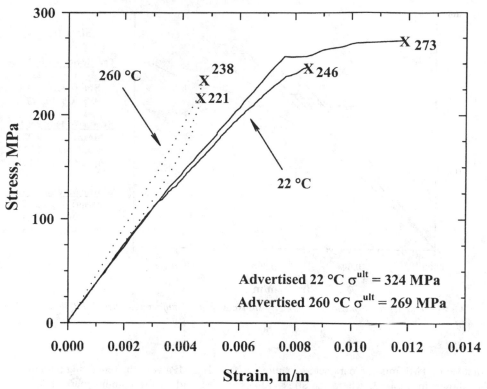

FIG. 5—*Static tensile behavior of as-manufactured T650-35/PMR-15 SMC at 22 and 260°C.*

the exploratory creep deformation section that time dependent deformation can be initiated at markedly low temperature values.

Given the noted "viscous" response at 260°C, a slightly lower temperature was examined for potential use as T^{Max} in the TMF cycle with the intent to avoid this type of deformation behavior as being typical for the mission cycle. However, the 260°C target was felt to be a minimum for representing the T^{RL} condition. Shown in Fig. 6 is the static tensile response at the slightly lower temperature of 232°C. Having established, within the first few tests, that the stiffening effect was not manifested at this temperature, a full complement of tests was conducted (i.e., 6 repeats) to establish statistically meaningful static properties; these properties are also shown in Fig. 6. Note there is a relatively large deviation in properties as one might expect with a SMC, particularly in view of the fiber orientation distribution information presented earlier. This variation was especially significant with regards to ε^f, where the standard deviation (Σ) was found to be approximately 25% of the mean value. At a minimum, this suggests that the panel-to-panel variation in properties can be quite large.

Determination of TMF Capabilities and TMF Deformation

With T^{Max} specified to be 232°C and T^{RL} established at 260°C, the remaining key parameter to be determined was σ^{Max} (recall that σ^{RL} is specified at $1.05\sigma^{Max}$). As an initial estimate, the maximum fatigue stress was taken as ($\sigma^{ult} - 3\Sigma$) where σ^{ult} is taken at T^{Max}. Thus, the

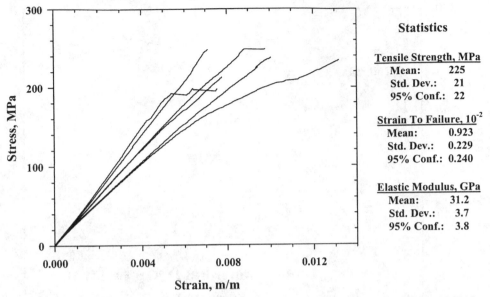

Statistics

<u>Tensile Strength, MPa</u>
Mean: 225
Std. Dev.:   21
95% Conf.:   22

<u>Strain To Failure, 10^{-2}</u>
Mean: 0.923
Std. Dev.: 0.229
95% Conf.: 0.240

<u>Elastic Modulus, GPa</u>
Mean: 31.2
Std. Dev.:   3.7
95% Conf.:   3.8

FIG. 6—*Static tensile behavior and property statistics of as-manufactured T650-35/PMR-15 SMC at 232°C.*

first set of TMF mission parameters features $\sigma^{Max} = 162$ MPa with the remaining parameters, as shown in Table 3, where the stress values correspond to the column marked ($\sigma^{Max} = \sigma^{ult} - 3\Sigma$). The specific idle parameters were selected as representative of the application without regard to the maximum and redline values. Also note that a σ^{Max} of 162 MPa represents a value corresponding to 72% of σ^{ult}, a seemingly modest level in view of the fact that T^{Max} was more than 100°C below T_g. A representative "cycle 1" deformation response for this TMF mission cycle is given in Fig. 7, where the strain plotted is the total component (i.e., thermal and mechanical). As shown in Fig. 7, the TMF cycle induces a relatively complex deformation behavior. Each of the secondary segments (idle to maximum) prior to the redline cycle is discernible as the material experienced a notable amount of creep deformation at maximum conditions. This effect is seen to decrease with progressive secondary segments as the material approaches a stabilized deformation response. When the redline condition was reached, the creep response was significantly revived, and then followed by a nominally elastic response corresponding to the remaining secondary segments of the mission cycle.

TABLE 3—*Temperature and stress values for TMF mission cycle.*

Engine Condition	Temperature, °C	Stress, MPa ($\sigma^{MAX} = \sigma^{ult} - 3\Sigma$)	Stress, MPa ($\sigma^{MAX} = \sigma^{ult} - 4\Sigma$)
Shutdown	26	0	0
Idle	123	36	36
Maximum	232	162	143
Redline	260	170	150

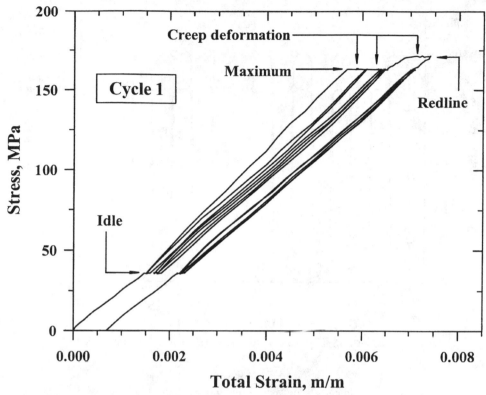

FIG. 7—*Stress/strain deformation behavior of T650-35/PMR-15 with TMF mission cycle parameters established by* $\sigma^{Max} = \sigma^{ult} - 3\Sigma$.

One obvious effect was the amount of strain recovery experienced within the cycle. For the test shown in Fig. 7, a total creep strain of 0.14% accumulated at maximum and redline conditions. However, after unloading from the final secondary segment, only 0.7% (i.e., half) remained. Further, if time permitted, more strain would likely have been recovered. Unfortunately, the redline condition of $\sigma^{Max} = 170$ MPa at 260°C proved to be too demanding to allow the full mission deformation response to stabilize. Note that the SMC mean ε^f (see Fig. 6) at 232°C is only slightly greater than that experienced after just one mission cycle. With progressive missions, a cyclic strain accumulation effect (comparable to creep ratchetting in metallic materials [16]) ensued leading to complete fracture of the sample. Although this test is revealing regarding material response under severe TMF conditions, the result prevented an important research objective: to establish residual properties subsequent to 100 h of TMF mission cycling. Therefore, TMF cycle revisions were necessary.

Shown in Fig. 8 is the final modified TMF mission cycle which was used for all residual properties. This cycle was based upon stress parameters defined by $\sigma^{Max} = \sigma^{ult} - 4\Sigma$. The values for σ^{Max} and σ^{RL} were reduced to 143 and 150 MPa, respectively, with all other parameters held constant (see Table 3). This σ^{Max} value represents 64% of σ^{ult} at 232°C. Representative deformation responses of the SMC to this mission cycle are given in Fig. 9, where both cycles 1 and 120 are shown. Several dramatic changes in the deformation behavior are noted when compared to that observed in Fig. 7. First, the stress/strain response

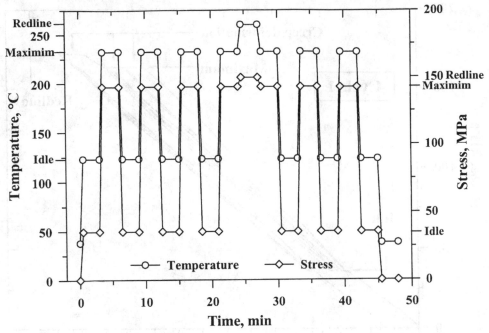

FIG. 8—*Finalized TMF mission cycle; parameters are based upon* $\sigma^{Max} = \sigma^{ult} - 4\Sigma$.

is seen to be essentially linear elastic, with only modest time dependent effects observable within a given cycle. Further, the viscous effects appear to be well stabilized within a given mission cycle with the redline condition having a seemingly inconsequential effect on the overall deformation behavior. Note that the material continues to experience a minor degree of permanent strain ratchetting on the order of 0.05 to 0.1% strain, which tends to accumulate during the early cycles. Having determined that the T650-35/PMR-15 SMC was capable of sustaining the new mission cycle loading for 100 h, the full complement of tests was conducted to the 50 and 100 h states of conditioning and then checked for residual properties.

Residual Properties

As indicated earlier, progressive fatigue damage accumulation was tracked through monitoring elastic stiffness degradation and damage tolerance was quantified on the basis of the static tensile property retention. Shown in Fig. 10 are the elastic stiffness values measured as a function of accumulated TMF missions. These values were measured isothermally at 232°C (i.e., T^{Max}) by applying a small elastic load (35 MPa) immediately after each mission cycle. Given the relatively large deviation in elastic stiffness (see Fig. 6), to facilitate specimen-to-specimen comparison, this property was normalized with respect to the original value measured prior to testing. The one major conclusion drawn from this data is that the TMF mission cycling had little to no effect on elastic stiffness. Of the 12 tests shown, six were cycled to 60 and six were cycled to 120 mission cycles, the data divide essentially equally above and below the original value, indicating the lack of a clear or overriding trend. Further, the data from Fig. 10 indicates that the modest changes observed are incurred early

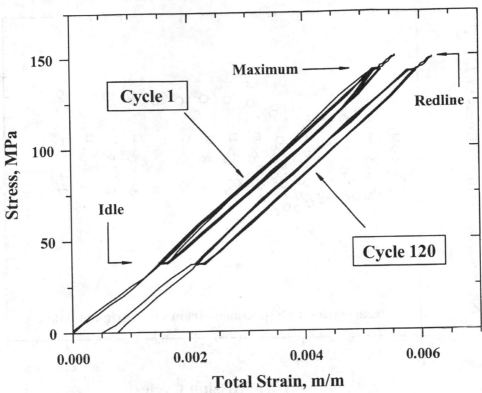

FIG. 9—*Stress/strain deformation behavior of T650-35/PMR-15 with TMF mission cycle parameters established by* $\sigma^{Max} = \sigma^{ult} - 4\Sigma$.

in the cyclic conditioning (prior to ~20 TMF missions) with only minimal exceptions. Subsequent to these early changes, the material tends to be cyclically neutral with respect to stiffness changes. The early changes may be indicative of a slight degree of fiber straightening, corresponding to the observed strain ratchetting and/or substructural damage (discussed later). Also note, in Fig. 10, identical symbols represent specimens taken from the same original panel: one tested to 60 missions and the other tested to 120 missions. No panel-specific patterns corresponding to elastic stiffness degradation were observed.

Shown in Fig. 11 are residual static tensile properties at 232°C after 50 and 100 h of mission cycling (six tests for each condition); the 0 h results from Fig. 6 are also shown for comparison. Though the figure is relatively crowded and difficult to distinguish single tests, it serves to show the significant spread and/or grouping associated with the data. Also given in Fig. 11 are the means ($\overline{X}$) and standard deviations (Σ), for ultimate strength, strain to failure, and elastic stiffness. By comparing the $\overline{X}$ and Σ values for the two populations (i.e., post 50 and 100 h TMF conditions) through an analysis of variance to the 0 TMF h material behavior, it was determined that there were no significant differences at the 95% confidence level in any of the three properties. The only property which exhibits a potential difference is the Σ of elastic stiffness for the 100 h TMF condition, which was significant at 92.5% confidence level, but not at the 95% level. This increased Σ over that exhibited prior to TMF cycling is consistent with the trend revealed in Fig. 10, where some of the material tends to

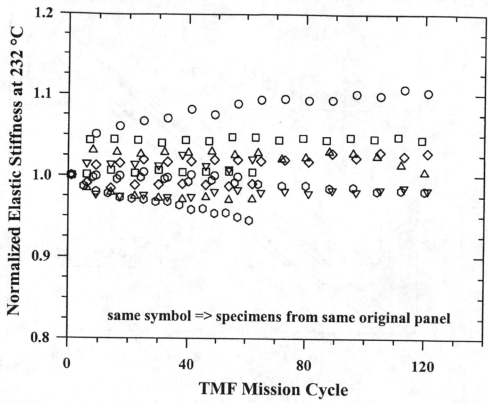

FIG. 10—*Normalized elastic stiffness measurements taken isothermally at 232°C during TMF mission cycling.*

stiffen slightly and some tends to become more compliant, but the overall mean remains constant. Similar to the progressive stiffness changes discussed above, panel-specific trends associated with residual static tensile properties were examined and none were evident.

Another residual property examined was T_g—to check for potential aging. It is well known that elevated temperature exposures can advance the cure states of polyimides, potentially introducing changes in mechanical properties. The result indicated no change in T_g after 100 h of TMF mission cycling, which was anticipated given that the total time spent at 232 and 260°C was 35 and 5 h, respectively. This observations is consistent with studies conducted on PMR-15 composites where volume changes were not found to occur after 100 h at 316°C [17] and T_g was found to be stable up to 2000 h of exposure at 260°C [18]. Thus, given the modest exposure times in the present study, it was assumed that the material did not incur any noteworthy aging.

Microstructural Examination

Microstructural examinations were conducted on several specimens subjected to TMF mission cycling. Although there was only minimal indication of damage corresponding to the macroscopic property degradation, there was clear evidence of highly localized damage at the microstructural level which was not evident in the untested control samples. Damage

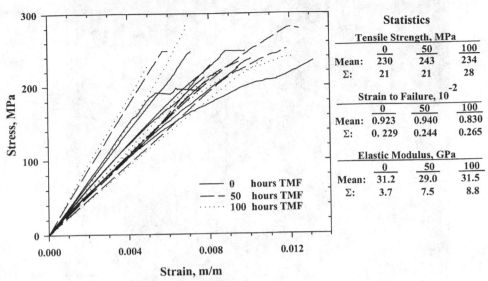

FIG. 11—*Residual static tensile properties at 232°C subsequent to 50 and 100 h of TMF mission cycling.*

was predominantly associated with fiber/matrix interface de-bonding at fibers oriented at angles of $90 \pm 40°$, where the specimen longitudinal axis coincides with 0 or 180°, that is, the loading direction. At points where such de-bonded fibers intersected with other fibers or fiber bundles having a different orientation, the cracks appear to have propagated from one bundle to the next, but generally remained confined to fibers oriented in the range specified above. This common pattern is well illustrated in Fig. 12a where a surface-visible crack is seen in the presence of at least two fiber orientations (upper portion of picture). The crack likely initiated at the fiber interface oriented at approximately 105°, then connected with the fibers oriented at approximately 60° and proceeded to cause a localized interface de-bond along this orientation, albeit, in much less aggressive fashion. As would be expected, whenever a crack at a fiber intersection such as this is noted, the more dominant crack is generally associated with the fiber orientation closest to 90°. Note that there is a third dominant fiber bundle orientation visible in the lower portion of the photograph where a starting crack is also associated; the orientation is approximately 135°.

In general, it appears that when the crack front encountered a bundle with a predominant longitudinal (PL) orientation, the fiber bundle effectively bridged the crack, leading to cases where the crack propagated "around" the bundle to another having a predominantly transverse (PT) orientation. This "around" crack path was seen to be associated either with the cut end or the outer diameter of the PL bundle. An example of the first is shown in Fig. 12b looking into the specimen thickness, where PT oriented fibers appear with a near circular cross-section, and PL oriented fibers appear elliptical. The crack shown here, connected to the surface, appears to propagate transverse to the loading direction through the PT bundles, but takes a path around the cut end of the PL bundle. Thus, in general, a through-thickness view of the SMC such as that shown in Fig. 12c revealed the vast majority of transverse cracks in PT bundles. Further, there is an indication of damage progression from the 50 to the 100 h TMF state, associated with higher transverse crack densities in the PT bundles, though the number of samples viewed (two for each case) was not sufficient to make a

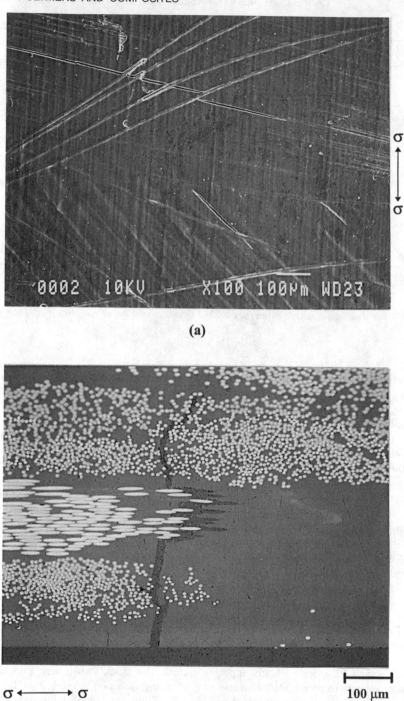

FIG. 12—*SEM (a) and optical (b) & (c) microscopy revealing TMF damage in the SMC. (a) surface cracking associated with de-bonds; (b) view into specimen thickness revealing a transverse crack proceeding around the end of a predominantly longitudinal bundle; (c) view into specimen thickness showing transverse cracking associated with transversely oriented bundles.*

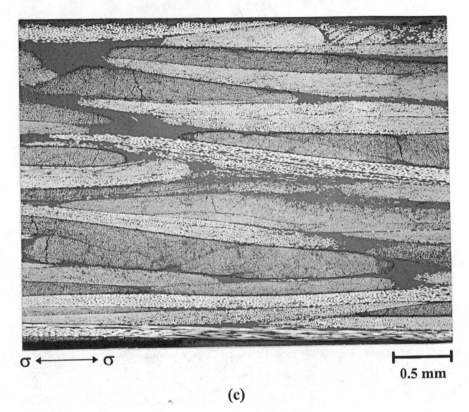

σ ←——→ σ

0.5 mm

(c)

FIG. 12—*Continued*

quantitative assessment. These microstructures do suggest, however, that given sufficient TMF cycling (significantly more than 100 h), the damage processes would eventually cause degradations in the macroscopic properties.

Fracture surfaces of residual strength specimens for all conditions (0, 50, and 100 h of TMF) appeared similar and are well represented by the fractography in Fig. 13. The fracture surface was generally transverse to the loading direction and revealed features dominated by pull-out of PL fibers/fiber bundles and separating of PT fiber bundles. The pull-out did not appear to be influenced by ply to ply interfaces, which were generally indistinguishable, both on the fracture surfaces and on the polished mounts. Thus, there was no indication of ply-to-ply delamination. A number of bundles having PL orientations reveal some degree of fiber fractures, but this effect appeared secondary to the pull-out feature. The general features were consistent with those cited by Beaumont and Schultz [19] for a room temperature fatigue failure of SMC-65.

Creep Response

During the process of determining the SMC's structural capabilities under TMF mission cycle loading, it became obvious that the more aggressive parameters investigated, which led to "premature" specimen fracture, induced a primary failure mode associated with creep deformation. This was facilitated by the fact that the TMF mission cycle featured a series

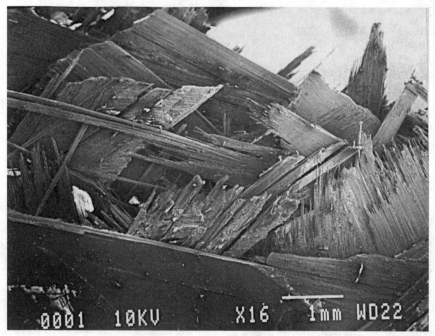

FIG. 13—*Typical fracture surface of T650-35/PMR-15 SMC revealing extensive fiber and fiber bundle pull-out.*

of elevated temperature stress holds (2.5 min each), allowing for creep. This "excessive" creep was unanticipated given that the mission maximum and redline temperatures (232 and 260°C, respectively) were significantly below the average T_g of approximately 340°C, raising concerns associated with the time dependent deformation response.

Recently, some emphasis has been placed on characterizing the creep response of carbon fiber/polyimide PMCs because of long-term durability issues [17,20,21], however, the work has been dominated by examining the effects of thermal aging at temperatures much closer to the material's T_g. Further, such investigations have dealt with continuous fiber reinforced materials, the much more common application. For PMR-15 SMC, the following questions needed to be addressed: At what stress/temperature threshold levels does creep occur? Further, do behaviors determined from routine dynamic mechanical loadings (e.g., storage and loss modulus, T_g, . . . etc.) give insight into thresholds for time dependent responses? A concise exploratory examination was conducted to address these critical issues.

Thermomechanical Creep Initiation—A novel thermomechanical creep deformation test was conducted where the temperature was ramped at 5°C/min to the redline temperature with an applied static load. The temperature ramp rate was felt to be sufficiently slow so as to allow for a quasi static thermal equilibrium through the thickness of the sample. A preliminary test was conducted under zero load to assess the free thermal expansion of the SMC during the temperature ramp. The specimen was then loaded at the redline stress level and subjected to the same temperature excursion. The goal of this test was to identify the temperature at which creep initiated. The results of this test are shown in Fig. 14 for two different conditions consisting of the two redline levels discussed previously. The creep strain plotted was reduced by subtracting off the time independent elastic and thermal strain components.

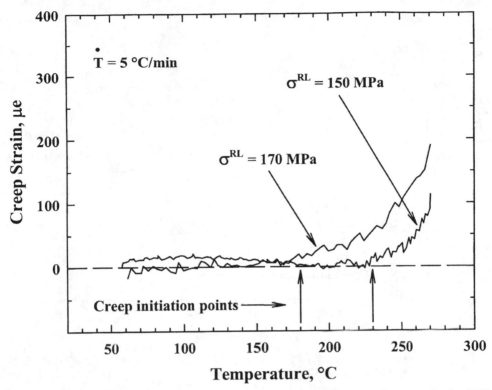

FIG. 14—*Thermomechanical creep threshold of T650-35/PMR-15 SMC at redline stress levels of 170 MPa ($\sigma^{ult} - 3\Sigma$ parameters) and 150 MPa ($\sigma^{ult} - 4\Sigma$ parameters).*

As seen in Fig. 14, with the nominal σ^{RL} of 150 MPa applied, the creep threshold was determined to be approximately 230°C. Further, the experiment revealed that the creep response at the redline temperature of 260°C, remains relatively modest. Note that 230°C essentially corresponds to the TMF mission cycle T^{Max}, at which point the stress is slightly less. These results correspond well to the deformation behaviors observed during the (-4Σ) TMF mission cycling, indicating that the creep experienced at maximum and redline conditions should be minimal. The creep threshold temperature drops quickly, however, when the creep stress level is set to the more demanding σ^{RL} of 170 MPa. Under these conditions creep deformation is found to initiate at approximately 180°C which is well below both the T^{Max} and T^{RL}. Again, the results here are consistent with the excessive creep observed during the exploratory phase of establishing TMF capabilities; recall Fig. 7. The data suggest that if a σ^{RL} of 170 MPa is to be tolerated, then the T^{RL} need be restricted to approximately 230°C. It is important to note that the creep deformations experienced were strictly associated with a viscoelastic response (no microstructural damage) and verified to be fully reversible/recoverable.

To gain further insight into the creep threshold temperature of the SMC, a comparable test was conducted on the neat PMR-15 resin with an identical post-cure cycle. The results are given in Fig. 15. The thermomechanical threshold stress for temperatures below 260°C was found to be ~21 MPa, which is seen to initiate creep at 240°C. By examining the tensile strength of the neat PMR-15, it was determined that this stress level corresponds to ~56%

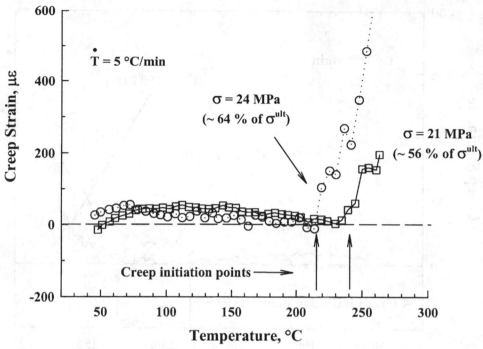

FIG. 15—*Thermomechanical creep threshold of neat PMR-15 at stress levels of 21 and 24 MPa (~56 and 64% of the 260°C σ^{ult}, respectively).*

of the material σ^{ult} at 260°C. A stress/temperature viscoelastic threshold "lower-bound" was estimated for neat PMR-15 by Kamvouris et al. [11] to be approximately 204°C, with stress levels up to 50% of σ^{ult}. This was concluded because no creep strains were observed over time periods approaching 720 h. The data presented in Fig. 15 agrees and further defines this stress/temperature creep threshold, suggesting that for a temperature of 204°C as proposed by Kamvouris et al., a stress level approaching 70% of σ^{ult} would be needed to induce time dependent deformation. Having established this threshold, a stress level corresponding to 64% of the 260°C σ^{ult} (i.e., 24 MPa) was applied to examine a macroscopic stress level "comparable" to that being applied to the SMC (i.e., 150 MPa). The results indicated that the neat PMR-15 resin experienced creep initiation at approximately 215°C. It should be noted that the "comparable" stress comparison between the neat resin and SMC materials is not to suggest an identical stress state. Unlike the neat resin, the localized matrix in the SMC is subjected to a highly complex multiaxial stress state. Further, use of the respective σ^{ult} values as the normalizing factors reflects only modest mechanistic relevancy, as these parameters are dictated by significantly different failure mechanisms for the two classes of materials. However, it remains noteworthy that the PMR-15 resin behavior clearly indicates that at mid-stress levels (50 to 70% σ^{ult}), time dependent deformation is likely to occur at temperatures as low as 200°C: a temperature which is nominally 150°C below commonly cited T_g values. The creep response measured on the neat matrix confirms the low temperature creep deformation exhibited by the SMC, where the characteristic creep behavior is dictated by the properties of the PMR-15. Also, if the possibility of interface damage associated with PT fibers is considered, both the creep threshold and creep rates may be detrimentally affected, giving rise to properties that are even less desirable than those of the matrix alone. Such an effect of "structural weakening" is relatively common in cases of transversely reinforced systems, and has been noted with specific reference to elevated temperature creep behavior of polyimide based composites [22].

Dynamic Mechanical Response—A final aim of the exploratory SMC creep investigation was to relate the thermomechanical creep thresholds to dynamic mechanical characterizations routinely performed on polymers and their composites to determine T_g. Specifically, does the dynamic mechanical response provide quantitative insight to the thermomechanical creep threshold? It is well known in the polymer science field that temperature dependent dynamically measured properties (e.g., storage modulus, loss modulus, T_g . . . etc.) are not time independent properties, but rather vary as functions of both temperature rate and loading rate. This fact, however, is often overlooked or at least minimized by mechanics researchers, who generally tend to treat and report material properties, such as T_g, as unique, time independent values (like melting points). Such issues of time dependency are central to the consideration of a thermomechanical creep threshold.

A series of dynamic mechanical characterization curves for the SMC and neat PMR-15 are shown in Figs. 16 and 17, respectively. The loading rate dependency was examined over the range from 0.1 to 10 rad/sec using the RMS 800 (Rheometrics™). The temperature sweep rate was maintained constant at 5°C/min for all of the tests and the T_g values were determined by the intercept method from the storage modulus as before. All of the samples used for this examination were taken from the same panel with the goal of minimizing material variability, and highlight the effects of time dependency. The T_gs of both the SMC and neat PMR-15 were found to vary considerably with loading frequency. The shear loss modulus, G″, representative of the imaginary part of the complex modulus, is usually dis-

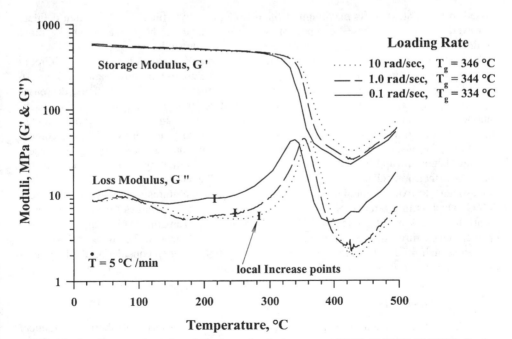

FIG. 16—*Loading rate dependent dynamic mechanical response of T650-35/PMR-15 SMC indicating where G″ trends suggest rate dependency at temperatures near 200°C.*

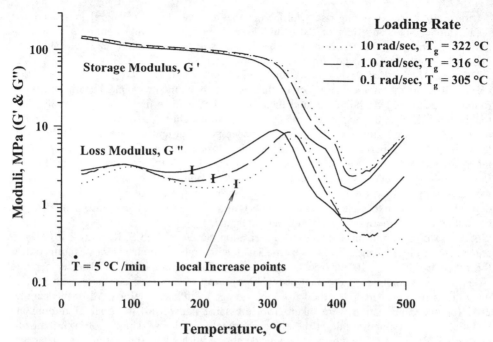

FIG. 17—*Loading rate dependent dynamic mechanical response of neat PMR-15 showing relatively low temperatures for rate dependency based upon G″ trends.*

cussed in the context of its maximum at the α transition, by which the T_g is often defined. However, this "out-of-phase" component is also indicative of sub-T_g transitions, or more subtle viscous effects pertinent to the discussion of time dependency.

As the temperature progresses from ambient conditions, G″ experiences a local minimum and then increases as the α transition is approached. The significant observation in Figs. 16 and 17 is that the approximate temperature at which the increase in G″ occurs, decreases with decreasing loading rate, showing a typical positive strain rate dependence. For clarification, each of the local increase points are designated with an "I" on the plots. Though this observation is not uncommon, such behavior is generally not discussed in the context of defining the onset of time dependent behavior. Note that as the loading rate is decreased, the state begins to approximate a static load condition, facilitating a comparison with the static load thermomechanical creep threshold tests. The G″ trends indicate that it is not unreasonable to expect time dependent deformation at temperatures as low as approximately 200°C, which is in the range of 150°C below commonly cited T_g values for PMR-15 and its composites. This result compares well with the findings from the exploratory thermomechanical creep threshold experiments. The fact that time dependent behavior can be experienced at such modest temperatures needs to be highlighted when considering the long term use of these materials in deformation critical applications.

Summary/Conclusions

A detailed experimental investigation was conducted to characterize the thermomechanical fatigue (TMF) durability and damage behavior of the carbon fiber/polyimide sheet molding compound (SMC), T650-35/PMR-15. High performance SMCs present alternatives to pre-preg forms with great potential for low cost component production. The TMF loading spectrum was proposed by Allison Advanced Development Company to be representative of a gas turbine engine compressor application where the SMC will be used for an inner vane endwall. The fiber distribution orientation was characterized through detailed quantitative image analyses, revealing a non-randomized fiber distribution orientation. Mechanical damage progression was tracked macroscopically on the basis of property changes, in addition to examining other properties such as glass transition temperature, T_g, and dynamic mechanical properties. Damage tolerance was quantified through residual static tensile properties after a prescribed number of TMF missions. Detailed microstructural examinations were conducted using optical and scanning electron microscopy to characterize the local damage. The SMC was further evaluated through a series of exploratory thermomechanical creep tests designed to determine stress/temperature creep thresholds. Similar thermomechanical tests on the neat PMR-15 resin permitted a comparison of the creep characteristics to the PMR-15 SMC.

The imposed TMF missions were found to have only a modest impact on inducing fatigue-type damage when limiting the exposure to 100 h. Microstructural damage was observed after 50 and 100 h of TMF cycling which consisted primarily of fiber debonding and transverse cracking local to various fiber bundles. However, no statistically significant degradations occurred in the elastic stiffness, or the residual properties of strain to failure and ultimate tensile strength. The TMF loadings did, however, promote creep damage and excessive strain accumulation which led to rupture in more aggressive stress scenarios. The creep behavior was found to occur in some cases at temperatures more than 150°C below glass transition temperatures commonly cited for PMR-15 composites. Thermomechanical exploratory creep tests revealed that below 260°C (the redline temperature investigated) the SMC undergoes time-dependent deformation at the stress/temperature threshold level of 150 MPa/240°C. Stress increases above this level allowed for lower temperature thresholds in the range of 180°C. Finally, upon examining relatively slow loading rates during dynamic mechanical testing, trends revealed by the loss modulus (G'') were found to serve as good indicators of the creep threshold temperatures.

Acknowledgments

The authors would like to thank Mr. Chris Burke, Mr. Ralph Corner, Ms. Linda Inghram, Ms. Linda McCorkle, and Mr. Dan Scheiman for their expert technical assistance in the various laboratory facilities at NASA LeRC; Mr. Dennis Keller for assistance with the statistical planning aspects; Dr. Gary Roberts for helpful technical discussions and supplying the neat PMR-15 resin, and Mr. Kevin Kannmacher and Mr. Wayne Maple of AADC for fabricating and supplying the SMC panels.

Appendix A

Procedure for Quantitative Image Analysis of the Fiber Orientation Distribution

Metallographic images were captured in gray scale from a video monitor/computer which is connected to a camera in the Nikon metallograph. A magnification of 50× was used to capture images across the entire cross section. The images were then imported to NIH Image Analysis program where the images were "seen as" 320 × 240 pixels (video squares) each with a level of gray between 0 and 255 (0 = white, 255 = black).

The images were calibrated against a known standard so that measurements could be made in millimeters.

The angle tool within the image analysis program was used to measure the predominant orientation of fibers within a fiber bundle region. The freehand tool was then used to outline this region to measure its geometric area. A MS Excel spreadsheet was used to record all measurements.

All fiber bundle regions at each cross section were analyzed. The total area of the fiber bundle regions within each orientation category was calculated. The fiber volume percent within the fiber bundles was taken into account to calculate the overall fiber volume orientation distribution.

References

[1] Wilson, D., "Polyimides as Resin Matrices for Advanced Composites," *Polyimides,* D. Wilson, H. D. Stenzenberger, and P. M. Hergenrother, Eds., Chapman and Hall, 1990, pp. 187–226.

[2] Stevens, T. "PMR-15 is A-OK," *Materials Engineering,* Oct. 1990, pp. 34–38.

[3] Allen, P. and Childs, B., "SMC: A Cost Effective Alternative to Prepreg Technology," *38th International SAMPE Symposium,* 1993, pp. 533–546.

[4] Hoff, S. M., "Applying Advanced Materials to Turboshaft Engines," *Aerospace Engineering,* Vol. 15, No. 2, 1995, pp. 27–30.

[5] *Investigation of Low Cost High Temperature PMC Components,* NASA Contract NAS3-97015, 1997.

[6] Hahn, H. T. and Kim, R. Y., "Fatigue Behavior of Composite Laminate," *Journal of Composite Materials,* Vol. 10, 1976, pp. 156–180.

[7] Talreja, R., "Stiffness Based Fatigue Damage Characterization of Fibrous Composites," *Fatigue of Composite Materials,* Technomic Publishing Company, Lancaster, Pennsylvania, 1987, pp. 73–81.

[8] Yang, J. N. and Liu, M. D., "Residual Strength Degradation Model and Theory of Periodic Proof Tests for Graphic/Epoxy Laminates," *Journal of Composite Materials,* Vol. 11, 1977, pp. 176–203.

[9] Reifsnider, K. L. and Stinchcomb, W. W., "A Critical Element Model of the Residual Strength and Life of Fatigue-Loaded Composite Coupons," *Composite Materials: Fatigue and Fracture, ASTM STP 907,* H. T. Hahn, Ed., ASTM, West Conshohocken, PA, 1986, pp. 298–303.

[10] Reardon, J. P. and Thorpe, J. D., U.S. Patent No. 5126085 920630, 30 June 1992.

[11] Kamvouris, J. E., Roberts, G. D., Pereira, J. M., and Rabzak, C., "Physical and Chemical Aging Effects in PMR-15 Neat Resin," *High Temperature and Environmental Effects on Polymeric Composites: 2nd Volume, ASTM STP 1302,* Thomas S. Gates and Abdul-Hamid Zureick, Eds., ASTM, West Conshohocken, PA, 1997, pp. 243–258.

[12] Roth, D. J., Baaklini, G. Y., Sutter, J. K., Bodis, J. R., Leonhardt, T., and Crane, E. A., "NDE Methods Necessary for Accurate Characterization of Polymer Matrix Composite Uniformity," *Advanced High Temperature Engine Materials Technology Program,* NASA CP 10146, 1994, paper 11.

[13] Castelli, M. G., "A Summary of Damage Mechanisms and Mechanical Property Degradation in Titanium Matrix Composites Subjected to TMF Loadings," *Thermal-Mechanical Fatigue of Aircraft Engine Materials,* AGARD Conference Proceedings 569, March 1996, pp. 12:1–12.

[14] Gyekenyesi, A. L., Castelli, M. G., Ellis, J. R., and Burke, C. B., "A Study of Elevated Temperature Testing Techniques for the Fatigue Behavior of PMCs: Application to T650-35/AMB-21," NASA TM-106927, July 1995.

[15] *Dexter Composites Division Data Sheet for HyComp M-300 Series Sheet Molding Compound,* Dexter Corporation (now HyComp, Inc.), Cleveland, Ohio.

[16] Skrzypek, J. J., *Plasticity and Creep: Theory, Examples, and Problems,* R. B. Hetnarski, Ed., CRC Press, Inc., Boca Raton, FL, 1993, p. 128.

[17] Skontorp, A. and Wang, S. S., "High-Temperature Aging, and Associated Microstructural and Property Changes in Carbon-Fiber Reinforced Polyimide Composites," *Proceedings of the 9th Technical Conference of the American Society of Composites,* Technomic Pub., 1994, pp. 1203–1212.

[18] Bowles, K. J., Roberts, G. D., and Kamvouris, J. E., "Long-Term Isothermal Aging Effects on Carbon Fabric-Reinforced PMR-15 Composites: Compression Strength," *High Temperature and*

Environmental Effects on Polymeric Composites: 2nd Volume, ASTM STP 1302, Thomas S. Gates and Abdul-Hamid Zureick, Eds., ASTM, West Conshohocken, PA, 1997, pp. 175–190.

[19] Beaumont, P. W. R. and Schultz, J. M., "Fractography," *Failure Analysis of Composite Materials: Delaware Composites Design Encyclopedia,* Vol. 4, 1990, pp. 134–135.

[20] Pasricha, A., Dillard, D. A., and Tuttle, M. E., "Effect of Physical Aging and Variable Stress History on the Creep Response of Polymeric Composites," *Mechanics of Plastics and Plastic Composites,* ASME, MD-Vol. 68/AMD-Vol. 215, 1995, pp. 283–299.

[21] Brinson, L. C. and Gates, T. S., "Effects of Physical Aging on Long-Term Creep of Polymers and Polymer Matrix Composites," *International Journal of Solids and Structures,* Vol. 32, No. 6, 1995, pp. 827–846.

[22] Rodeffer, C. D., Maybach, A. P., and Ogale, A. A., "Influence of Thermal Aging on the Transverse Tensile Creep Response of a Carbon Fiber/Thermoplastic Polyimide Composite," *Journal of Advanced Materials,* Jan. 1996, pp. 46–51.

James S. Loverich,[1] Blair E. Russell,[1] Scott W. Case,[1] and Kenneth L. Reifsnider[1]

Life Prediction of PPS Composites Subjected to Cyclic Loading at Elevated Temperatures

---

REFERENCE: Loverich, J. S., Russell, B. E., Case, S. W., and Reifsnider, K. L., **"Life Prediction of PPS Composites Subjected to Cyclic Loading at Elevated Temperatures,"** *Time Dependent and Nonlinear Effects in Polymers and Composites, ASTM STP 1357,* R. A. Schapery and C. T. Sun, Eds., American Society for Testing and Materials, West Conshohocken, PA, 2000, pp. 310–317.

ABSTRACT: Combinations of failure mechanisms are frequently encountered in the life prediction of composite materials. A life prediction methodology is developed and applied to one such failure mechanism combination. This method uses experimental data and analytical tools to predict the long-term behavior of a composite under service conditions. The prediction scheme is based on the assumption that damage accumulation progressively reduces the remaining strength of a composite. An overview of the fundamental concepts of the life prediction method is presented. The method is used to model the elevated temperature fatigue behavior of a unidirectional AS-4 carbon fiber/PolyPhenylene Sulfide (PPS) matrix composite material. The nonlinear combined effects of time at elevated temperature and fatigue are taken into account by considering elevated temperature tensile rupture and room temperature fatigue behavior. The life prediction for the combined loading is compared to 90°C tensile-tensile fatigue data. This comparison shows good correlation between the prediction and data and demonstrates the method's effectiveness in life prediction modeling.

KEYWORDS: life prediction, fatigue, elevated temperature, polymer matrix composites, PolyPhenylene Sulfide (PPS), tensile rupture, micromechanics

Life prediction of composite materials subjected to combined loading is a common occurrence in design situations. Combined loads may be comprised of quasi-static mechanical loads (tension, compression, shear, etc.), mechanical fatigue, elevated temperatures, thermal cycling and chemical degradation. It is usually not possible to include all of these factors and their variations in a single test. Therefore, it is necessary to have a method for combining the applicable conditions to predict the life of composites.

The analysis and experiments of the current study were conducted to characterize and predict the life of a unidirectional AS-4 carbon fiber/PolyPhenylene Sulfide (PPS) matrix composite material. This material's intended application is the helical tensile strength layers, referred to as the tensile armor, in flexible risers used by the offshore oil industry. These layers support the hanging weight of the riser as well as the axial load caused by internal pressure. A typical flexible riser design is shown in Fig. 1. In order to reduce the weight of flexible risers, replacement of the standard steel tensile armor with composite tensile armor

---

[1] Graduate Student, Research Associate, Assistant Professor, and Professor, respectively, Virginia Tech, Engineering Science and Mechanics, 120 Patton Hall, Blacksburg, VA 24061.

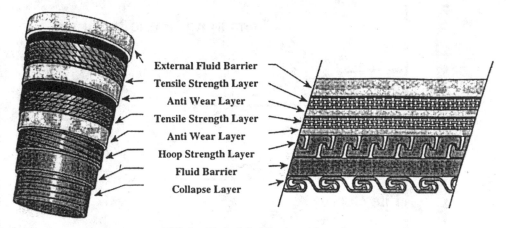

FIG. 1—*Typical flexible riser schematic.*

is being considered. A weight reduction of 30% and increased corrosion resistance is expected for a composite armored riser designed to the specifications of a comparable conventional flexible pipe.

In order to use this material for this application, the service life prediction must include the effects of fatigue and elevated temperatures up to 90°C. The service life model for the material under consideration was constructed using the methodology proposed by Reifsnider et al. [1]. This methodology has been used for a variety of material systems in many service environments and has been integrated into a performance simulation code, MRLife, developed by the Materials Response Group at Virginia Tech [2–4]. This methodology uses experimental data and analytical tools to predict the long-term behavior of a composite. The method is verified for the current material via comparison of experimental data to the predicted life.

Life Prediction Overview

The present life prediction scheme is based on damage accumulation concepts applied to composite materials. The basic principles of this scheme are described as follows. We begin our analysis by postulating that remaining strength may be used as a damage metric [2–5]. We next assume that the remaining strength may be determined (or predicted) as a function of load level and some form of generalized time. For a given load level, a particular fraction of life corresponds to a certain reduction in remaining strength. We claim that a particular fraction of life at a second load level is equivalent to the first if and only if it gives the same reduction in remaining strength, as illustrated in Fig. 2. In the case of Fig. 2, time t_1 at an applied stress level S_a^1 is equivalent to time t_2 at stress level S_a^2 because it gives the same remaining strength. In addition, the remaining life at the second load level is given by the amount of generalized time required to reduce the remaining strength to the applied load level. In this way, the effect of several increments of loading may be accounted for by adding their respective reductions in remaining strength. For the general case, the strength reduction curves may be nonlinear, so the remaining strength and life calculations are path dependent.

Our next step in the analysis is to postulate that normalized remaining strength (our damage metric) is an internal state variable for a damaged material system. This normalized

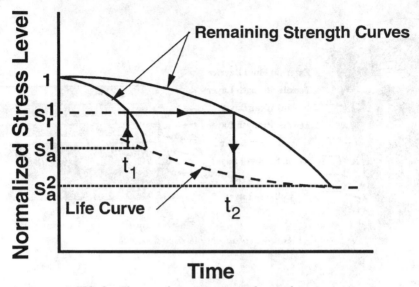

FIG. 2—*The use of remaining strength as a damage metric.*

remaining strength is based on the selection of an appropriate failure criterion (such as maximum stress or Tsai-Hill) which is a scalar combination of the principal material strengths and applied stresses in the critical element. In this way, we are able to consider a single quantity rather than the individual components of the strength tensor. We denote this failure function by *Fa.* We next construct a second state variable, the continuity [6], defined as $(1 - Fa)$ and denote it by ψ. We shall attempt to define our remaining strength and life in terms of ψ. To do so, we assume that the kinetics are defined by a specific damage accumulation process for a particular failure mode, and assign appropriate rate equations to each of the processes that may be present.

For the present case, we assume that a common kinetic equation (a power law) describes the damage accumulation in our material such that

$$\frac{d\psi}{d\tau} = -A\psi j\tau^{j-1} \tag{1}$$

where

ψ = continuity
τ = generalized time variable
A = constant
j = material parameter

The generalized time, τ, is defined by

$$\tau = \frac{t}{\hat{\tau}} \tag{2}$$

where

t = time
$\hat{\tau}$ = characteristic time

The characteristic time could be a creep rupture life, a creep time constant, or a fatigue life, in which case

$$\tau = \frac{n}{N} \tag{3}$$

where

n = number of fatigue cycles
N = number of cycles to failure for the applied loading condition

Next, we rearrange Eq 1 and integrate so that we have

$$\int_{\chi_0}^{\chi_i} d\psi = -A \int_{\tau_1}^{\tau_2} \psi j \tau^{j-1} d\tau \tag{5}$$

If we integrate the left-hand side and substitute $(1 - Fa)$ for ψ we have

$$\psi_2 - \psi_1 = Fa_1 - Fa_2 = \Delta Fa \tag{6}$$

Next, we define our normalized remaining strength, Fr so that

$$Fr = 1 - \Delta Fa = 1 - A \int_0^\tau (1 - Fa)j\tau^{j-1} d\tau \tag{7}$$

For loading cases in which Fa is constant, Eq 7 may be integrated to obtain

$$Fr = 1 - A(1 - Fa)\tau^j \tag{8}$$

The point at which Fr is equal to Fa defines material failure. Based upon the definition of τ, for the constant amplitude case this condition should occur when τ is equal to unity, so that A is also equal to unity. The resulting evolution equation for the remaining strength is then given by

$$Fr = 1 - \Delta Fa = 1 - \int_0^\tau (1 - Fa)j\tau^{j-1} d\tau \tag{9}$$

The material parameter, j, is related to the shape of the remaining strength curve [3]. The parameter, j, can be determined from remaining strength data for a given failure mechanism. In our case, $j = 1.2$.

Equation 9, which is a direct result of the kinetic equation given in Eq 5 has been used with a great deal of success for polymer composites [2–4] under fatigue loading conditions. Recently, an approach has been developed to predict the lifetime under combined fatigue and rupture conditions [7]. This approach is used in the present work.

Method Validation

We will now apply the damage accumulation scheme previously described to predict the elevated temperature fatigue behavior of the material under study and then compare these predictions to experimental data.

Tensile Rupture

The first step in implementing the method for this task is to characterize the time dependent behavior of the failure process. In our case this is tensile rupture at an elevated temperature of 90°C.

An equation presented by Kachanov [6] of the form

$$t_{rupture} = \frac{1}{(n + 1) \cdot B \cdot Fa^n} \tag{10}$$

where

$Fa = \sigma/\sigma_{ult}$ = applied stress/ultimate strength
$t_{rupture}$ = time to rupture hours
B, n = constants determined through curve fitting

is used to describe the tensile rupture behavior. Setting σ equal to S_r (the remaining strength fraction) Eq 10 can be rewritten in the form

$$Fa = \left[\frac{1}{(n + 1) \cdot B \cdot t_{rupture}} \right]^{\frac{1}{n}} \tag{11}$$

Equation 10 is then fit, using a least squares method, to 90°C tensile rupture data. The values of B and n were determined to be 0.0382 and 26.0, respectively (for the time to failure given in seconds). The tensile rupture curve fit and data are shown in Fig. 3.

Fatigue

Next, the fatigue effect of the combined loading was characterized. The room temperature fatigue data were fit with a conventional S-N curve of the form

$$\frac{S}{S_{ULT}} = A_n + B_n (\log N)^{P_n} \tag{12}$$

where

S/S_{ult} = normalized maximum stress
N = number of cycles to failure
A_n, B_n, P_n = material constants

The data and curve are shown in Fig. 4. The constants were found to be: $A_n = 0.9207$, $B_n = -0.0189$, and $P_n = 1$.

Finally, the effects of time at elevated temperature and fatigue were combined by defining a characteristic time as described previously in Ref 6. The prediction for the combined

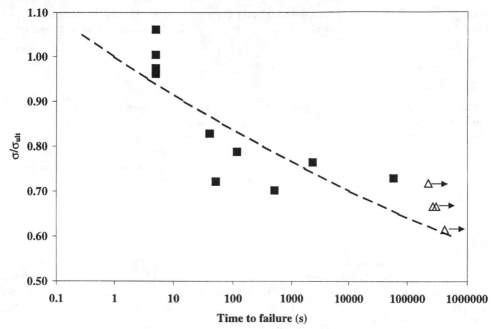

FIG. 3—*Tensile rupture curve fit and data.*

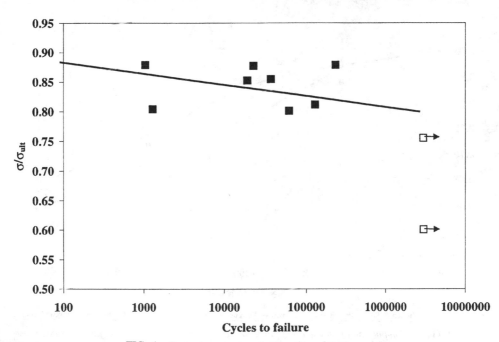

FIG. 4—*Room temperature fatigue fit and S-N curve fit.*

loading and the 90°C fatigue data shows a good correlation (Fig. 5). This verifies the technique for predicting the combined effects of time and cyclic processes for the material and conditions under study.

R-Ratio Influence

For the present case, tension-tension ($R = 0.1$) fatigue of a unidirectional composite at elevated temperature, this method resulted in a good prediction of life under present combined loading. It is expected that for tension-tension fatigue loading with other R-ratios, the present characterization of the fatigue effect on life will produce similarly good predictions. This should be confirmed through experiment. As the failure mechanism is expected to be different for tension-compression or compression-compression fatigue this failure mechanism would need to be characterized in order to represent the fatigue effect in this method.

Conclusion

The life prediction model presented is based on an iterative strength reduction scheme which allows for the consideration of several types of combined loads. This paper studied only environmental conditions and cyclic loading; however, any number of other load types may be able to be combined using a similar technique. Comparison of experimental data with the life prediction results displayed the model's effectiveness for the material and loading case under study.

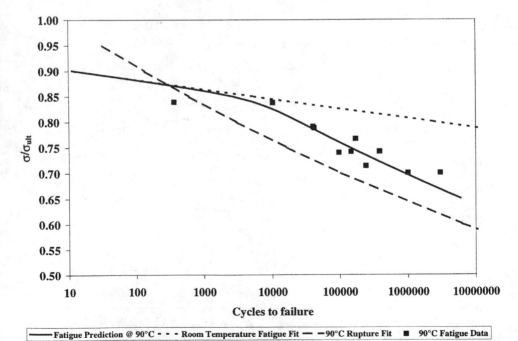

FIG. 5—*Comparison of elevated temperature fatigue prediction and 90°C fatigue data.*

References

[*1*] Reifsnider, K. L. and Stinchomb, W. W., "A Critical Element Model of the Residual Strength and Life of Fatigue Loaded Coupons," *Composite Materials: Fatigue and Fracture, ASTM STP 907,* H. T. Hahn, Ed., American Society for Testing and Materials, West Conshohocken, PA, 1986, pp. 298–303.

[2] Case, S. W. and Reifsnider, K. L., *MRLife11™ A Strength and Life Prediction Code for Laminated Composite Materials,* Materials Response Group, Virginia Tech, 1998.

[*3*] Reifsnider, K. L., "Use of Mechanistic Life Prediction Methods for the Design of Damage-Tolerant Composite Material Systems," *Advances in Fatigue Lifetime Predictive Techniques: Second Volume, ASTM STP 1211,* American Society for Testing and Materials, West Conshohocken, PA, 1993, pp. 3–18.

[*4*] Reifsnider, K., Case, S., and Iyengar, N., "Recent Advances in Composite Damage Mechanics," *Proceedings of Conference on Spacecraft Structures, Materials & Mechanical Testing* (ESA SP-386), Noordwijk, The Netherlands, June 1996, pp. 483–490.

[5] Schaff, J., "Life Prediction Methodology for Composite Laminates, Constant Amplitude and Two-Stress Level Fatigue" WL-TR-94-4046, Wright Laboratory Air Force Material Command, September 1994.

[6] Kachanov, L. M., *Introduction to Continuum Damage Mechanics.* Martinus Nijhoff Publishers, Boston, 1986.

[*7*] Case, S., Iyengar, N., and Reifsnider, K., "Life Prediction Tool for Ceramic Matrix Composites at Elevated Temperatures," *Composite Materials: Fatigue and Fracture, Seventh Volume, ASTM STP 1330,* R. B. Bucinell, Ed., American Society for Testing and Materials, West Conshohocken, PA, 1998, pp. 165–178.

J. R. Reeder,[1] *D. H. Allen,*[2] *and W. L. Bradley*[2]

Accelerated Strength Testing of Thermoplastic Composites

REFERENCE: Reeder, J. R., Allen, D. H., and Bradley, W. L., **"Accelerated Strength Testing of Thermoplastic Composites,"** *Time Dependent and Nonlinear Effects in Polymers and Composites, ASTM STP 1357,* R. A. Schapery and C. T. Sun, Eds., American Society for Testing and Materials, West Conshohocken, PA, 2000, 318–337.

ABSTRACT: Constant ramp strength tests on unidirectional thermoplastic composite specimens oriented in the 90° direction were conducted at constant temperatures ranging from 149°C to 232°C. Ramp rates spanning 5 orders of magnitude were tested so that failures occurred in the range from 0.5 s to 24 h (0.5 to 100 000 MPa/sec). Below 204°C, time-temperature superposition held allowing strength at longer times to be estimated from strength tests at shorter times but higher temperatures. The data indicated that a 50% drop in strength might be expected for this material when the test time is increased by 9 orders of magnitude. The shift factors derived from compliance data applied well to the strength results.

To explain the link between compliance and strength, a viscoelastic fracture model was investigated. The model, which used compliance as input, was found to fit the strength data only if the critical fracture energy was allowed to vary with stress rate. This variation in the critical parameter severely limits its use in developing a robust time-dependent strength model. The significance of this research is therefore seen as providing both the indication that a more versatile acceleration method for strength can be developed and the evidence that such a method is needed.

KEYWORDS: accelerated testing, superposition, polymer matrix composites, strength

Advanced polymer matrix composites are being developed for a number of elevated temperature applications. As polymeric composites are used in applications that are closer to the polymer's glass transition temperature, the compliance and strength of the material will change with time. The duration of tests that can conveniently be conducted in a laboratory will be orders of magnitude shorter than the necessary life span of composite structure. To avoid failures due to unexpected strength loss after long periods of time, it is imperative that accelerated tests be developed to determine long-term strength properties.

The time-temperature-superposition (TTSP) technique is a common method used to determine long-term compliance properties from shorter term tests [1,2]. Accelerated strength testing is not as well established. Miyano, et al. [3,4], suggested that it may be possible to accelerate the strength testing of advanced composite materials using time-temperature-superposition and shift factors obtained from creep compliance tests.

The eventual goal of this research is to develop models that are able to predict damage that will initiate and grow in materials that are in service for decades. These models will have to be able to model full thermo-mechanical fatigue accounting for viscoelastic deformation, plastic deformation, physical and chemical aging, and environmental effects. One

---

[1] NASA Langley Research Center, Hampton, VA 23681.
[2] Texas A&M University, College Station, TX 77853.

step toward the final goal would be to accurately predict the long-term static strength of composite materials using much shorter-term strength tests. The development of such an accelerated test for strength was the objective of this research project. To accomplish this objective, tests were conducted to verify that the time-temperature superposition technique can be used with strength data and that the shift rates for strength are the same as those measured from creep compliance tests. The link between compliance and strength was then investigated in an attempt to better understand the failure mechanism so that a more robust accelerated test methodology might be developed.

Transverse tension strength of a IM7/K3B composite was tested over a wide range of time scales and at several temperatures. At each temperature, strength was determined as a function of stress rate to show the magnitude of the time-scale effect. Once the failure curves were defined at different temperatures, they were shifted to form a master curve. The formation of a master curve allowed temperature to be used as an accelerator because an elevation in temperature affected strength in a similar manner to a known increase in time scale. The shift factors were then compared to see if they were the same as the shift factors derived from compliance data.

The reason why the superposition of compliance and strength might be tied together was also investigated. A viscoelastic fracture mechanics model was formed which defined a critical fracture parameter based on viscoelastic compliance properties. This model was derived from a work-of-fracture model proposed by Schapery [5]. The results of the fracture model were compared to the strength results. For the fracture model to be an appropriate model for these strength tests, the strength property must be controlled by a fracture type phenomenon where cracks initiate and grow from pre-existing flaws. A model of this type, with failure based on compliance data, would not only make long-term strength prediction possible, but would also dramatically reduce the number of strength tests needed to characterize a material's strength.

Experimental Testing

Transverse strength tests were performed on IM7/K3B composite specimens at temperatures between 149°C and 232°C over a range of time scales so that time to failure ranged from 0.5 s to 24 h. During the tests, load, displacement, and strain readings were recorded. This section describes the material, test specimen, experimental apparatus, and test procedure.

Material and Test Specimen

The test material was IM7/K3B, a composite made of IM7 graphite fibers in a thermoplastic resin designated K3B. The glass transition temperature (T_g) of K3B has been reported to be 236°C [6]. The high T_g of the K3B resin makes IM7/K3B a candidate material in elevated temperature applications.

From the IM7/K3B prepreg material, four 8-ply unidirectional laminates measuring 30.5 cm × 61 cm were prepared with the fibers running in the 61-cm direction. After these panels were cured using the manufacturer's recommended cure cycle, each 1.1-mm thick panel was cut into 30 test specimens measuring 3.8 cm × 15 cm (see Fig. 1a). The specimens were cut so that the fibers ran transverse to the loading direction. Three strain gauges were attached to each specimen as shown in Fig. 1b and were arranged so that the bending front-to-back and side-to-side could be measured as well as average strain. The high temperature strain gauges designated WK-00-250BG-350 were bonded to the specimen with a high temperature strain gauge adhesive. Loading tabs were applied to each end of the specimen to reduce stress concentrations due to gripping. To reduce stress concentrations at the end of the tab,

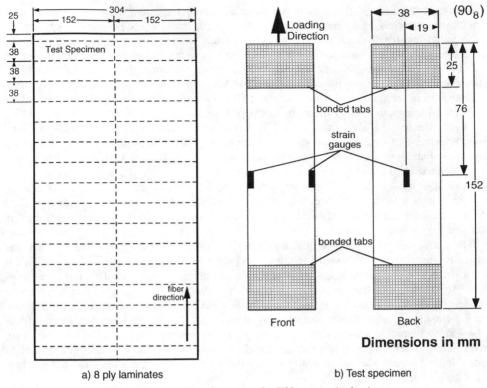

a) 8 ply laminates   b) Test specimen

FIG. 1—*Test specimen layout on the K3B composite laminates.*

a relatively compliant tabbing material made of an open weave sand paper infused with an epoxy resin was used.

The specimens were dried with a multistage drying cycle lasting 24 h with a peak temperature of 150°C. They were then aged at 204°C for 4000 min. (66 h) which was significantly longer than the duration of any of the tests to be conducted. This aging cycle was conducted because the K3B resin material has been shown to physically age [6]. Physical aging in polymers occurs due to excess free volume being trapped in a polymer when cooled below T_g. As the polymer then moves toward equilibrium, the physical properties change. The change in the polymer matrix causes the properties of the composite to change. Although aging of this material was unavoidable, it was not the focus of this investigation. By pre-aging the specimens, the effect of the aging that occurred during the subsequent tests was reduced, thus reducing skew in the data for tests of different durations. To prevent moisture from re-entering the specimens, they were stored at 107°C for less than one week until they could be tested.

Test Apparatus

The transverse strength tests were performed in a hydraulic load machine. The test setup is presented in Fig. 2. The test machine was equipped with an environmental chamber which surrounded the test section and high-temperature hydraulic wedge grips.

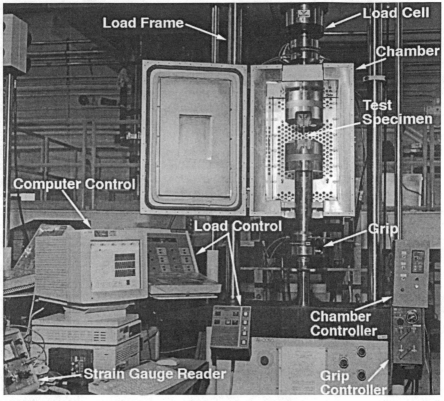

FIG. 2—*The test apparatus used during transverse tension tests.*

In addition to load and position, both temperature and strain were measured during tests. The temperatures, both inside and outside the chamber, were monitored using type T thermocouples. Each strain gauge was connected in a quarter bridge configuration and monitored using a bridge amplifier and meter.

All instruments were connected to the test computer. The computer collected data from the thermocouple reader, the hydraulic load frame controller and the strain gauge readers. The computer read, processed, plotted, and stored the data, and sent appropriate control commands to the load frame based on the processed data.

Test Procedure

Since the test specimens were cut from several panels which could be slightly different due to variations in processing, the specimen for each test was chosen using a random number generator so that the data would not be biased by such effects. The width and thickness of each specimen was measured to the nearest 2.5 μm using micrometers. Measurements were taken at three different positions spaced along the length of the specimen. The average of the measurements was recorded for the thickness (b) and width (w). Strain gauges were then applied followed by the ends tabs, as described earlier. The specimens were aged and stored until they could be tested. Prior to performing a test, the test chamber was heated to the test temperature and allowed to reach equilibrium temperature. This could take many hours because of the large mass of the grips. The specimen was then aligned in the grips and allowed to heat up to test temperature while hanging free from the lower grip. Thermocouples were located near each grip and at the center of the specimen to monitor temperature. Once at temperature, the strain gauges were zeroed and the bottom grip was closed. To minimize stress concentrations due to gripping, the tabs were allowed to extend out of the grips by one-fourth of their length, and the grip pressure was kept as low as possible without incurring significant slip.

Transverse strength tests were performed at four temperatures: 149, 177, 204, and 232°C. At each temperature, constant ramp rate tests were performed with a range of stress rates that produced failure times from half a second to approximately 24 h. The five ramp rates were 94 500, 4730, 189, 9.45, and 0.473 MPa/sec. These rates were chosen to be fairly evenly distributed when plotted on a logarithmic scale. At each stress rate and temperature, three or more duplicate tests were performed.

It was found that the zero point of the load cell could change by as much as 222 N due to changes in load cell temperature which changed with ambient lab temperature and with environmental chamber heating. The change in indicated load with load cell temperature was measured and found to be a linear function. By programming the test computer to correct the measured load for load cell temperature, the zero point was held constant within 10 N (less than 5% of minimum failure load). Because the load cell temperature changed very slowly, the correction was only made during tests lasting an hour or longer. The measurement of ultimate load was even more precise because a zero load reading for each test was taken just after failure. This zero point should be the same as that at failure since the time for drift in the signal was extremely small.

During a test, load, position, and the three strain gauge readings were recorded along with temperature data. At least 100 data points were recorded while each specimen was being loaded to failure.

After each test, average specimen strain as well as the difference in strain readings front-to-back and side-to-side were calculated and plotted as a function of load. The side-to-side and front-to-back bending strains generally stayed below 200 $\mu\varepsilon$ which is less than 5% of the average strain at failure and indicates that the specimens were well aligned. The specimen strength was determined by dividing the maximum load value by the specimen area (the product of average width and average thickness for each specimen).

Analysis

This section describes three different analytical models. The first model produces an expression for the compliance of the K3B matrix material as a function of time and temperature. This will be used as input for the subsequent models. The second model predicts composite compliance transverse to the fiber direction and is used in the results section to verify the accuracy of the matrix model. The third model is a viscoelastic fracture model which relates the matrix compliance properties to composite failure. This model attempts to explain the link between the time-temperature effects of compliance and strength and therefore is an attempt at developing a robust accelerated test method.

Matrix Compliance Model

The model for the compliance of the matrix (D_m) is based on creep data for the K3B material collected during a study on physical aging [6]. The data from the reference were in the form of momentary master curves shown as dashed curves in Fig. 3. Each momentary master curve was constructed with data from several creep tests conducted at different aging times, but with the data adjusted to represent compliance at a constant reference aging time. The reference time used in producing these momentary master curves was 2 h.

To make predictions of compliance for longer testing times than could be actually be performed, the time temperature superposition (TTSP) technique was used. In this technique the dashed curves are shifted horizontally to form a TTSP master curve as shown by the solid line. For this master curve, all the data were shifted to a reference temperature of 200°C. A horizontal shift in data has the effect of shifting the data in time, and the distance that each curve must be shifted is recorded as the shift factor (A_T), for that temperature. The master curve was fitted with a stretched exponential function which has the form

$$D_m = D_0 \, e^{(t/\tau)^\beta} \tag{1}$$

The curve fit parameters: initial compliance (D_0), time constant (τ), and stretch factor (β) have values of 0.355 1/GPa, 31 500 sec, and 0.31, respectively. To use the master curve at a temperature other than the reference temperature the curve must be shifted horizontally using an appropriate shift factor. The relationship between temperature and shift factor is derived from the shift factors determined in developing the master curve. Figure 4 shows that the relationship between A_T and temperature is well modeled by an Arrhenius style equation [7] of the form

$$\log A_T = \eta\left(\frac{1}{T_K} - \frac{1}{T_0}\right) \tag{2}$$

with a shift rate (η) of 16 500°K (corresponding to an activation energy of 316 kJ/mole).

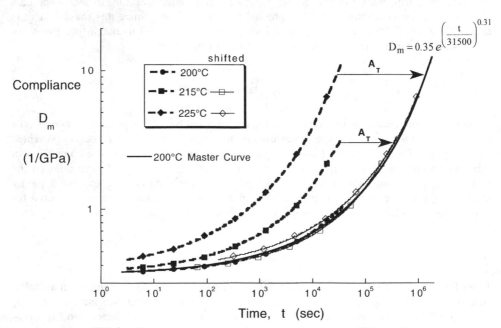

FIG. 3—*Time-temperature compliance master curve for K3B matrix.*

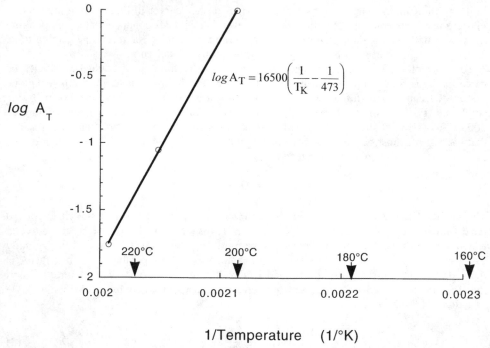

$$\log A_T = 16500\left(\frac{1}{T_K} - \frac{1}{473}\right)$$

1/Temperature   (1/°K)

FIG. 4—*Shift factors from K3B master curve.*

The reference temperature of the master curve (T_0) and the temperature to which the data should be shifted (T_K) must both be expressed in absolute units. Since this master curve is at 200°C, $T_0 = 473$°K. Equations 1 and 2 can then be combined to obtain the following expression for compliance as a function of time and temperature.

$$D_m = D_0 \, e^{\left(\frac{t}{\tau 10^{-\eta(1/T_K - 1/T_0)}}\right)^\beta} \tag{3}$$

One more correction was needed before this model could be used to predict the compliance of the matrix of the composite specimens used in this study. The compliance curves presented in Fig. 3 were for resin material with a constant aging time of 2 h (t_{ref}). The composites in this study were aged for 66 h at 204°C. The additional aging time will have affected the compliance properties so it must be accounted for. Aging time can be accounted for by shifting the master curve just as the TTSP technique shifts the master curve to account for temperature. The following equation relates the shift factor for aging (a_T) to the aging time (t_{age}).

$$\log a_T = \mu(\log t_{age} - \log t_{ref}) \tag{4}$$

From Ref 6 the shift rate (μ) for aging at 204°C was found to be 0.93. Aging 66 h at 204°C therefore caused an aging shift factor (a_T) of 26.0. To shift the reference master curve, a_T was multiplied by the time constant (τ). Incorporating the aging shift factor into Eq 3 produced the following compliance model for the K3B matrix using the numerical values found in Table 1.

$$D_m = D_0 \, e^{\left(\frac{t}{a_T \pi 10^{-\eta(1/t_K - 1/T_0)}}\right)^{\beta}}$$

(5)

Composite Compliance Model

The composite compliance model was used to relate the matrix compliance model to the composite stress-strain response which was measured during this study. The matrix compliance was related to the compliance of the composite through a simple series model.

$$D_c = v_f D_f + v_m D_m$$

(6)

where D and v were respectively the compliance and volume fraction of the composite, fiber, and matrix. Because the matrix is much more compliant than the fiber in this type of graphite-polymer composite, the fiber was assumed to be rigid ($D_f = 0$). Equation 6 therefore reduced to

$$D_c = v_m D_m$$

(7)

This, the simplest of composite micromechanics models, will be shown to be adequate for the purposes of this research. Equation 7 indicates that a simple scaling factor should be able to scale the matrix compliance to that of the composite, and further that the scaling factor would have physical significance as the matrix volume fraction.

Figure 5 shows the stress-strain response of a transverse tension composite specimen tested at 149°C and 2220 N/sec and the prediction of the matrix compliance model for a neat resin specimen tested under the same conditions. At this low temperature and high stress rate, the material behaved essentially linear elastically, producing a linear stress-strain curve. The figure shows that a scaling factor of 0.33 was needed to scale the predicted neat resin response to the measured response of the composite thus creating a composite compliance model. The composite compliance was then used to model the composite material at test conditions where the material exhibited viscoelasticity.

A hereditary integral [8], given as Eq 8, was used to predict the strain response due to a given loading history based on the material's viscoelastic compliance.

TABLE 1—*Matrix compliance model property values.*

Parameter		Value
D_0	Initial compliance	0.355 1/GPa
a_T	Aging shift factor	26.0
τ	Time constant	31 500 sec
η	TTSP shift rate	16 500°K
T_0	Reference temperature	473°K
β	Exponential stretch factor	0.31

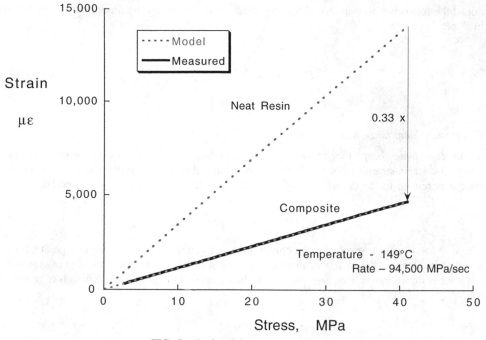

FIG. 5—*Scaling of matrix compliance.*

$$\varepsilon(t) = \int_0^t D(t - \psi) \frac{d\sigma(\psi)}{d\psi} \, d\psi \tag{8}$$

Stress and strain are given by σ and ε, respectively and ψ is a dummy time variable. In Eq 8, D can be the compliance of the matrix or the composite depending on which material response is to be modeled.

Viscoelastic Fracture Model

To help explain the transverse strength results which will be presented, a fracture model was considered. The use of a fracture model assumed that specimen failure was precipitated by a pre-existing flaw which initiated a crack that grew quickly to failure. Many different parameters have been used to describe when a crack will extend. The three most notable are K, G, and J [9]. The stress intensity factor (K) is a scaling factor to describe the magnitude of the stress field around a crack tip. The strain energy release rate (G) describes the amount of energy made available to the crack tip from external forces and internal strain energy when the crack length is extended. The J-integral is an alternative method of evaluating the energy released as a crack grows which can account for plasticity in the body. The J-integral is calculated by a path independent integral of traction, displacement and strain energy, around the crack tip. For linear elastic problems, the three parameters are related to each other by material properties and therefore predictions made from them are equivalent. Fracture in a viscoelastic body is more complicated.

The G parameter assumes that the energy supplied to a cracked body is readily available to extend the crack. This is not the case in a viscoelastic body where some of the absorbed energy may require significant amounts of time to be released. The W_f fracture parameter describes only the amount of energy that can be released by a viscoelastic body at a given point in time. Schapery [5] used a path independent integral and the elastic-viscoelastic correspondence principle to develop the expression for W_f given in Eq 9.

$$W_f = E \int_{t_0}^{t_i} D'(t_i - \Psi) \frac{dJ^E}{d\Psi} d\Psi \qquad (9)$$

where J^E is the J-integral for an elastic body similar to the viscoelastic body of interest but with a constant modulus, E. The E value is actually arbitrary because it eventually cancels out of the equation. The effective compliance, D', is a complex function of the orthotropic compliance of the composite [10]. For the polymer matrix composite used in this study, where the viscoelastic matrix is soft in comparison to the elastic fiber, D' will be assumed to be proportional to the compliance of the matrix as indicated in Eq 10.

$$D'(t) = C_1 D_m(t) \qquad (10)$$

The integral is from the initiation of loading (t_0) to the time when the crack begins to grow (t_i). Since J^E is for the reference elastic system, it is equivalent to G^E, the strain energy release rate of the reference system and related to K through the reference modulus and the Poisson's ration (ν) as shown in Eq 11. The stress intensity factor is only a function of the applied stress (σ) and the geometry parameters such as crack size and specimen dimensions which are represented by a constant (C_2).

$$J^E = G^E = \frac{1 - \nu^2}{E} K^2 = \frac{1 - \nu^2}{E} (C_2 \sigma)^2 \qquad (11)$$

Substituting Eq 11 into the Eq 9 yields

$$W_f = (1 - \nu^2) C_1 C_2^2 \int_{t_0}^{t_i} D_m(t_i - \Psi) \frac{d(\sigma^2)}{d\Psi} d\Psi \qquad (12)$$

For an elastic body, W_f can be shown to be equal to J^E and G^E since D, which is not a function of time, can be moved out of the integral in Eq 9 and since $1/D = E$.

Although it is not necessary that W_f be constant with time and temperature, if one does make that assumption, prediction of failure as a function of time and temperature becomes very straight forward. One could test a specimen geometry for which the strain energy release rate constant (C_2) is known. From Eq 12, the critical value of W_f could then be calculated using the time to failure and the other test parameters. The critical W_f could then be used to calculate the time to failure of any other loading history and geometry by inverting Eq 12. Of course, the C_2 parameter would need to be known for the new geometry.

In the current application, the size and shape of the critical flaw was not known resulting in an unknown value of C_2. To use the model for this case, C_2, W_f, and material properties that were assumed constant in the current study were grouped together to define one unknown constant. Once this constant was determined by a single strength test, the fracture model was used to predict strength after other loading and temperature histories. Although the unknown C_2 limits the model predictions to similar geometries, the model was still quite useful because failure predictions due to a wide range of loading and temperature histories were possible.

The experiments in this study were performed with constant stress rates (R); therefore

$$\sigma = Rt \tag{13}$$

This expression was then substituted into Eq 12, to obtain

$$W_f = (1 - v^2)C_1C_2^2 \int_{t_0}^{t_i} D_m(t_i - \Psi) \frac{d(R^2 \Psi^2)}{d\Psi} d\Psi$$

$$= 2(1 - v^2)C_1C_2^2R^2 \int_{t_0}^{t_i} \Psi D_m(t_i - \Psi) d\Psi \tag{14}$$

Equation 15 was derived by assuming that W_f is constant and collecting the terms that are constant for this study:

$$R^2 \int_{t_0}^{t_i} \Psi D_m(t_i - \Psi) \, d\Psi = \frac{W_f}{2(1 - v^2)C_1C_2^2} = \text{Constant} \tag{15}$$

Once the critical value of this constant was determined from a "reference" strength test, predictions of failure could be made for other temperature and load histories. Figure 6 shows predictions of this model created by inverting Eq 15 to predict times to failure at different stress rates and temperatures, using Eq 5 as the expression for matrix compliance (D_m) and

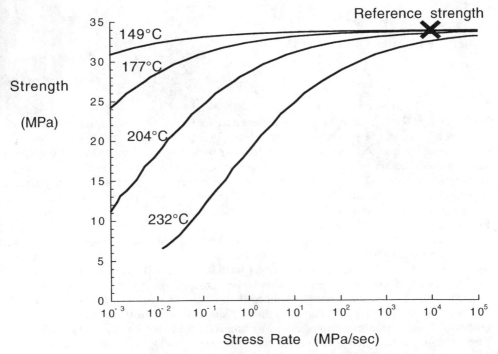

FIG. 6—*Constant* W_f *contours of strength predictions.*

the experimentally measured failure stress at 149°C and 94 500 MPa/sec as the reference strength. Although Fig. 6 only shows predictions for constant stress rate tests, failures due to other loading histories could have been predicted. A mathematical manipulation computer program was used to create the contour plots in Fig. 6 because inverting the integral in Eq 15 to obtain a closed form expression for t_i as a function of R and D_m was not possible. Figure 6 actually shows predictions of strength (the product of t_i and R) versus stress rate to emulate how experimental results might actually be presented.

Figure 6 shows that the model predicts strength decreasing with longer test times or slower ramp rates. One way of thinking about the results is to think of a failure occurring at a given stress level. If the time to reach that stress level were to be increased, the time for viscoelastic deformation would also be increased, therefore producing greater deformation. The model assumes that the work to failure is constant; therefore, if the deformation increases, the critical stress would have to decrease since work is the product of stress and deformation.

Without assuming that W_f remains constant with time and temperature, the predictive capabilities of the model become much more limited. Instead of being able to predict a wide range of failure events with just one test, many would be needed. Once the failure envelope for how W_f changes with histories of load and temperature is defined for one specimen geometry, it could then be used to predict the failure of other geometries. Since the variety of load and temperature histories to failure are unlimited, defining the failure envelope would be problematic without further understanding of how W_f should change. The W_f model would still be quite useful for making predictions of failure for some complicated structure based on a coupon test conducted such that the history of W_f for the critical flaw was duplicated in an accelerated mode by increasing the temperature.

Results and Discussion

This section presents the results of the transverse strength tests conducted over a range of temperatures and stress rates and compares the results to predictions from the various models.

Strength Results

Strength tests were performed at four different temperatures and at stress rates that varied by 5 orders of magnitude. The results of these tests are plotted in Fig. 7. Numerical values for all data points are listed in Ref *11*. There is a large amount of scatter in the data which blurs the effect of stress rate and temperature, but scatter is a common problem for this type of transverse tension specimen [*12*]. The scatter is believed to be due to the premature failure of some of the specimens. The strength of a test specimen may be artificially low for many reasons: specimen misalignment, damage caused during specimen machining, stress concentration due to gripping, etc. Because there are many reasons that would cause a specimen strength to be abnormally low and no anticipated reason that would cause a specimen to show an artificially high strength, only the highest three strength values at each test condition were used in the rest of this study. The strength results are replotted in Fig. 8 with the lower strength values filtered out. It is obvious that not only is the scatter greatly reduced, but trends in the data can be much more readily seen. Most of the strength results that were filtered out, were from specimens that failed at the grip line while others were from specimens with noticeable edge defects, but not all the rejected data showed signs of problems. Of the data that were kept, some specimens failed at the grip line which can be a sign of early failure due to stress concentrations but can also naturally occur due to random strength variations along the length of the specimen. In Fig. 8, the trends in the data are marked by second order polynomial curves fitted through the data by a least squares procedure. At each

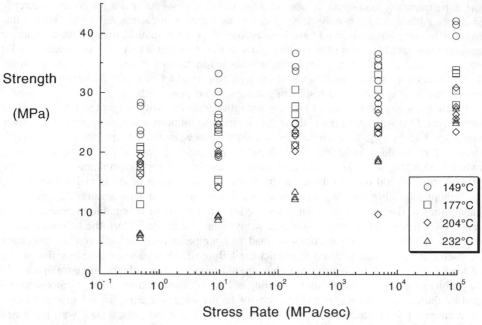

FIG. 7—*Results of transverse tension tests on K3B/IM7.*

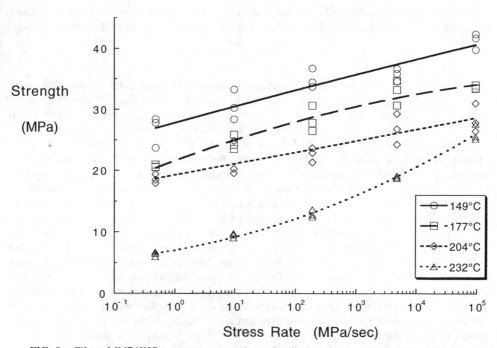

FIG. 8—*Filtered IM7/K3B transverse strength results (highest 3 strengths at each condition).*

temperature, the strength decreased by at least 25% as the test time was increased by 5 orders of magnitude. At 232°C, which was the temperature at which the strength was the most sensitive to stress rate, the strength dropped by 66%. The highest strength was always at the highest stress rate and therefore the shortest test. The strength results also showed a significant temperature dependence. Each increment of 28°C reduced the strength at a given stress rate by an average of 6.2 MPa. The form of the curves are such that they might be shifted to form a master curve. Figure 9 was created by shifting the results presented in Fig. 8 using the same shift constant ($\eta = 16\ 500$ °K) that created the matrix compliance master curve shown in Fig. 3. In Fig. 9, strength is plotted against "temperature reduced" stress rate indicating that data at different temperatures were shifted in time. The shift factors, A_T, are given by Eq 2. However, for the strength data, the reference temperature, T_0, was chosen as 422°K (149°C) corresponding to the minimum temperature tested, as opposed to $T_0 = $ 473°K which had been used with compliance data. Since the shift factor at 149°C is 1, the plot can be thought of as the master curve at 149°C plotted against stress rate. In Fig. 9, only the averages of the high 3 replicate tests are shown. The light lines are the best fit lines from the previous figure shifted in time. The strength results up to 204°C do appear to form a fairly well behaved master curve as shown by the heavy line. At 232°C, the strength dropped off faster than would have been predicted by the master curve fitted through the remaining data.

The formation of the master curve indicates that an increase in temperature creates an effect that is comparable to decreasing the stress rate and therefore increasing the time scale of a test. The formation of a master curve allows the prediction of the 149°C strength tested at a stress rate of 10^{-5} MPa/sec which would require approximately 50 years to conduct. If

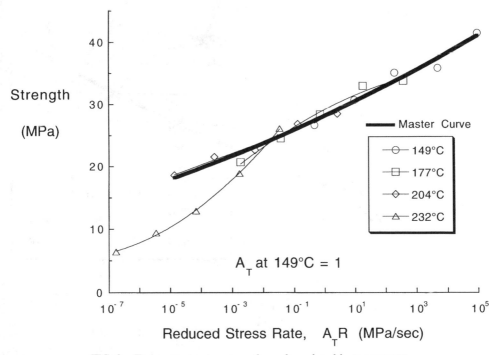

FIG. 9—*Transverse tension strength results reduced by temperature.*

the 232°C data had also followed the master curve, 149°C tests conducted at 10^{-7} MPa/sec could have been predicted which would have required 5000 years to conduct! The current master curve allows predictions of 177°C tests lasting half a year while 50 year tests could have been predicted if the 232°C tests had also fallen along the master curve.

Although this way of making predictions of strength over long time scales is significant, the technique by itself only allows for accelerating similar load histories at a constant temperature. Additional knowledge would be needed to, for example, use results to predict creep test data where the stress is held constant until specimen failure or predict the strength of a constant ramp tests during which the temperature changed. To better understand how the entire history of load and temperature leading up to failure effects strength, the cause of failure was explored. The development of such understanding could significantly expand the usefulness of strength master curves such as the one presented in Fig. 9. Because the shift constant for strength is the same as the shift constant for compliance, it is suspected that the two time dependent effects may be due to the same phenomenon. The change in strength with time may even be caused directly by the change in compliance. The W_f fracture parameter relates compliance to failure.

Comparison to W_f Fracture Model

Predictions of the W_f fracture model are shown in Fig. 10. The constant W_f parameter was fit to the data at two different reference strength values. The experimental data in Fig. 10 is plotted against temperature reduced stress rate just as it was presented in Fig. 9. It is clear from the graph that this simple fracture model does not capture the failure characteristics of the material, and that using a different constant value of W_f would not help significantly.

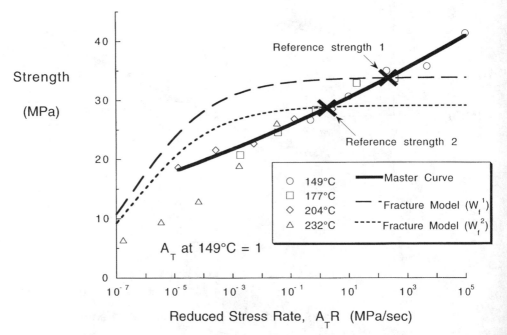

FIG. 10—*Fracture model fitted to strength results reduced by temperature.*

Verification of Compliance Model

One possible cause of the discrepancy with the W_f model predictions was that it relied on matrix properties based on neat resin tests. The properties of neat resin can be significantly different from the matrix due to differences in curing when fibers are present or differences in chemistry due to additives in the prepregging process. To test for this difference, the stress-strain response measured during the strength tests were compared to predictions from the composite compliance model. Figure 11 shows these results for the slowest ramp rate at each temperature. The model seems to predict the experimental results quite well at 204°C and below, but at 232°C, the measured compliance is much greater than predicted. The discrepancy might actually be worse than is indicated. At 232°C, the specimen are shown to be extremely compliant. When measuring strain of such a compliant specimen with a strain gauge, the strain gauge can stiffen the specimen therefore causing the measured compliance to be less than the true response. The failure of the compliance model at 232°C is not surprising since it is based on neat resin results with a peak temperature of 225°C. Extrapolating compliance results near the glass transition temperature where deformation properties are changing very rapidly can cause the large modeling errors observed in this case. The magnitude of the change in K3B properties can be seen in Ref *13* where changes in modulus were measured near T_g. Changing material properties near the glass transition temperature would also be a reason why the 232°C strength data does not fall on the master curve created by the strength values at the lower temperatures.

Figure 12 shows the composite compliance model prediction for the 204°C tests at all the different stress rates tested. The agreement between the predicted and the measured response indicates that the model captures the time-dependent deformation at this temperature. The

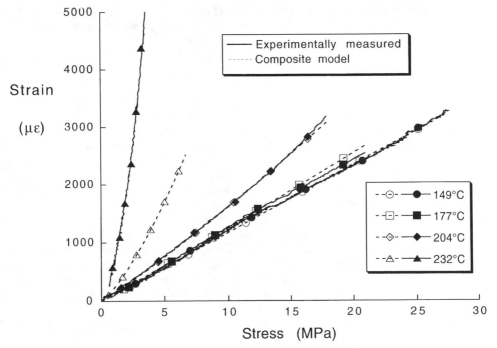

FIG. 11—*Strain response of slow ramp tests (0.473 MPa/sec).*

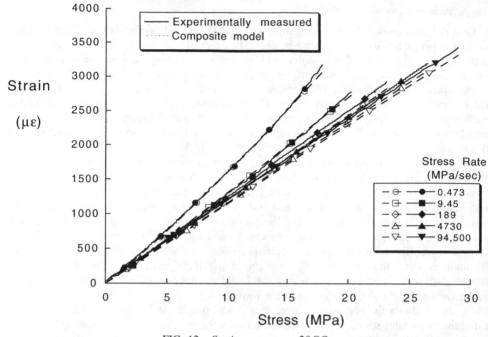

FIG. 12—*Strain response at 204°C.*

compliance model of the composite, and therefore of the matrix, appears to perform well up to 204°C so this should not be the reason for the poor agreement between the fracture model strength predictions and the experimentally measured master curve.

Variable W_f *Model Results*

The fracture model predictions shown in Fig. 10 were based on the assumption that the W_f parameter was not a function of time and temperature. Although there was a significant modeling advantage to making this assumption, it did not produce predictions that matched the material response. To investigate how W_f might have changed with time and temperature, a W_f value was calculated from the average strength at each test condition. These W_f values were plotted on Fig. 13 normalized to the peak value. Only normalized values were possible because the value of C_2 was not known. Figure 13 indicates that the W_f parameter did change significantly with temperature reduced stress rate. On this figure where the data were shifted using the shift factors from compliance tests, all data fell along a master curve except for the 232°C data. Fitting a line through the remaining data on this log-log plot produces an expression for the W_f master curve. At 232°C, the W_f values shown are in error because they are based on the compliance model which was shown to greatly underestimated the actual deformation of the material at this temperature. If the material compliance were correctly modeled at 232°C, these W_f values would be higher bringing them closer to the master curve.

The fact that W_f appears to be a function of rate may not be a complete surprise if one considers the Dugdale assumption [*14*] that

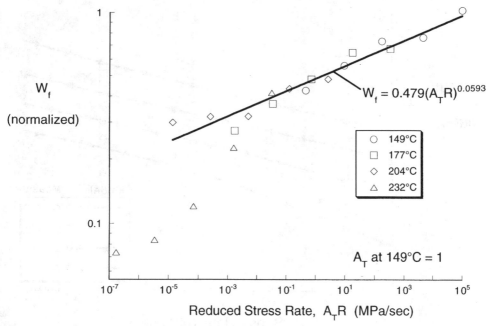

FIG. 13—W_f *results from strength tests reduced.*

$$G_{Ic} \cong \sigma_F \delta_{CTOD} \tag{16}$$

The displacement at the crack tip just prior to crack growth (δ_{CTOD}), has been shown to remain constant over a range of temperatures for certain classes of materials [15]. Assuming TTSP, δ_{CTOD} would be constant with time as well. The plastic flow stress (σ_F) for the material at the crack tip might be assumed to increase with increasing stress rates since the creep rupture stress of polymers increases with shorter times to failure [16]. A rate dependence in σ_F would produce a rate dependence in G_{Ic} and therefore in W_f which would explain the rate dependence seen in Fig. 13.

Because W_f changed with time and temperature, the predictive capabilities of the fracture model were severely limited, but some predictions were possible because W_f varied in a way that could be modeled. Using the expression for how W_f changes with temperature reduced stress rate, predictions were made with the fracture model. Figure 14 shows that the variable W_f model performed well in modeling the experimental results up to 204°C, but at 232°C where the model is suspect as discussed previously, the model over predicted strength. The fact that the model fits the data up to 204°C should not be a surprise since the curve was in fact fitted to the data, just in a transformed space.

If W_f is shown to be the controlling parameter for these failures, then this model will allow these results to be used to predict failures of other configurations where fracture is the controlling mechanisms. Unfortunately because W_f proved to be a function of time and temperature, predictions of tests with other types of load-temperature histories could not be predicted without further understanding of how the W_f parameter changes.

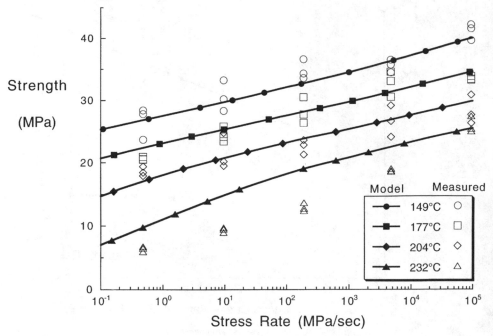

FIG. 14—*Strength predictions assuming varying* W_f.

Concluding Remarks

The use of time-temperature superposition was shown to apply to the transverse tensile strength of a thermoplastic composite material using the same shift factors as measured for compliance. This was demonstrated with transverse tension strength data for IM7/K3B at four temperatures from 149°C to 232°C and at five different stress rates that produced failure in the range from 0.5 s to 24 h. This accelerated test method applied well up to 204°C where a master curve could be formed from the experimental data. The master curve showed that a composite might suffer losses of 50% in strength when the time scale is increased by 9 orders of magnitude. Results at 232°C, however, did not fit on the master curve.

To explain why the same shifts rates might apply to both strength and compliance, a viscoelastic fracture (W_f) model was investigated based on the hypothesis that transverse strength is controlled by the work of fracture for crack initiation from a critical flaw. An expression for W_f based on the viscoelastic J-integral theory and the elastic-viscoelastic correspondence principle was used to relate viscoelastic compliance to strength. Had the W_f parameter remained constant with time and temperature, the time-dependent nature of strength would have been controlled solely by the time-dependent compliance properties used as input to the model. The model was found to fit the strength data only if W_f was allowed to vary as a power law in temperature reduced stress rate. The varying W_f parameter, may limit the ability to extrapolate from one set of experimental results to predictions of failures after arbitrary load and temperature histories.

The significance of this research is, therefore, seen as providing an indication that a more versatile acceleration method for strength may be possible while also providing the evidence that such a method is needed. The promise of an accelerated test method comes from the strength data forming a master curve. The need for an accelerated test is demonstrated by the significant decrease in strength over the five orders of magnitude in stress rate which were tested.

References

[1] Ferry, J. D., *Viscoelastic Properties of Polymers,* Wiley, New York, 1970.
[2] Scott, D. W., Lai, J. S., and Zureick, A.-H. "Creep Behavior of Fiber-Reinforced Polymeric Composites: A Review of the Technical Literature," *Journal of Reinforced Plastics and Composites,* Vol. 14, 1995, pp. 588–617.
[3] Miyano, Y., Kanemitsu, M., Kunio, T., and Kuhn, H. A., "Role of Matrix Resin on Fracture Strengths of Unidirectional CFRP," *Journal of Composite Materials,* Vol. 20, 1986, pp. 520–538.
[4] Mohri, M., Miyano, Y., and Suzuki, M., "Time-Temperature Dependence on Flexural Strength of Pitch-Based Carbon Fiber Unidirectional CFRP Laminates," *Composites: Design, Manufacture, and Application,* S. W. Tsai and G. S. Springer, Eds., *Proceedings* of ICCM VIII, SAMPE, Honolulu, 1991, pp. 3-B-1 to 9.
[5] Schapery, R. A., "Correspondence Principles and a Generalized J Integral for Large Deformation and Fracture Analysis of Viscoelastic Media," *International Journal of Fracture,* Vol. 25, 1984, pp. 195–223.
[6] Feldman, M., "The Effects of Elevated Temperature on the Physical Aging of a High Temperature Thermoplastic Resin and Composite," Master's Thesis in Engineering Mechanics, Old Dominion University, Norfolk, VA, 1996.
[7] Gol'dman, A. Y., "Prediction of the Deformation Properties of Polymeric and Composite Materials," M. Shelef and R. A. Dickie, Eds., American Chemical Society, Washington, D.C., 1994.
[8] Schapery, R. A., "Viscoelastic Behavior and Analysis of Composite Materials," Ch. 4 of *Mechanics of Composite Materials,* Vol. 2, Sendeckyj, Ed., Academic Press, New York, 1974.
[9] Broek, D., "The Energy Principle," Ch. 5 *of Elementary Engineering Fracture Mechanics,* 4th ed., Kluwer Academic Publishers, Boston, 1991, pp. 123–149.
[10] Brockway, G. S. and Schapery, R. A., "Some Viscoelastic Crack Growth Relations for Orthotropic and Prestrained Media," *Engineering Fracture Mechanics,* Vol. 10, 1978, pp. 453–468.
[11] Reeder, J. R., "Prediction of Long-Term Strength of Thermoplastic Composites Using Time-Temperature Superpositon," Ph.D. Dissertation in Mechanical Engineering, Texas A&M University, College Station, TX, 1998.
[12] O'Brien, T. K. and Salpekar, S. A., "Scale Effects on the Transverse Tensile Strength of Graphite/ Epoxy Composites," *Composite Materials: Testing and Design, ASTM STP 1206,* Vol. 11, E. T. Camponeschi, Jr., Ed., American Society for Testing and Materials, West Conshohocken, PA, 1993.
[13] Lear, C. M., "Material Characterization of Thermoplastic Candidate Matrix Materials for Advanced Composites," Master's Thesis in Engineering Mechanics, Ohio State University, Columbus, OH, 1994.
[14] Brown, W. F. Jr. and Strawley, J. E., Eds., *Plane Strain Crack Toughness Testing of High Strength Metallic Materials, ASTM STP 410,* American Society of Testing and Materials, Philadelphia, 1966.
[15] Kinloch, A. J. and Young, S., *Flow and Fracture of Polymers,* Elsevier, New York, 1985.
[16] Rosato, D. V., DiMattia, D. P., and Rosato, D. V., "Plastics: Design Criteria," Ch. 3 of *Designing with Plastics and Composites: A Handbook,* Van Nostrand Reinhold, New York, 1991, pp. 125–252.

Samit Roy[1]

Hygrothermal Modeling of Polymers and Polymer Matrix Composites

REFERENCE: Roy, S., **"Hygrothermal Modeling of Polymers and Polymer Matrix Composites,"** *Time Dependent and Nonlinear Effects in Polymers and Composites, ASTM STP 1357,* R. A. Schapery and C. T. Sun, Eds., American Society for Testing and Materials, West Conshohocken, PA, 2000, pp. 338–352.

ABSTRACT: The theory of irreversible thermodynamics is applied to derive governing equations for history-dependent diffusion in polymers and polymer matrix composites from first principles. A special form for Gibbs free energy is introduced using stress, temperatures, and moisture concentration as independent state variables. The resulting governing equations are capable of modeling the effect of interactions between complex stress, temperature, and moisture histories on the diffusion process within an orthotropic material. Since the mathematically complex nature of the governing equations precludes a closed-form solution, a variational formulation is used to derive the weak form of the nonlinear governing equations which are then solved using the finite element method. For model validation, the model predictions are compared with published experimental data for the special case of isothermal diffusion in an unstressed Graphite-Epoxy symmetric angle-ply laminate.

KEYWORDS: non-Fickian, anomalous, diffusion, entropy, polymers, composites, two-stage, finite-element, modeling

It is now widely recognized that moisture plays a significant role in influencing the mechanical behavior, and therefore, long-term durability of polymers and polymer matrix composites (PMC). Numerous diffusion models have been proposed over the years for modeling hygrothermal effects in PMC The one most frequently used by researchers is the one-dimensional Fickian model due to its simplicity and mathematical tractability. Unfortunately, this model tends to overestimate the moisture absorption in panels for short diffusion times, even when it is modified to take into account edge effects [1]. Some researchers have suggested that the deviation can be explained by a two-stage Fickian process [2,3]. Others claim that the diffusion process in a PMC is really non-Fickian [4,5]. The applicability of Fick's law for a given material system under a specified loading cannot be guessed "a priori" but must be determined from moisture absorption/desorption test data.

Frisch [6] and Crank [7] were among the first researchers to recognize and attribute non-Fickian moisture transport in resins to time-dependent molecular mechanisms within a polymer. To model this phenomenon, Weitsman [8] applied basic principles of continuum mechanics and irreversible thermodynamics to derive governing equations and boundary conditions for coupled stress-assisted diffusion in elastic and viscoelastic materials. Following an approach originally proposed by Biot [9] and Schapery [10], viscoelasticity was intro-

[1] Assistant Professor, Department of Mechanical Engineering, University of Missouri-Rolla, Rolla, MO 65401.

duced by means of scalar-valued internal state variables, also referred to as hidden coordinates. The results of the analysis allowed an insight into the mechanism that causes a time-drift toward equilibrium at the boundary of a viscoelastic material subjected to a constant chemical potential of the ambient vapor. The governing equations also indicated that the saturation levels vary quadratically with stress and that they can be expressed in terms of the dilatational and the deviatoric stress invariant. Subsequently, Weitsman [11] extended the rigorous phenomenological model to incorporate polymer "free volume" as a thermodynamic internal state variable in order to include the effect of physical aging on moisture diffusion. The resulting governing equations were not amendable to closed-form solution due to mathematical complexity although it was observed that the diffusion equation followed a time-retardation process analogous to mechanical viscoelastic response, and that it exhibited an aging behavior characteristic of glassy polymers. Employing a similar phenomenological approach, Weitsman [12] developed a model for coupled damage and moisture transport in an elastic, transversely isotropic, fiber reinforced polymer composite. The damage entity was represented as a skew-symmetric tensor and was included in the model as an internal state variable.

The objective of this paper is to present a thermodynamically consistent phenomenological approach for modeling non-Fickian diffusion in viscoelastic polymers and polymer composites. The model is developed using thermodynamic internal state variables that includes hidden coordinates to represent the various degrees of freedom associated with the configurations of long-chain polymer molecules. The law of conservation of mass and the second law of thermodynamics in the form of the Clausius-Duhem inequality are invoked to obtain governing equations and boundary conditions for stress-accelerated, temperature-dependent moisture transport in an anisotropic viscoelastic material. The governing equations are subsequently cast in a weak form using variational formulation and numerical solutions are obtained using the finite element method for the special case of viscoelastic diffusion in an unstressed laminate. Although there are numerous physical and chemical mechanisms that are deemed responsible for non-Fickian transport kinetics [13], the focus of the current work is restricted to the development of a phenomenological model for a class of non-Fickian moisture transport that is directly attributable to the intrinsic viscoelastic constitutive behavior of polymers. While it is feasible to include aging and damage in the formulation, the present derivation is restricted to an un-aged and undamaged material to maintain solution tractability.

Governing Equations for Diffusion in Viscoelastic Media

Following Schapery's approach [10], a material element with fixed mass is taken to represent the thermodynamic system. The maximum number of variables required to define its state is assumed to be the N generalized coordinates q_i, absolute temperature, T, and vapor mass concentration M. A generalized force Q_i conjugate to the generalized coordinate q_i is also defined based on virtual work. In addition, the generalized coordinates are divided into two groups: hidden and observed. Hidden coordinates are defined by the condition that their conjugate forces are always zero. Since the hidden coordinates do not appear explicitly in the final diffusion equations, a precise physical interpretation is not actually necessary. Nevertheless, they may be interpreted as entities that define the configuration of chain-like molecules in polymers.

Expanding Gibbs free energy per unit mass of a polymer with respect to hidden coordinates only, and assuming that terms beyond second-order in these coordinates can be neglected gives

$$G = G_R + \hat{C}_r q_r + \frac{1}{2} \hat{d}_{rs} q_r q_s \qquad r, s = k + 1, N \ (r \text{ summed}) \tag{1}$$

where

$$G_R = C_R(M,T) + C_m (M,T)Q_m + \frac{1}{2} d_{mn}(M,T)Q_m Q_n \qquad m, n = 1, k$$

$$\hat{C}_r = C_r(M,T) + d_{rm}(M,T)Q_m \tag{2}$$

$$\hat{d}_{rs} = \hat{d}_{rs}(Q_m, M, T)$$

where:

> M : Diffusing vapor mass per unit volume of solid (i.e., vapor mass concentration).
> T : Spatially uniform temperature at time t.
> Q_m : m^{th} component of the k generalized forces associated with observed coordinates.
> q_r : r^{th} component of the $(N\text{-}k)$ hidden coordinates.

Let the polymeric solid absorb vapor through its exposed boundary. Conservation of mass within a volume element in a polymer in which vapor diffusion is occurring yields

$$\frac{\partial M}{\partial t} = - \vec{\nabla} \cdot \vec{f} = - f_{i,i} \qquad i = 1,3 \tag{3}$$

where, following Weitsman [11], diffusing vapor flux, f_i, in the absence of temperature gradient is defined by

$$f_i = -D_i \frac{\partial \mu^\psi}{\partial X_i} \tag{4}$$

In Eq 4, D_i is the effective diffusivity of the vapor in the polymer in the ith lamina principal direction, and μ^ψ is the chemical potential of the vapor at time ψ. Note that repeated subscripts does not imply summation in Eq 4. Combining Eqs 3 and 4 gives the governing equation for non-Fickian or anomalous diffusion in an orthotropic media

$$\frac{\partial M}{\partial t} = \frac{\partial}{\partial X_i} \left(D_i \frac{\partial \mu^\psi}{\partial X_i} \right) \tag{5}$$

where, the subscript i is summed from 1 to 3.

Since the reduced entropy inequality must hold for all thermodynamic processes, hence the chemical potential per unit volume of the solid is given by [11]

$$\mu^\psi = \rho_s \left(\frac{\partial G}{\partial M} \right) \tag{6}$$

where, ρ_s is the mass density of the dry polymeric solid.

Combining (1), (2), and (6), and retaining only first-order terms in the hidden coordinates q_r

$$\mu^\psi = \rho_s \left\{ \frac{\partial C_R}{\partial M} + \frac{\partial C_m}{\partial M} Q_m + \frac{1}{2} \frac{\partial d_{mn}}{\partial M} Q_m Q_n + \left(\frac{\partial C_r}{\partial M} + \frac{\partial d_{rm}}{\partial M} Q_m \right) q_r \right\} \tag{7}$$

Expressing C_m and d_{rm} in terms of thermal (α_m) and hygroscopic (β_m) dilatational coefficients as proposed by Weitsman [11] yields

$$C_m(M,T) = \alpha_m \Delta T + \beta_m M \tag{8}$$

$$d_{rm}(M,T) = \left(\frac{\partial \alpha_m}{\partial q_r} \right) \Delta T + \left(\frac{\partial \beta_m}{\partial q_r} \right) M = \alpha_{rm} \Delta T + \beta_{rm} M$$

Substituting (8) in (7) yields

$$\mu^\psi = \rho_s \left\{ \frac{\partial C_R}{\partial M} + \left(\frac{\partial C_r}{\partial M} \right) q_r + \left[\left(\frac{\partial \alpha_m}{\partial M} + \frac{\partial \alpha_{rm}}{\partial M} q_r \right) \Delta T \right. \right.$$

$$\left. \left. + \left(\frac{\partial (M\beta_m)}{\partial M} + \frac{\partial (M\beta_{rm})}{\partial M} q_r \right) \right] Q_m + \frac{1}{2} \frac{\partial d_{mn}}{\partial M} Q_m Q_n \right\} \tag{9}$$

Solving thermodynamic governing equations for the hidden coordinates [10], they may be expressed as

$$q_r = -\left(\frac{\hat{C}_r}{d_r^R a_G} \right) \left(1 - e^{-\psi/\tau_r} \right) = -K_r \left(1 - e^{-\psi/\tau_r} \right) \tag{10}$$

where, ψ is the reduced time, and reduced time increment $d\psi = a_G(Q_m, T)dt/a_D(Q_m,M,T)$ and, retardation time $\tau_r = b_r^R/d_r^R$, (r not summed). The quantities a_D and a_G are entropy production coefficients analogous to those defined by Schapery [10]. Substituting (10) in (9), gives

$$\mu^\psi = \rho_s \left\{ \frac{\partial C_R}{\partial M} - \left(\frac{\partial C_r}{\partial M} \right) K_r \left(1 - e^{-\psi/\tau_r} \right) + \left[\left(\frac{\partial \alpha_m}{\partial M} \right) - \left(\frac{\partial \alpha_{rm}}{\partial M} \right) K_r \left(1 - e^{-\psi/\tau_r} \right) \right) \Delta T \right.$$

$$\left. + \left(\frac{\partial (M\beta_m)}{\partial M} \right) - \frac{\partial (M\beta_{rm})}{\partial M} K_r \left(1 - e^{-\psi/\tau_r} \right) \right] Q_m + \frac{1}{2} \frac{\partial d_{mn}}{\partial M} Q_m Q_n \right\} \tag{11}$$

If it is assumed that the dilatational coefficients α_m, β_m are independent of moisture concentration [11], Eq 11 reduces to

$$\mu^\psi = \rho_s \left\{ \frac{\partial C_R}{\partial M} - \left(\frac{\partial C_r}{\partial M} \right) K_r \left(1 - e^{-\psi/\tau_r} \right) \right.$$

$$\left. + \left[\beta_m - \beta_{rm} K_r \left(1 - e^{-\psi/\tau_r} \right) \right] Q_m + \frac{1}{2} \frac{\partial d_{mn}}{\partial M} Q_m Q_n \right\} \tag{12}$$

Substituting for K_r from Eq 10 and $\hat{C}_r$ from Eq 2, into Eq 12 gives

$$\mu^{\psi} = \rho_s \left\{ \frac{\partial C_R}{\partial M} - \left(\frac{\partial C_r}{\partial M} \right) \left(\frac{C_r}{d_r^R a_G} \right) \left(1 - e^{-\psi/\tau_r} \right) \right.$$

$$+ \beta_m Q_m - \left(\frac{\partial C_r}{\partial M} \right) \left(\frac{d_{rm}}{d_r^R a_G} \right) \left(1 = e^{-\psi/\tau_r} \right) Q_m$$

$$- \beta_{rm} \left(\frac{C_r}{d_r^R a_G} \right) \left(1 - e^{-\psi/\tau_r} \right) Q_m + \frac{1}{2} \frac{\partial d_{mn}}{\partial M} Q_m Q_n$$

$$- \beta_{rm} \left(\frac{d_{rn}}{d_r^R a_G} \right) \left(1 - e^{-\psi/\tau_r} \right) Q_m Q_n \right\} \tag{13}$$

Following Schapery [10] and assuming $C_r(M,T) = a_H(M,T)C_r^R$, and substituting (8) in (13)

$$\mu^{\psi} = \rho_s \left\{ \frac{\partial C_R}{\partial M} - \frac{(C_r^R)^2}{d_r^R} \left(\frac{\partial a_H}{\partial M} \right) \left(1 - e^{-\psi/\tau_r} \right) \left(\frac{a_H}{a_G} \right) + \beta_m Q_m \right.$$

$$- \left(\frac{C_r^R}{d_r^R} \right) \left(\frac{\partial a_H}{\partial M} \right) \alpha_{rm} \left(1 - e^{-\psi/\tau_r} \right) \frac{Q_m \Delta T}{a_G} - \left(\frac{C_r^R}{d_r^R} \right) \left(\frac{\partial a_H}{\partial M} \right) \beta_{rm} \left(1 - e^{-\psi/\tau_r} \right) \frac{Q_m M}{a_G}$$

$$- \beta_{rm} \left(\frac{C_r^R}{d_r^R} \right) \left(1 - e^{-\psi/\tau_r} \right) \frac{a_H Q_m}{a_G} + \frac{1}{2} \left(\frac{\partial d_{mn}}{\partial M} \right) Q_m Q_n$$

$$- \left(\frac{\beta_{rm} \alpha_m}{d_r^R} \right) \left(1 - e^{-\psi/\tau_r} \right) \frac{Q_m Q_n \Delta T}{a_G} - \left(\frac{\beta_{rm} \beta_{rn}}{d_r^R} \right) \left(1 - e^{-\psi/\tau_r} \right) \frac{Q_m Q_n M}{a_G} \right\} \tag{14}$$

In the presence of time varying forces, concentrations, and temperatures, Eq 14 takes the form of a convolution integral

$$\mu^t = \rho_s \left\{ \frac{\partial C_R}{\partial M} - \int_0^t \Delta A(\psi - \psi') \frac{d(a_H/a_G)}{d\tau} d\tau + \beta_m Q_m \right.$$

$$- \int_0^t \Delta B_m(\psi - \psi') \frac{d(Q_m \Delta T/a_G)}{d\tau} d\tau$$

$$- \int_0^t \Delta E_m(\psi - \psi') \frac{d(Q_m M/a_G)}{d\tau} d\tau - \int_0^t \Delta F_m(\psi - \psi') \frac{d(a_H Q_m/a_G)}{d\tau} d\tau$$

$$+ \frac{1}{2} \left(\frac{\partial d_{mn}}{\partial M} \right) Q_m Q_n - \int_0^t \Delta G_{mn}(\psi - \psi') \frac{d(Q_m Q_n \Delta T/a_G)}{d\tau} d\tau$$

$$- \int_0^t \Delta H_{mn}(\psi - \psi') \frac{d(Q_m Q_n M/a_G)}{d\tau} d\tau \right\} \qquad m, n = 1,6 \tag{15}$$

where

$$\Delta A(\psi) = \sum_r \left\{ \frac{(C_r^R)^2}{d_r^R} \left(\frac{\partial a_H}{\partial M}\right) \left(1 - e^{-\psi/\tau_r}\right) \right\}$$

$$\Delta B_m(\psi) = \sum_r \left\{ \left(\frac{C_r^R}{d_r^R}\right) \left(\frac{\partial a_H}{\partial M}\right) \alpha_{rm} \left(1 - e^{-\psi/\tau_r}\right) \right\}$$

$$\Delta E_m(\psi) = \sum_r \left\{ \left(\frac{C_r^R}{d_r^R}\right) \left(\frac{\partial a_H}{\partial M}\right) \beta_{rm} \left(1 - e^{-\psi/\tau_r}\right) \right\}$$

$$\Delta F_m(\psi) = \sum_r \left\{ \left(\frac{C_r^R}{d_r^R}\right) \beta_{rm} \left(1 - e^{-\psi/\tau_r}\right) \right\}$$

$$\Delta G_{mn}(\psi) = \sum_r \left\{ \left(\frac{\beta_{rm}\alpha_{rn}}{d_r^R}\right) \left(1 - e^{-\psi/\tau_r}\right) \right\}$$

$$\Delta H_{mn}(\psi) = \sum_r \left\{ \left(\frac{\beta_{rm}\beta_{rn}}{d_r^R}\right) \left(1 - e^{-\psi/\tau_r}\right) \right\}$$

$$\text{and,} \quad \psi^t = \int_0^t \left(\frac{a_G}{a_D}\right) d\xi = \int_0^t \frac{d\xi}{a_\sigma}$$

where, the shift-factor $a_\sigma = a_D/a_G$.

Special case: isothermal isotropic diffusion under zero stress
  For this case, Eq 15 reduces to

$$\mu^t = \rho_s \left[\frac{\partial C_R}{\partial M} - \int_0^t \Delta A(\psi^t - \psi^\tau) \frac{d(a_H/a_G)}{d\tau} d\tau \right] \qquad (16)$$

where:

$$\Delta A(\psi^t - \psi^\tau) = \left(\frac{\partial a_H}{\partial M}\right) \sum_r \left\{ \Delta A_r \left(1 - e^{-(\psi^t-\psi^\tau)/\tau_r}\right) \right\} \qquad (17)$$

Substituting (17) in (16) gives

$$\mu^t = \rho_s \left[\frac{\partial C_R}{\partial M} - \int_0^t \left\{ \sum_r \Delta A_r \left(1 - e^{-(\psi^t-\psi^\tau)/\tau_r}\right) \left(\frac{\partial a_H}{\partial M}\right) \frac{d(a_H/a_G)}{d\tau} \right\} d\tau \right]$$

$$= \rho_s \left[\frac{\partial C_R}{\partial M} - \int_0^t \left\{ \sum_r \Delta A_r \left(1 - e^{-\hat\psi/\tau_r}\right) \frac{d(hM)}{d\tau} \right\} d\tau \right] \qquad (18)$$

where:

$$\hat\psi = \psi^t - \psi^\tau.$$

It can be shown that

$$\frac{\partial a_H}{\partial M} \frac{\partial(a_H/a_G)}{\partial \tau} = \left(\frac{\partial a_H}{\partial M}\right)\left(\frac{\partial(a_H/a_G)}{\partial M}\right)\left(\frac{dM}{d\tau}\right) = \frac{d(hM)}{d\tau} = \frac{d(hM)}{dM}\left(\frac{dM}{d\tau}\right)$$

giving

$$\frac{d(hM)}{dM} = \frac{\partial a_H}{\partial M}\left\{\frac{\partial(a_H/a_G)}{\partial M}\right\}$$

and hence, nonlinear diffusion parameter

$$h(M) = \frac{1}{M}\int\left(\frac{\partial a_H}{\partial M}\right)\left(\frac{\partial(a_H/a_G)}{\partial M}\right)dM$$

Writing $H(M) = h(M)M$, Eq 18 becomes

$$\mu^t = \rho_s\left[\frac{\partial C_R}{\partial M} - \int_0^t\left\{\sum_r \Delta A_r\left(1 - e^{-\psi/\tau_r}\right)\frac{dH}{d\tau}\right\}d\tau\right] \tag{19}$$

Substituting Eq 19 in Eq 5 yields the governing equation for anomalous diffusion in an isotropic viscoelastic material

$$\frac{\partial M^t}{\partial t} = \frac{\partial}{\partial X_i}\left\{\rho_s D\left(\frac{\partial^2 C_R}{\partial M^2}\right)\frac{\partial M}{\partial X_i} - \rho_s D\frac{\partial}{\partial X_i}\left[\int_0^t\left\{\sum_r \Delta A_r\left(1 - e^{-\psi/\tau_r}\right)\frac{dH}{d\tau}\right\}d\tau\right]\right\} \tag{20}$$

The integral on the R.H.S. of Eq 20 may be simplified as

$$\int_0^t\left\{\sum_r\Delta A_r\left(1 - e^{-\psi/\tau_r}\right)\frac{dH}{d\tau}\right\}d\tau = \sum_r\Delta A_r(H^t - H^0) - \int_0^t\sum_r\Delta A_r e^{-\psi/\tau_r}\frac{dH}{d\tau}d\tau \tag{21}$$

Writing $\lambda_r = 1/\tau_r$, the second term on the R.H.S. of (21) can be re-written as

$$\int_0^t e^{-\lambda_r(\psi^t - \psi^\tau)}\frac{dH}{d\tau}d\tau = \int_0^{t-\Delta t}e^{-\lambda_r(\psi^t - \psi^\tau)}\frac{dH}{d\tau}d\tau + \int_{t-\Delta t}^t e^{-\lambda_r(\psi^t - \psi^\tau)}\frac{dH}{d\tau}d\tau$$

$$= e^{-\lambda_r\Delta\psi^t}q_r^{t-\Delta t} + (H^t - H^{t-\Delta t})\beta_r^t \tag{22}$$

Where, the hereditary integral q_r^t is obtained from the recurrence formula

$$q_r^t = e^{-\lambda_r\Delta\psi^t}q_r^{t-\Delta t} + (h^t M^t - h^{t-\Delta t}M^{t-\Delta t})\left(\frac{1 - e^{-\lambda_r\Delta\psi^t}}{\lambda_r\Delta\psi^t}\right) \tag{23}$$

and the function β_r^t is given by,

$$\beta_r^t = \frac{1 - e^{-\lambda_r \Delta \psi^t}}{\lambda_r \Delta \psi^t}, \quad \text{where } \Delta \psi = \psi^t - \psi^{t-\Delta t}$$

Substituting Eqs 21 and 22 in 20

$$\frac{\partial M^t}{\partial t} = \frac{\partial}{\partial X_i} \left\{ \rho_s D \left(\frac{\partial^2 C_R}{\partial M^2} \right) \frac{\partial M}{\partial X_i} - \rho_s D \left[\sum_r \Delta A_r (1 - \beta_r^t) \frac{\partial H}{\partial X_i} - \frac{\partial Q^t}{\partial X_i} - R_i^t \right] \right\} \quad (24)$$

where, the herditary diffusion terms

$$Q^t = \sum_r \Delta A_r \{ e^{-\lambda_r \Delta \psi^t} q_r^{t-\Delta t} - \beta_r^t H^{t-\Delta t} \}$$

$$R_i^t = \sum_r \Delta A_r \frac{\partial \beta_r^t}{\partial X_i} H^t$$

If it is assumed that power law approximations could be used for the unknown moisture-dependent nonlinearizing parameters in Eq 24, that is

$$\rho_s D \left(\frac{\partial^2 C_R}{\partial M^2} \right) = D_0 (1 + a_1(T) \overline{M}^n)$$

and

$$h(\overline{M}) = a_2(T) + \hat{a}_3(T) \overline{M}^m = a_2(T) + \left(\frac{a_3(T)}{m + 1} \right) \overline{M}^m \quad (25)$$

where

$$\overline{M} = \frac{M}{M_m}, \quad \text{and} \quad \hat{a}_3(T) = \frac{a_3(T)}{m + 1}$$

Note that M_m is a normalizing parameter that could be the maximum (saturation) moisture concentration of the polymer under reference conditions. Substituting (25) in (24), and using the chain rule gives

$$\frac{\partial M^t}{\partial t} = \frac{\partial}{\partial X_i} \left\{ D_0 (1 + a_1 \overline{M}^n) \frac{\partial M^t}{\partial X_i} - \rho_s D \left[\sum_r \Delta A_r (1 - \beta_r^t) \frac{\partial H^t}{\partial M} \frac{\partial M^t}{\partial X_i} - \frac{\partial Q^t}{\partial X_i} - R_i^t \right] \right\} \quad (26)$$

But

$$\frac{\partial H^t}{\partial M} = \frac{1}{M_m} \frac{\partial}{\partial \overline{M}} \{ M_m (a_2(T) + \hat{a}_3(T) \overline{M}^m) \overline{M} \}$$

$$= a_2(T) + (m + 1) \hat{a}_3(T) \overline{M}_m$$

$$= a_2(T) + a_3(T) \overline{M}_m \quad (27)$$

Substituting (27) in (26), gives

$$\frac{\partial M^t}{\partial t} = \frac{\partial}{\partial X_i}\left[D^t \frac{\partial M^t}{\partial X_i}\right] + \frac{\partial}{\partial X_i}\left[\rho_s D \frac{\partial Q^t}{\partial X_i}\right] + \frac{\partial}{\partial X_i}[\rho_s D R_i^t] \tag{28}$$

where, the instantaneous diffusivity D^t is given by

$$D^t = D_0(1 + a_1(T)\overline{M}^n) - (a_2(T) + a_3(T)\overline{M}^m)\sum_r \Delta D_r(1 - \beta_r^t) \tag{29}$$

Also, for the special case of a spatially uniform shift factor, the gradient of the hereditary diffusion term, Q^t, reduces to

$$\rho_s D \frac{\partial Q^t}{\partial X_i} = \rho_s D \sum_r \Delta A_r \left\{ e^{-\lambda_r \Delta\psi^t} \frac{\partial q_r^{t-\Delta t}}{\partial X_i} - \beta_r^t \frac{\partial H^{t-\Delta t}}{\partial X_i} \right\}$$

$$= \sum_r \Delta D_r \left\{ e^{-\lambda_r \Delta\psi^t} \frac{\partial q_r^{t-\Delta t}}{\partial X_i} - \beta_r^t(a_2 + a_3\overline{M}^m)^{t-\Delta t} \frac{\partial M^{t-\Delta t}}{\partial X_i} \right\} \tag{30}$$

It should be noted that when $\Delta\Psi^t \to 0$, then $\beta_r^t \to 1$, and $D^t \to D_0$. Therefore D_0 can be obtained from the initial Fickian diffusivity of vapor in a polymer. The remaining model parameters that need to be determined by material characterization are the Prony series coefficients for transient diffusivity, ΔD_r, the reciprocal retardation time parameters, λ_r, the parameters a_1, a_2, a_3, m, n, and the shift factors a_G and a_σ.

Variational Formulation

The variational (weak) form of Eq 28 is given by

$$\int_{V^{(e)}} \left[u \cdot \frac{\partial M^t}{\partial t} + D^t \frac{\partial u}{\partial X_i}\frac{\partial M^t}{\partial X_i} + \bar{\rho}\frac{\partial u}{\partial X_i}\cdot\frac{\partial Q^t}{\partial X_i} - u\frac{\partial R_i^t}{\partial X_i} \right] dV$$

$$- \int_{A^{(e)}} \left[u\left(D^t\frac{\partial M^t}{\partial X_i} + \bar{\rho}\frac{\partial Q^t}{\partial X_i}\right)n_i \right] dA = 0$$

Based on the variational statement, the diffusion boundary conditions can now be identified as

$$\left(D^t\frac{\partial M^t}{\partial X_i} + \bar{\rho}\frac{\partial Q^t}{\partial X_i}\right)n_i + \hat{q} = 0 \qquad \text{on } A_1^{(e)} \quad \text{(specified solvent flux)}$$

$$M = \hat{M} \qquad \text{on } A_2^{(e)} \quad \text{(specified concentration)}$$

where, $A_1^{(e)} + A_2^{(e)} = A^{(e)}$.

and n_i are the components of the unit outward normal at the boundary. Thus

$$\int_{V^{(e)}} \left[u\cdot\frac{\partial M^t}{\partial t} + D^t\frac{\partial u}{\partial X_i}\frac{\partial M^t}{\partial X_i} + \bar{\rho}\frac{\partial u}{\partial X_i}\frac{\partial Q^t}{\partial X_i} \right] dV = -\int_{A_1^{(e)}} u\hat{q}dA + \int_{V^{(e)}} u\cdot\frac{\partial R_1^t}{\partial X_i}dV \tag{31}$$

Finite Element Approximation

Approximating the concentration within an element using standard isoparametric finite element interpolation functions N_j, yields

$$M^t = \sum_{j=1}^{N} N_j M_j$$

$$u = N_k$$

After finite element discretization and using matrix notation, Eq 31 becomes

$$[T^{(e)}] \left\{ \dot{M} \right\} + [K^{(e)}]\{M\} = \{F^{(e)}\} \tag{32}$$

where the superscript (e) is used to denote that the equations are satisfied over each element and

$$T_{jk}^e = \int_{V^{(e)}} (N_j N_k) dV$$

$$K_{jk}^e = \int_{V^{(e)}} \left\{ D^t \frac{\partial N_j}{\partial X_i} \frac{\partial N_k}{\partial X_i} \right\} dV$$

$$F_j^e = -\int_{V^{(e)}} \bar{\rho} \frac{\partial N_j}{\partial X_i} \frac{\partial Q^t}{\partial X_i} dV - \int_{A_1^{(e)}} N_j \hat{q} dA + \int_{V^{(e)}} N_j \frac{\partial R_i^t}{\partial X_i} dV, \quad i = 1,3 \quad \text{and } j,k = 1,N$$

where, N is the number of nodes per element.

The time derivative $\{\dot{M}\}$ is approximated using a theta-family of approximations yielding, for discrete times t_n and t_{n+1},

$$[A^{(e)}]\{M\}_{n+1} + [B^{(e)}]\{M\}_n = \{P^{(e)}\}_n \tag{33}$$

where:

$$[A^{(e)}] = [T^{(e)}] + \theta \Delta t_{n+1}[K^{(e)}]$$
$$[B^{(e)}] = [T^{(e)}] + (1 - \theta)\Delta t_{n+1}[K^{(e)}]$$
$$\{P^{(e)}\} = \Delta t_{n+1}(\theta\{F^{(e)}\}_{n+1} + (1 - \theta)\{F^{(e)}\}_n)$$

Equation 33 is solved iteratively using a value of $\theta = 0.5$, which corresponds to the Crank-Nicholson scheme and is unconditionally stable. Note that for $n = 1$, the value of the starting concentration in Eq 33 is known from initial conditions.

Model Verification Results

The finite element governing equations derived in the preceding sections were programmed into a finite element test-bed code (NOVA-3D) that was developed by the author. The model predictions were verified by comparing with data from moisture uptake experiments performed by Blikstad et al. [14] on a T300/1034 graphite/epoxy symmetric angle-ply laminate subjected to 98% relative humidity at 50°C. There were several compelling reasons for selecting these data for model validation. First, the test data for this case exhibits a distinct two-stage absorption behavior that is presumably caused by a free energy mechanism that is analogous to the one that gives rise to time-dependent mechanical response in a polymer even when exposed to constant ambient environment [11]. Secondly, much of the transient diffusivity data required as input to the proposed model could be extracted from the time-varying moisture boundary condition data published by Cai and Weitsman [15] for non-Fickian moisture uptake in a T300/1034 laminate.

Cai and Weitsman [15] developed a simplified methodology within the framework of linear Fickian diffusion with time varying boundary conditions that would allow reduction of non-Fickian moisture weight gain data in a manner which enables the evaluation of the diffusion coefficient and through-thickness concentration profiles. Based on Frisch's [6] assumption of uniform chemical potential for sorption equilibrium at the boundaries, they employed a seven term Prony series to model the simulated time-varying moisture boundary condition, given by

$$C(\pm L, t) = C_0 + \sum_{r=1}^{7} C_r(1 - e^{-\lambda_r t}) \tag{34}$$

In order to convert the time-varying moisture boundary condition data into a Prony series for instantaneous diffusivity as mandated by Eq 29 of the current model, a simple scaling procedure was employed. In this procedure, each Prony coefficient in Eq 34 was normalized by the initial coefficient, C_0, and then multiplied by the initial diffusivity, D_0, for the T300/1034 laminate as tabulated in [14]. The resulting Prony series coefficients for instantaneous diffusivity, as well as the corresponding retardation time, τ_r, are listed in Table 1.

Figure 1 shows the two-dimensional finite element mesh that was used to model the cross-section of a 20-mm wide, 1 mm thick specimen of T300/1034 symmetric angle-ply laminate, similar to the one used in the moisture uptake experiments [14]. A unit specimen depth was assumed for modeling purposes. Based on data published by Blikstad [14], a uniform boundary moisture concentration of $M = 2.12 \times 10^{-8}$ Kg/m^3 was applied along the exposed boundaries of the specimen and is depicted in Fig. 1. In contrast with the methodology proposed by Cai and Weitsman [15], the current procedure assumed that the boundary moisture concentration remained constant for the entire duration of the analysis, although the chemical potential on the boundary could vary with time in accordance with Eq 19. A uniform initial concentration of $M_i = 0.65 \times 10^{-8}$ Kg/m^3, obtained from test data [14]

TABLE 1—*Diffusion data for T300/1034 angle ply laminate.*

Prony Term	Prony Coefficient D_r(mm^2/sec)	Retardation Time, τ_r(sec)
0	5.22×10^{-8}	—
1	0.7353×10^{-8}	36×10^3
2	0.1635×10^{-8}	180×10^3
3	-0.6197×10^{-8}	36×10^4
4	2.047×10^{-8}	180×10^4
5	-4.116×10^{-8}	36×10^5
6	13.299×10^{-8}	180×10^5
7	-11.797×10^{-8}	36×10^6

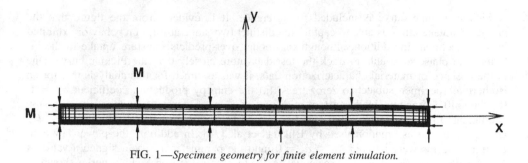

FIG. 1—*Specimen geometry for finite element simulation.*

assuming a density of 1580 Kg/m³ for dry T300/1034 composite, was applied as initial condition.

Figure 2 shows the moisture weight gain percentage predicted by the model compared with actual test data for the case of a specimen subjected to 98% relative humidity at 50°C.

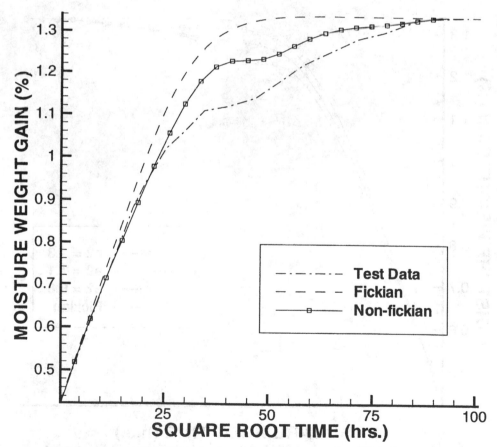

FIG. 2—*Comparison of finite element results with test data for T300/1034 laminate.*

A Fickian uptake curve is included for reference. It is evident from the figure that the predicted uptake curve is able to capture the distinct two-stage absorption behavior exhibited by the specimen. In addition, although the model over-predicts moisture uptake in the intermediate phase, it is able to track the test data more closely than the Fickian curve. Due to the absence of material characterization data, it was assumed for this analysis that for an isothermal specimen subject to zero stress: (a) the entropy production coefficient $a_G = 1$, (b) the shift factor remained constant with time, i.e., a_σ = constant, and, (c) the diffusion coefficients in the laminate global directions could be approximated by assuming $D_x = D_y$, analogous to the assumption used by Blikstad et al. [14]. In addition, the power-law concentration terms were ignored ($a_1 = a_3 = 0$) and the parameter a_2 was assigned a value of 1.3, and the shift factor a_σ was set equal to ⅙, to obtain the non-Fickian curve shown in Fig. 2.

A parametric sensitivity study was conducted to investigate the sensitivity of the solution to changes in the parameter, a_2, and the shift factor, a_σ. Figure 3 shows the variation in moisture uptake profiles with changes in a_2. Starting with a value of $a_2 = 0$ corresponding to the Fickian uptake curve, the predicted uptake curve progressively transitions to the classic two-stage non-Fickian form as the magnitude of a_2 is increased. A value of $a_\sigma = $ ⅙ was

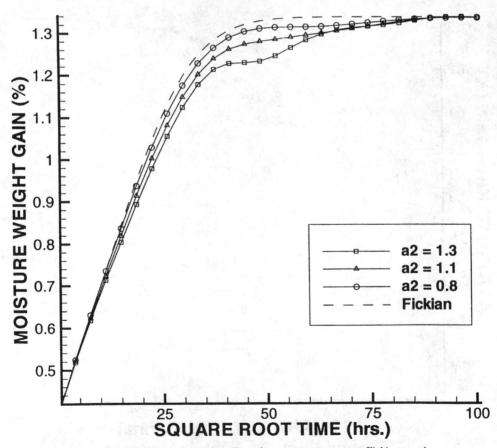

FIG. 3—*Sensitivity study on the effect of parameter* a₂ *on non-Fickian uptake.*

used for this set of analyses. Figure 4 shows that decreasing the magnitude of the shift factor a_σ results in a slightly lower "shoulder" of the uptake curve that is followed by a steeper approach to final saturation. A value of $a_2 = 1.2$ was used for this set of analyses.

Conclusions

Theory of irreversible thermodynamics was applied to derive governing equations for history-dependent diffusion in polymers and polymer matrix composites from first principles. A special form for Gibbs free energy is introduced using stress, temperature, and moisture concentration as independent state variables. Since the mathematically complex nature of the governing equations were not amenable to a closed-form solution, a variational formulation was used to derive the weak form of the nonlinear governing equations that were then solved using the finite element method. For model validation, the model predictions were compared with published experimental data for the special case of isothermal diffusion in an unstressed, symmetric Graphite/Epoxy angle-ply laminate. Effects of residual stresses on diffusion were ignored in the preliminary validation. The model was capable of reasonably simulating two-stage absorption behavior in a polymer composite even though some of the material param-

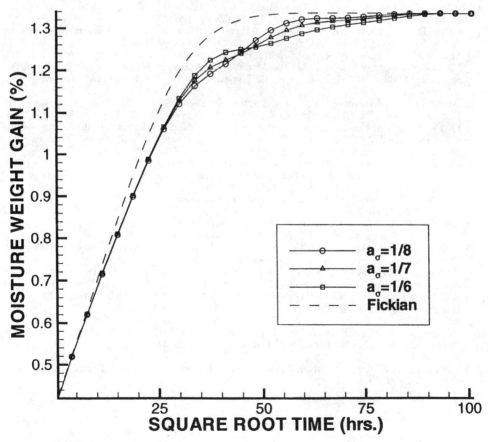

FIG. 4—*Sensitivity study on the effect of shift factor* a_σ *on non-Fickian uptake.*

eters were unknown, and therefore, had to be estimated. It is feasible that a more accurate prediction may be obtained through: (a) a comprehensive experimental characterization of the concentration dependence of the shift factor (a_σ) and the entropy production coefficient (a_G), and (b) retaining the power-law moisture concentration term in Eq 27. Moisture uptake experiments for additional verification of the model are currently underway.

Acknowledgment

The author would like to express his gratitude to Professor Ken Liechti at the University of Texas-Austin, to Lockheed-Martin, and to the Automotive Composites Consortium for funding this research. The author would like to thank Mr. Vikas Gupta, Mr. Weiqun Xu, and Mr. Sreesh Inguva for their help in preparing the manuscript.

References

[1] Shen, C. H. and Springer, G. S., "Effects of Moisture and Temperature on the Tensile Strength of Composite Materials," *Environmental Effects on Composite Materials,* G. S. Springer, Ed., Lancaster, PA, Technomic Publishing Co., Inc., 1981, pp. 79–93.

[2] Gurtin, M. E. and Yatomi, C., "On a Model for Two Phase Diffusion in Composite Materials," *Journal of Composite Materials,* Vol. 13, April 1979, pp. 126–130.

[3] Carter, H. G. and Kibler, K. G., "Langmuir-Type Model for Anomalous Diffusion in Composite Resins," *Journal of Composite Materials,* Vol. 12, April 1978, pp. 118–130.

[4] Shirrell, C. D., Leisler, W. H., and Sandow, F. A., "Moisture-Induced Surface Damage in T300/ 5208 Graphite/Epoxy Laminates," in *Nondestructive Evaluation and Flaw Criticality for Composite Materials, ASTM STP 696,* R. B. Pipes, Ed., American Society for Testing and Materials, 1979, pp. 209–222.

[5] Weitsman, Y., "Moisture in Composites: Sorption and Damage," *Fatigue of Composite Materials,* K. L. Reifsnider, Eds., Elsevier Science Publishers B. V., 1991, pp. 385–429.

[6] Frisch, H. J., "Irreversible Thermodynamics of Internally Relaxing Systems in the Vicinity of the Glass Transition," *Non-Equilibrium Thermodynamics, Variational Techniques, and Stability,* R. J. Dennelly, R. Herman, and I. Prigogine, Eds., University of Chicago Press, pp. 277–280.

[7] Crank, J., *The Mathematics of Diffusion.* Oxford University Press, 1975.

[8] Weitsman, Y., "Stress Assisted Diffusion in Elastic and Viscoelastic Materials," *Journal of Mechanics and Physics of Solids,* Vol. 35, No. 1, 1987, pp. 73–93.

[9] Biot, M. A., "Thermoelasticity and Irreversible Thermodynamics," *Journal of Applied Physics,* Vol. 27, No. 3, 1956, pp. 240–253.

[10] Schapery, R. A., "Further Development of a Thermodynamic Constitutive Theory: Stress Formulation," *A&S Report,* No. 69-2, Purdue University, West Lafayette, 1969.

[11] Weitsman, Y., "A Continuum Diffusion Model for Viscoelastic Materials," *Journal of Physical Chemistry,* Vol. 94, No. 2, 1990, pp. 961–968.

[12] Wietsman, Y., "Coupled Damage and Moisture Transport in Fiber-Reinforced Polymeric Composites," *International Journal of Solids and Structures,* Vol. 23, No. 7, 1987, pp. 1003–1025.

[13] Roy, S., Lefebvre, D. R., Dillard, D. A., and Reddy, J. N., "A Model for the Diffusion of Moisture in Adhesive Joints. Part III: Numerical Simulations," *Journal of Adhesion,* Vol. 27, 1989, pp. 41–62.

[14] Blikstad, M., Sjöblom, P. O. W., and Johannesson, T. R., "Long-Term Moisture Absorption in Graphite/Epoxy Angle-Ply Laminates," *Journal of Composite Materials,* Vol. 18, 1984, pp. 107–121.

[15] Cai, L. W. and Weitsman, Y., "Non-Fickian Moisture Diffusion in Polymeric Composites," *Journal of Composite Materials,* Vol. 28, No. 2, 1994, pp. 130–154.

Annette Roy,[1] Eléna Gontcharova-Bénard,[2] Jean-Louis Gacougnolle,[2] and Peter Davies[3]

Hygrothermal Effects on Failure Mechanisms of Composite/Steel Bonded Joints

REFERENCE: Roy, A., Gontcharova-Bénard, E., Gacougnolle, J-L., and Davies, P., "**Hygrothermal Effects on Failure Mechanisms of Composite/Steel Bonded Joints,**" *Time Dependent and Nonlinear Effects in Polymers and Composites, ASTM STP 1357*, R. A. Schapery and C. T. Sun, Eds., American Society for Testing and Materials, West Conshohocken, PA, 2000, pp. 353–371.

ABSTRACT: The aging in water of steel/composite glass polyester specimens bonded with an epoxy adhesive has been studied at 25° and 40°C. The damage processes are controlled by water diffusion. The main damage is due to the adhesive swelling but, when specimens are loaded during immersion, the stress gradient at the extremities of the joint accelerates the water diffusion and then the failure of specimens. A simple logarithmic relationship between the specimen life time and the applied load makes easy the life prediction.

KEYWORDS: glass polyester composites, stress gradient, aging, bonded joints, epoxy adhesive

Large amounts of glass fiber reinforced polyester composite are used in a wide range of marine structures, from pleasure boats to minesweepers [*1–3*]. In many of these applications, connections are required between the composite and adjacent metallic components. A large-scale example is the connection between the metal hull and the composite superstructure of the Lafayette frigate [*4*]. A weight-efficient way to assemble these very different materials is to bond them adhesively together. The adhesive may be either a polyester or, as in this study, an epoxy resin. When adhesive bonding is employed, the durability in water of both the adhesive and the various interfaces is critical and must be characterized. Much literature is available on the durability of adhesive joints, both in general textbooks on adhesives [*5,6*] and specific studies [*7,8*], and for the aerospace and automotive industries, in particular, extensive work has been performed. The behavior of adhesively bonded marine structures has received less attention as structural adhesive bonding is not widespread in this industry, but some studies are available [*9,10*].

This study [*11*] deals with the aging of composite/steel single lap joints in a marine environment. In order to predict the long term behavior of these joints in water, it is assumed

[1] Research Manager, CRITT Matériaux, Avenue Marcel Dassault, B.P. 115, 17303 Rochefort Cedex, France.

[2] Ph.D. Student and Chargé de Recherches au CNRS, respectively, Laboratoire de Mécanique et Physique des Matériaux, ENSMA, Site du Futuroscope, B.P. 109, 86960 Futuroscope Cedex, France.

[3] Senior Research Engineer, IFREMER, Laboratoire des Matériaux Marins, Centre de Brest, B.P. 70, 29280 Plouzané.

here that damage is controlled mainly by water diffusion in both composite and adhesive. Then, two parameters are varied to accelerate this aging: temperature and load. The water diffusion is also increased by substituting deionized water for sea water. A final accelerating factor is the absence of a gel-coat.

The aging of the polyester composite and the epoxy adhesive are first studied separately, then the aging of the assemblies in water is studied without load, and finally the behavior of the assemblies under load in water is described.

Materials and Experimental Conditions

The composite studied is a glass/polyester laminate made up of six "rovimat" layers; "rovimat" is used in Europe to name a non–symmetrical layer made up of a 0°/90° woven roving layer of areal weight 500 g/m² loosely stitched to a 300 g/m² random mat layer. These are impregnated with an isophthalic polyester matrix (Cray Valley S70390 TA). The laminate thickness is usually about 6.5 mm, it varies from 6.3 mm up to 7.4 mm, fiber content is around 50% by weight and the tensile modulus in the 0° direction is 18 GPa.

The composite panels were produced by contact moulding at IFREMER. They are not symmetrical about the mid-plane as they have woven rovings on one face and mat on the other. The composite was postcured for 24 h at 70°C.

The composite adherends are cut as rectangular coupons aligned in the 0° roving direction. They are bonded on the roving face, which is lightly abraded with 1000 grit paper and then degreased with alcohol. The geometry of the joints is shown in Fig. 1. Just before bonding, the steel substrate is sand blasted and degreased on the overlap surface. The free surface of the steel A 284 is coated with an epoxy paint which has given a good protection against corrosion for more than two years.

The specimens are prepared in a grooved fixture to obtain good axial alignment. The adherends are bonded with a structural epoxy adhesive (Ciba Geigy Redux 420 NA) and held under constant pressure for 24 h to obtain a nearly constant thickness 0.2 mm of the adhesive layer. Then the specimens are post-cured for 2 h at 70°C.

The glass transition temperature of the epoxy adhesive determined by differential scanning calorimetry (D.S.C.) at 10°C/min, begins at 51°C. Tensile tests made at 50°C on bonded specimens showed a drop of the mechanical properties due to an adhesive failure at the

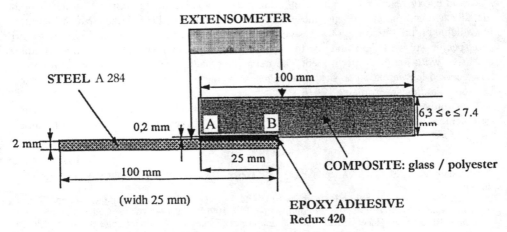

FIG. 1—*Geometry of steel/composite bonded joint.*

epoxy-metal interface. Based on these results the highest aging temperature was limited to 40°C.

Stress Analysis

The stress field in the metal/composite specimens subjected to a tensile force F was calculated by the finite element method using a two dimensional model. Each layer of roving and mat was described respectively by a line of eight-node quadrilateral elements. It has been shown that with this geometry of joint, the peel stresses due to the misalignment of applied forces are at the origin of the crack initiation [12–14]. The stress field of σ_{22} peel stresses between the points A and B at the interface rovimat-adhesive is shown in Fig. 2a. The stresses are concentrated at the extremities of the lap, where the gradient is very high in the thickness of the laminate. This gradient in the vicinity of the point A is illustrated in Fig. 2b.

Tensile Test Behavior of Unaged Joints

The specimens are tested in tension with an extensometer fixed to both sides of the joint as shown in Fig. 1. The elongation recorded by the extensometer is the addition of several strains and rotations, namely: i) the shear strains in the adhesive layer and in the adherends, ii) the tensile strains of the adherends, and iii) the difference between the rotations of the composite and the steel. This last parameter is difficult to estimate, it depends strongly on the exact position of the extensometer. The applied force F is plotted versus this apparent elongation ΔL in Fig. 3, for a test made at 0.5 mm/min.

The tensile behavior is typical of that of a ductile material: the elastic domain up to the threshold force F_s is followed by an apparent plastic domain characterized by an elongation to failure ε_r and a strain hardening. Below the threshold $F_s = 5000 \pm 500\ N$ the deformation is proportional to the applied force and an apparent rigidity of the specimen can be defined. This rigidity is obviously a simple function of the composite, steel, and adhesive moduli and also the overlap length. At 5000 N, the proportionality between force and deformation ceases due to the crack initiation in the composite near the point A (Fig. 1) where the peel stresses are at a maximum [12,15,16].

It should be emphasized that failure occurs neither in the adhesive nor at one of its interfaces but in the layer of the composite in contact with the adhesive. Often the crack initiates at a bubble of about 0.1 mm diameter, in the polyester near the composite/adhesive interface, and then propagates along and within the first layer of glass fibers. During this propagation, many fiber bridges are left between the two lips of the crack, they contribute to reduce the stress concentration at the crack tip and induce the apparent strain hardening of the specimen. The failure force F_r is 9000 $\pm$ 700 N. The scatter in F_s and F_r is due partly to the scatter of the laminate thickness caused by the manual fabrication method.

The aging of the specimen will be followed through the decrease of its rigidity, the threshold F_s, and the hardening beyond F_s.

Influence of Aging in Water on the Mechanical Properties of Adhesive and Composite

Aging of Adhesive

Cast tensile specimens of epoxy adhesive, of dimensions $3 \times 10 \times 60$ mm in the gauge length, were prepared. In such a thickness, the epoxy resin contains small air bubbles of 0.1 mm diameter. Samples were tested at 25° and 40°C at a strain rate of $6 \times 10^{-5}\ s^{-1}$. The

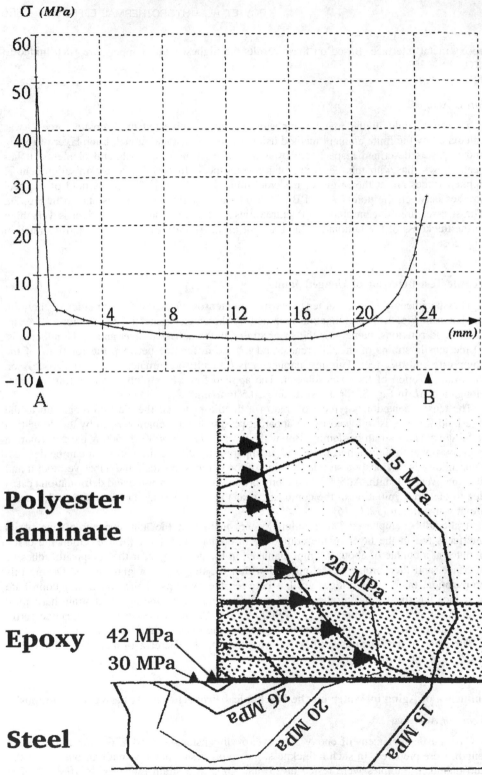

FIG. 2—*Stress field in the specimen under tension.*

Force, N

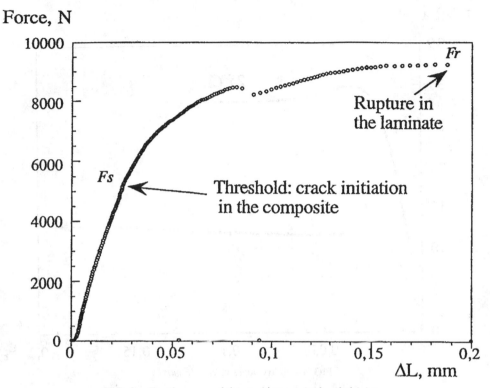

FIG. 3—*Tensile curve of the steel/composite bonded joint.*

tensile curves in Fig. 4 show an elongation at failure ε_r of 11% at 25°C and greater than 17% at 40°C, the modulus drops from 1.8 GPa down to 1 GPa when temperature is raised and the yield point from 30 MPa down to 12 MPa.

Five adhesive samples of dimensions $40 \times 40 \times 3$ mm were placed in water at 25 and 40°C. Their maximal weight and thickness were measured periodically, as shown in Figs. 5 and 6. The weight increase after 24 days of the epoxy adhesive is 4% at 25° and 40°C, while the increase in thickness is 0.8% at 25°C and 1.1% at 40°C. This difference between the amount of swelling at 25° and 40°C for the same content of water could be explained by the proximity of the glass transition at the higher temperature which increases the free volume.

As shown in Fig. 7, after an immersion of one month in water at 40°C tensile, failure occurs at $\varepsilon_r = 5\%$. The major part of the plastic strain has been lost, and the modulus has decreased by 20%. When the specimens are dried, their modulus is restored but the failure elongation is reduced to 3%. It is probable that the additives which plasticize the epoxy resin are extracted by water.

These tests have been performed on thick adhesive specimens, and water can diffuse easily into these specimens through all their surfaces. In the bonded specimens, water can enter the adhesive through the thin areas of the sides of the joint and indirectly through the composite. In order to isolate the influence of the diffusion through the sides of the joint, single lap shear specimens with both of the adherends made of steel have been tested. The failure force of these specimens is 16 000 N and their apparent elongation at failure is 0.18

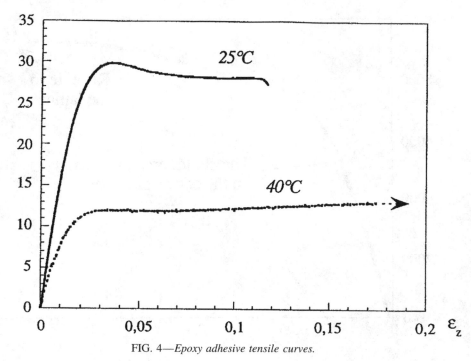

FIG. 4—*Epoxy adhesive tensile curves.*

mm (Fig. 8). After three months in water at 25°C, the failure force drops to 13 000 N for an elongation of 0.13 mm; after drying, this failure force is partly restored at 14 000 N. The influence of the loss of plasticity detected on the bulk specimens seems to be less dramatic on the bonded specimens than in the bulk specimens, which may be attributed to the predominant shear stress in the joint specimens.

Aging of Composite

The aging in water of the glass fiber reinforced polyester composite results from the aging of both the matrix polyester and of the fiber/polyester interfaces.

The maximum increase in weight of cast unreinforced polyester resin samples immersed in water is 0.6% at 25° and 40°C. The effect of the temperature on the total amount of absorbed water and the water diffusion kinetics is negligible. When water is removed from specimens by drying, the weight of the specimens is 0.2% lower than before immersion. This means that: i) the weight increase due to water absorption is at least 0.8% and, ii) short molecules have been extracted from the polyester resin.

As seen in Fig. 9, an immersion of one month in water does not modify the modulus of the polyester, but the failure stress is reduced from 55 MPa down to 45 MPa. This decrease will be crucial for the damage of bonded specimens subjected to a constant load in water.

The composite glass/polyester absorbs in weight about twice as much water after immersion for 24 days as the polyester alone, Fig. 10. If the weight of glass fibers is subtracted from the initial weight of specimens, the weight increase due to water absorbed in the matrix

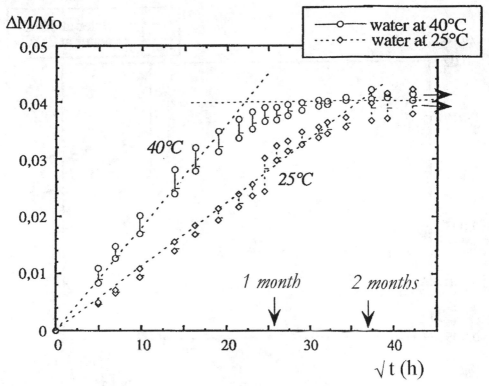

FIG. 5—*Water absorption in epoxy adhesive at 25° and 40°C.*

and glass/polyester interfaces is 2.2%; so these interfaces can absorb several percent of water. It must be underlined that the fiber-matrix region is 3 times larger in the mat fibers than in the roving fibers, therefore, it is to be expected that the mechanical degradation concerns much more the mat interfaces than the roving interfaces.

To quantify the influence of aging on the mechanical properties, tensile tests were performed, both in the plane of the laminate in the 0° direction and in the through-thickness direction. The latter were performed because of the presence of peel stresses at the ends of the joints in loaded specimens.

The degradation of the tensile properties during immersion in water at 40°C is shown in Fig. 11. The modulus drops from 18 GPa down to 10 GPa in proportion to the immersion duration. This drop seems to be related to the high amount of water absorbed in the interfaces glass/polyester, which weakens the glass/matrix bonding. The failure stress is reduced from 210 MPa down to 120 MPa by the nine months' immersion in water.

For tensile tests in the transverse direction, the specimens are 25 × 25 mm squares bonded to square steel blocks. Failure always occurs in a mat layer at a very low stress, only 7.5 MPa. The same tests were performed on specimens dried after aging. (It is not possible to test specimens containing water because the presence of water molecules within the polyester makes the adhesive inefficient). The results indicate only the irreversible damage due to water in the laminate. After three months in water at 25°C and drying, σ_R drops down to 5.2 MPa, after the same stay in water at 40°C, it drops as low as 2.8 MPa. This loss of transverse strength is attributed to the degradation of mat fiber interfaces. This interpretation

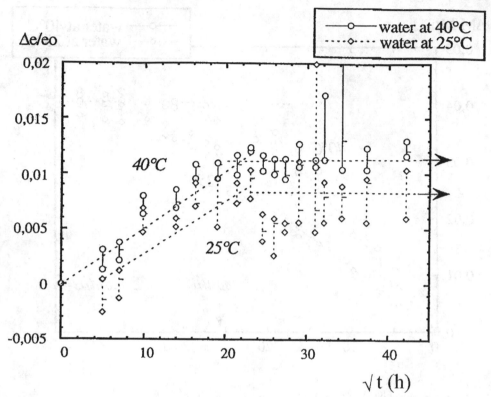

FIG. 6—*Increase of epoxy specimens thickness during their immersion in water at 25° and 40°C.*

is confirmed by shear tests with unidirectional (UD) and Rovimat laminates. The shear strength of UD specimens is not modified by a four months immersion at 40°C, whereas the shear strength of Rovimat laminates decreases from 35 MPa down to 23.5 MPa.

Aging in Water of Bonded Specimens

Bonded specimens are immerged in deionized water at 25° and 40°C. After 1, 3, 6, 9, and 12 months of immersion, three of them are tested in tension and three others are dried out and tested. Typical tensile curves obtained during the aging at 25° are reported in Fig. 12.

The apparent rigidity of the joints decreases by about 20%, this is in accordance with the decay of the tensile moduli of laminate and epoxy adhesive. In Fig. 12, the rupture modes are also mentioned. During the first three months in water the failure occurs, as in unaged specimens, along the first roving layer of laminate. After six months, the failure is partly adhesive at the steel/composite interface. The amount of this adhesive rupture increases with the immersion duration, it becomes predominant after a year in water but, at this moment, a specimen exhibits a rupture located in the first mat layer. This last failure mode is strange because the stresses are much lower in the mat layer than in the adhesive, nevertheless, it can be explained by the dramatic drop of interlaminar failure stress observed in that layer after aging in water.

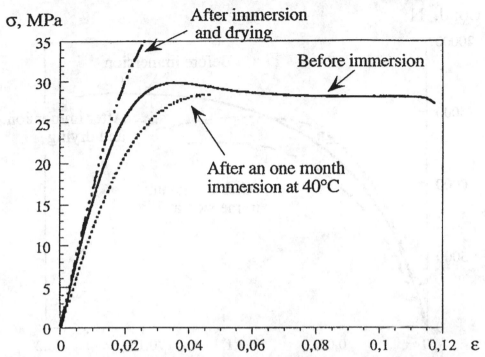

FIG. 7—*Epoxy resin tensile curves in the initial state, after one month in water at 40°C and after immersion and drying.*

The adhesive fracture at the epoxy/steel interface can be attributed to either the degradation of the steel surface by water molecules or the swelling of the adhesive due to water absorption. A cooperation between these processes can also be considered. The swelling generates shear stresses which, added to the applied stresses, can be at the origin of the adhesive failure. The influence of the adhesive swelling was isolated by drying the specimens out. Indeed, after drying, the specimens fail again by propagation of the crack along the first roving layer. Thus, the adhesive swelling seems to be a crucial parameter.

In Fig. 13 both the threshold F_s and the rupture force F_R are reported, the scatter of the results is represented by the extreme values of each set of measurements. F_s and F_R decrease during the first six months in water. The decay of the threshold F_s is directly related to the drop of the failure force of polyester aged in water. During six months, F_s and F_R decrease in the same way, they then remain constant. After one month in water, it is surprising to see that F_s is lowered by specimens drying. This is attributed to the increase of the tensile modulus of the aged epoxy resin by drying. When the adhesive modulus increases, the peel stresses in the laminate are also increased so that the crack initiation in the polyester occurs for a lower applied force.

The immersion in water at 40°C accelerates the aging rate. The plateau of F_s and F_R is reached after only three months. No new failure process can be detected but the failure in the mat layer is observed after nine months.

Aging Under Load in Water

To limit the scatter in the mechanical properties of the aged specimens, three specimens with the same laminate thickness were tested for each condition. An apparatus has been

Load, N

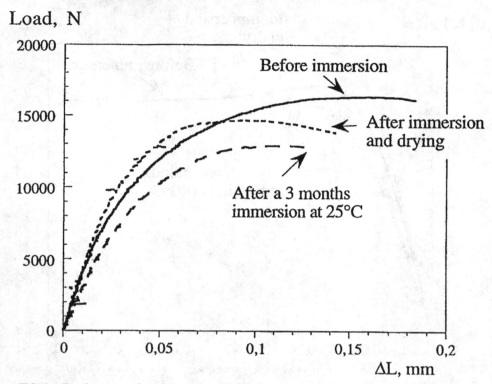

FIG. 8—*Tensile curves of steel/epoxy/steel bonded specimens in the initial state, after three months immersion at 25° and after drying at 40°C.*

designed to apply vertically the same load to three specimens placed in series, Fig. 14. Because of the out-of-alignment of the forces applied to each specimen, a couple is generated at each. In order to remove this couple, each grip is maintained in a vertical slide. This line of joints is placed within a transparent PMMA cylinder in which deionized water flows at 25° or 40°C, this flow makes the water temperature homogeneous throughout the cylinder. Several of these immersion columns were manufactured to multiply the number of specimens tested.

When a crack propagates within the laminate, its presence is revealed by whitening which is visible through the laminate. It is thus possible to follow the propagation of cracks in the laminate visually, from the outside of the PMMA cylinder. After the failure of the first specimen, the two remaining specimens are removed and tested to failure on a tensile test machine.

Before aging the specimens under load in water, their creep behavior was investigated in air. When the applied load F is higher than the threshold F_s, i.e., $F = 1.2$, $F_s = 6000$ N, a small crack of about 3 mm length is initiated in the laminate during loading of the specimen. This crack propagates slowly for several hours and stops completely at a length of about 6 mm. This arrest, like the strain hardening in a tensile test, can be attributed to the fiber bridging which increases as the crack propagates. When F is lower than F_s no crack can be detected even after two years at this force level. In summary, the application in air of a constant load $F \le 1.2$ F_s does not cause the failure of the assembly.

σ, MPa

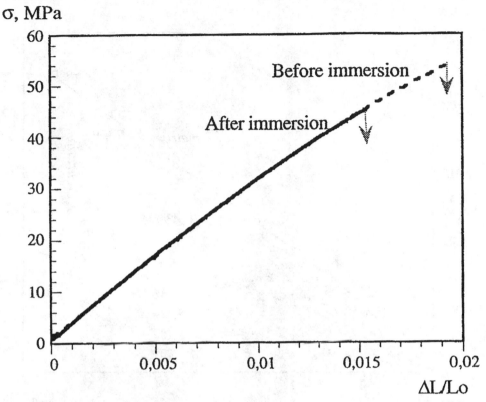

FIG. 9—*Polyester tensile curves before and after an immersion of one month in water at 25°C.*

In water, the results are totally different. The load of 1.2 F_s initiates the same small crack as in air but this crack propagates continuously and the specimen fails after about 6 h. Water diffuses quickly along the fibers within the laminate and weakens the fiber bridges, which are no longer able to blunt the crack. The rapid crack propagation implies a high acceleration of water diffusion along fibers, which can be attributed to the damaging loading of the fiber-matrix interfaces at the crack tip. Crack propagation and water diffusion are cooperative mechanisms.

When the applied load is lower than F_s, $F = 0.8\ F_s$, for example, the specimen failure occurs in several stages. After one day at 40°C, a crack is initiated at the usual site, in the first layer of the laminate, probably due to the aging of the polyester resin. The threshold decreases more than in the specimens aged in water without load, probably due to the high stress field at this extremity which increases diffusion and water content.

Then the crack propagates slowly during the next two days and stops at a length of about 5 mm. The failure occurs suddenly 10 days after the application of the load. A fractographic observation shows that this failure has three distinct areas. Under the initial crack and at the other extremity of the joint, as shown in Fig. 15, the failure is adhesive between epoxy and steel, in the middle of the joint, the failure is again in the first layer of the roving. The steel surface shows no sign of corrosion.

The failure modes by de-cohesion at the epoxy/steel interface are observed on all the specimens aged in water at 40° and 25°C under loads as low as 0.5 F_s. The process proposed

ΔM/Mo

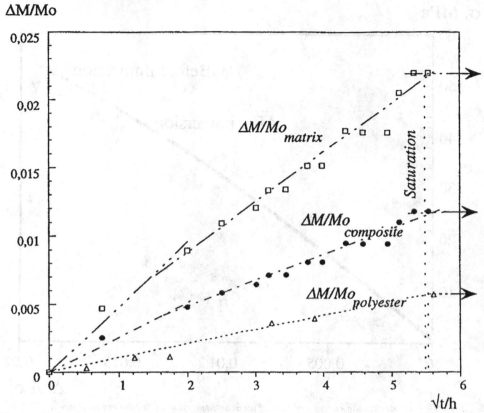

FIG. 10—*Water absoption kinetics at 25°C by polyester and glass/polyester laminate.*

for the successive stages of the failure is illustrated in Fig. 16. At the extremities of the lap (the shaded areas in Fig. 16*a*), the stresses are high in adhesive and laminate, therefore, the water diffusion is accelerated and maximum water content is enhanced in these materials. The polyester matrix is embrittled and a crack can be initiated in the laminate as in air, but under lower loads. During the crack propagation, fiber bridges are left between the two lips of the crack. Even though it has been shown above that the fiber interfaces in bridged regions are damaged by water, these interfaces can reduce the stresses at the crack tip and arrest the crack propagation.

Nevertheless, the stress field in the specimen in the vicinity of the crack tip has the same shape, with a reduced intensity, as the stress field in the point A before crack propagation Fig. 16*b*. The stress gradient is very high, especially for the peel stress σ_{22}, and its maximal value is located at the steel/epoxy interface. This stress gradient at points A and B accelerates and guides the water diffusion towards this interface. Then, during the crack arrest, the epoxy located just under the crack tip (point C) absorbs water very rapidly, swells, generates local shear stresses, and a crack is initiated at the steel/epoxy interface, Fig. 16*c*.

We suggest that this new crack propagates rapidly after its initiation. This assertion is based on the following reasons: i) the presence of the previous bridged crack in the polyester resin erases slightly the maximum of the stresses. When the interface crack starts to propagate, it leaves this protective influence, ii) the effective length of the adhesive lap is reduced

σ, MPa

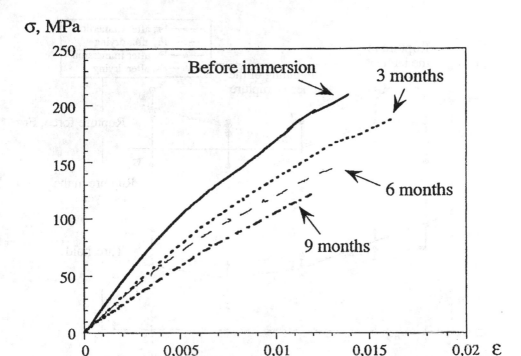

FIG. 11—*Tensile curves of composite glass/polyester after 3, 6, and 9 months of immersion at 40°C.*

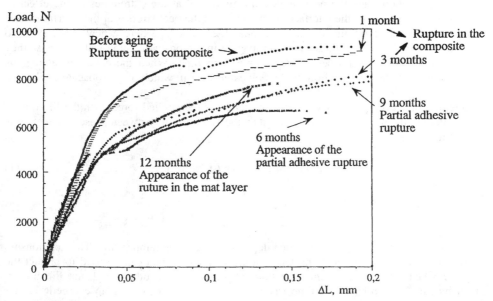

FIG. 12—*Tensile curves of steel/composite assemblies after 1, 3, 6, 9, and 12 months in water at 25°C.*

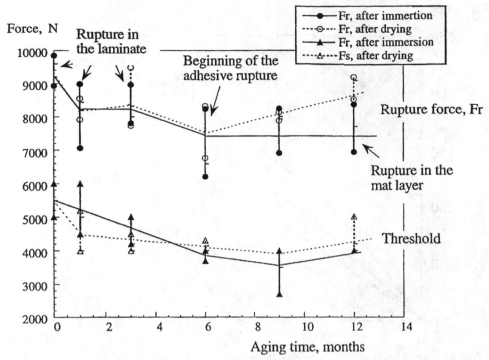

FIG. 13—*Evolution of the threshold* F_s *and the rupture force* F_R *of steel/composite assemblies during their immersion in water at 25°C. Dashed lines correspond to dried specimens.*

by this propagation, then the peel stresses are increased at the extremities of the effective joint. The propagation of the interface crack increases the peel stresses at the interface steel-epoxy at the other extremity of the joint, near the point B, where a new crack starts to propagate. Initially, both of these cracks propagate in regions containing water because they have been subjected to the peel stresses for a long time. When the cracks meet regions containing less water, they go back to the polyester layer in which they propagate in absence of water.

This interpretation is confirmed by the influence of the applied force F on the life time t_ℓ of specimens. In Fig. 17, it can be seen that the relationship between these parameters is logarithmic:

$$F/Fs = A(T) - B(T) \log (tc) \tag{1}$$

$$\text{or } t_\ell = 2.3 \exp \left[\frac{A(T) - F/F_S}{B(T)} \right] \tag{2}$$

$A(T)$ and $B(T)$ being parameters which depend on the aging temperature. This relationship shows that an increase of the aging temperature or applied load can be used to predict the long term behavior of steel/composite bonded joints. On the other hand, the life time t_ℓ given by Eq 2 is similar to the time t necessary for an atom to cross an energetic barrier ΔG in an Eyring diffusion process under a stress gradient:

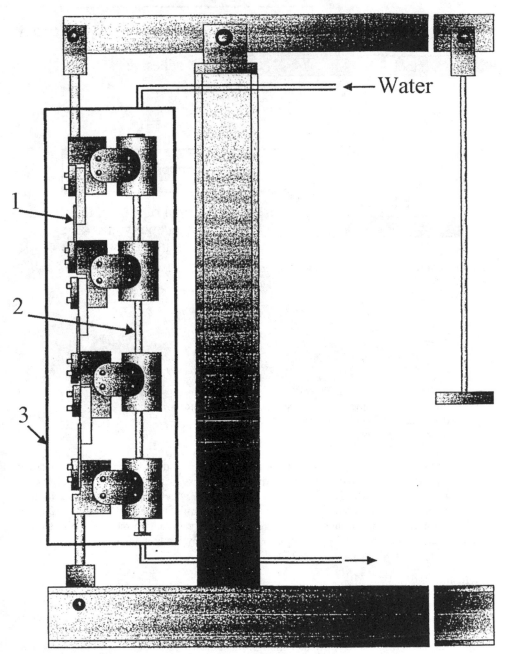

FIG. 14—*Creep apparatus: specimens (1) are fixed in grips which move in slides (2) within a PMMA cylinder (3).*

Initial crack in the composite

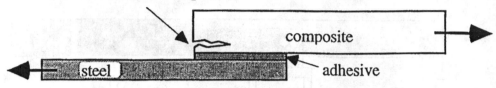

Crack in the composite

adhesive

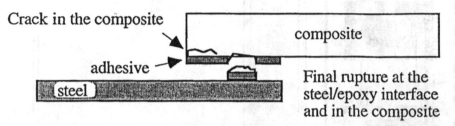

Final rupture at the
steel/epoxy interface
and in the composite

FIG. 15—*Schematic diagram of initial crack and failure in assembly aged in water at 25°C under a load of 0.8 F$_s$.*

$$t \approx \exp \frac{\Delta G - \delta G}{kT} \tag{3}$$

where

δG = the variation of the energetic barrier due to the stress gradient.
k = the Boltzmann constant.
T = the absolute temperature.

It can be assumed that the life time of bonded specimens under load in water is governed by the diffusion of water in a stress gradient. This process is the stress assisted diffusion, well known in the case of hydrogen in metals [17,18] and which received some attention in the context of stress corrosion of glass fiber composite laminates in the 1980's [19,20]. To model the life time in water quantitatively requires additional data on the water diffusion in these materials in a stress gradient. This aspect is currently being studied.

Conclusion

Prediction of the long term durability of adhesively bonded composite/metal joints is complex due to the interaction between competing damage mechanisms. The aging in water of steel/composite glass polyester specimens bonded with an epoxy adhesive is controlled by water diffusion through the polyester matrix, the adhesive, and the fiber/matrix interface. The mat fiber interfaces are seriously degraded by immersion, while the swelling of the epoxy adhesive induces a de-cohesion at the steel/epoxy interface. When a load is applied during immersion, the damage process is governed by diffusion of water in the stress gradient at the extremities of the joint. The aging processes at 40°C and at 25°C are identical but the

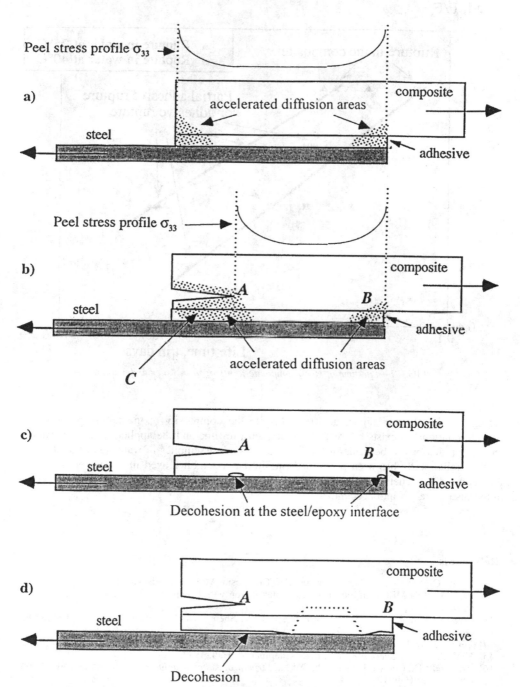

FIG. 16—*Failure process of steel/composite joint aged in water under load* F < F$_s$.

Load, F/F$_s$

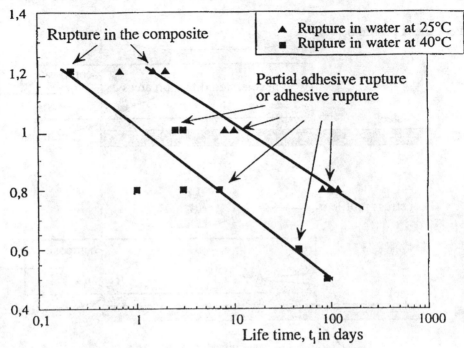

FIG. 17—*Life time of steel/composite joints aged in water under load.*

aging rate is 10 times higher at 40°C. Despite the complexity of the failure processes, a simple relationship exists between the specimen life time and the applied load. This should enable predictions to be made of the lifetime of such assemblies in water under load.

A parallel study has been performed on similar assemblies aged in sea water. This will be reported elsewhere, but the sea water environment adds a corrosion mechanism which must also be taken into account.

References

[1] Smith, C., "Design of Marine Structures in Composite Materials," Elsevier 1990.
[2] Proc 3rd IFREMER conf. on "Nautical Construction with Composite Materials," P. Davies and L. Lemoine, Eds., 1992.
[3] Composite Materials in Marine Structures, R. A. Shenoi and J. F. Wellicome, Eds., Cambridge Ocean Technology series, 1993.
[4] LeLan, J. Y., Parneix, P., and Gueguen, P. L., in Ref 2, p. 399.
[5] Kinloch, A. J., *Adhesion & Adhesives,* Chapman & Hall, 1987.
[6] Adams, R. D., Comyn, J., and Wake, W. C., *Structural Adhesive Joints in Engineering,* 2nd Edition, Chapman & Hall, 1997.
[7] Kinloch, A. J., Ed., "Durability of Structural Adhesives," *Applied Science,* 1983.
[8] Bowditch, M. R., *Int J. Adhesion & Adhesives,* Vol. 16, 1996, p. 73.
[9] Knox, E. M., Cowling, M. J., and Hashim, S. A., Proc SAEIV, Bristol UK, *Inst. of Materials,* 1995, p. 153.

[10] Stevenson, A., *Int. J. Adhesion & Adhesives,* Vol. 5, 1985, p. 81.

[11] Gontcharova-Bernard, E., "Vieillissement d'assemblages collés acier/composite dans l'eau et sous charge," Thèse de l'Université de Poitiers, 1997.

[12] Roy, A., Mabru, C., Gacougnolle, J. L., and Davies, P., "Damage Mechanisms in Composite/ Composite Bonded Joints Under Tensile Loading," *Applied Composite Materials,* Vol. 4, 1997, p. 95.

[13] Adams, R., "Strength Prediction for Lap Joints, Especially with Composite Adherends, A Review," *J. Adhesion,* Vol. 30, 1989, p. 219.

[14] Renton, J. W. and Vinson, J. R., "On the Behavior of Bonded Joints in Composite Material Structures," *Journal of Engineering Fracture Mechanics,* Vol. 7, 1975. p. 41.

[15] Garcia Lafuente, D., "Estudo de las unianes par adhesivo, de sola pamiento simple, de tipo acierocomposite, projet de fin d'études," Université de Saragosse, 1997.

[16] Roy, A., "Comportement mécanique en sollicitation monotone et cyclique d'assemblages collés composite/composite et composite/acier," Thèse de l'Université de Poitiers, 1994.

[17] Van Leuwen, H. P., *Engineering Fracture Mechanics,* Vol. 6, 1974, pp. 141–161.

[18] Bockris, J. O. M., Beck, W., Genshaw, M. A., Subramanyan, P. K., and Williams, F. S., *Acta Met,* Vol. 19, 1971, pp. 1209–1218.

[19] Hogg, P. J. and Hull, D., Chapter 2 "Developments in GRP Technology-1," B. Harris, Ed., *Applied Science,* 1983.

[20] Pritchard, G. and Speake, S. D., *Composites,* Vol. 19, No. 1, January 1988, pp. 29–35.

Author Index

A

Allen, D. H., 318

B

Benson, Dianne, 285
Bocchieri, Robert T., 238
Bowles, Kenneth J., 3
Bradley, W. L., 318
Brinson, L. Catherine, 141

C

Case, Scott W., 310
Castelli, Michael G., 285
Cerrada, Marria L., 47

D

Daniel, I. M., 223
Davies, Peter, 353

G

Gacougnolle, Jean-Louis,
    353
Gates, Thomas S., 141,
    160
Gontcharova-Bénard,
    Eléna, 353

H

Ho, Kwangsoo, 118
Holmes, Gale A., 98
Hsiao, H. M., 223
Hung, S.-C., 176
Hunston, Donald L., 98

K

Kandachar, Prabhu, 70
Krempl, Erhard, 118

L

Liechti, K. M., 176
Loverich, James S., 310

M

McDonough, Walter G., 98
McKenna, Gregory B., 18,
    47
McManus, Hugh L., 3

P

Peterson, Richard C., 98

R

Reeder, J. R., 318
Reifsnider, Kenneth L.,
    310
Roy, Annette, 353
Roy, Samit, 338
Russell, Blair E., 310

S

Schapery, Richard A., vii,
    238
Schutte, Carol L., 98
Simon, Sindee L., 18
Skrypnyk, Ihor D., 70, 83
Smit, Willem, 83
Spoormaker, Jan L., 70, 83
Sun, Chin-Teh, vii, 266
Sutter, James K., 285

T

Tsuji, Luis C., 3

V

Veazie, David R., 160

W

Whitley, Karen S., 141

Z

Zhu, Changming, 266

Subject Index

A

Accelerated strength testing, thermoplastic composites, 318

Aging
  composite/steel bonded joints, 353
  elevated temperature stress relaxation, 141
  thermo-oxidative, 3

AS4/PEEK, nonlinear multiaxial behavior, 176

B

Bonded joints, failure mechanisms, 353

C

Carbon fiber-reinforced composites, nonlinear viscoelastic behavior, 238

Coefficient of thermal expansion, 3

Cole-Cole process, 47

Composite/steel bonded joints, failure mechanisms, 353

Compression
  effects on composite time-dependent behavior, 160
  with shear, 176
  testing, composites with fiber waviness, 223

Constitutive equations, 118, 238

Cox model, 98

Creep
  effects on composite time-dependent behavior, 160
  elevated temperature stress relaxation, 141
  loadings, 285
  overstress model, 118
  PEN, 47
  prediction, 83
  short- and long-term, 70

Cyclic loading, polyphenylene sulfide, 310

D

Damage tolerance, polyimide chapped fiber composite, 285

Degradation, PMR-15 resin, 3

Differential scanning calorimetry, temperature modulated, 18

D (continued)

Diffusion, history-dependent, 338

Diglycidyl ether of biphenol-A/meta phenylenediamine epoxy resin matrix, 98

Discretization schemes, 83

Durability, polyimide chapped fiber composite, 285

Dynamic response, composites with fiber waviness, 223

E

E-glass fiber, interfacial shear strength, 98

Enthalpy recovery, time dependent, 18

Epoxy adhesive, 353

Epoxy resin, interfacial shear strength, 98

Equilibrium stress, 266

F

Failure mechanisms, hygrothermal effects, 353

Fiber distribution, 285

Fiber-reinforced composites
  laminates, time-dependent behavior, 160
  nonlinear multiaxial behavior, 176

Fiber waviness, 223

Finite-element analysis, 338
  constitutive model implementation, 83

Four-point-bend tests, 3

G

Gibbs free energy, 338

Glass fiber-reinforced composites, nonlinear viscoelastic behavior, 238

Glass polyester composites, failure mechanisms, 353

Glass transition, 18

Graphite fiber, durability and damage tolerance, 285

H

Henriksen scheme, discretization of hereditary integral, 83

Hygrothermal effects, failure mechanisms, 353

Hygrothermal modeling, polymers, 338